Progress in
Microemulsions

Progress in Microemulsions

Edited by

S. Martellucci
The Second University of Rome
Rome, Italy

and

A. N. Chester
Hughes Research Laboratories
Malibu, California

Plenum Press • New York and London

Library of Congress Cataloging in Publication Data

Progress in microemulsions / edited by S. Martellucci and A. N. Chester.
     p.   cm. — (Ettore Majorana international science series. Physical sciences; v. 41)
   "Based on the proceedings of the International School of Quantum Electronics Workshop on Progress in Microemulsions, held October 26–November 1, 1985, in Erice, Italy"—T.p. verso.
   Includes bibliographical references and index.
   ISBN 0-306-43212-9
   1. Emulsions—Congresses. I. Martellucci, S. II. Chester, A. N. III. International School of Quantum Electronics Workshop on Progress in Microemulsions (1985: Erice, Sicily) IV. Series.
TP156.E6P76 1989   89-33430
541.3′4514—dc20   CIP

Based on the proceedings of the International School of
Quantum Electronics Workshop on Progress in Microemulsions,
held October 26–November 1, 1985, in Erice, Italy

© 1989 Plenum Press, New York
A Division of Plenum Publishing Corporation
233 Spring Street, New York, N.Y. 10013

Printed in the United States of America

PREFACE

     The current state of the art of various aspects of microemulsion
systems is reflected in this volume. Major topics discussed include:
general background on solubilized systems, phase diagrams and phase equilibria, bicontinuous microemulsions, Winsor's phases, theories and models
of complex self association structures, crytical behaviour, phase transitions in lyotropic liquid crystals.

     I hope that this book will serve its intended objective of reflecting
our current understanding of microemulsions both in theory and practice,
and that it will be useful to researchers, both novices as well as experts,
as a valuable reference source.

     I feel indebted to the people of the Ettore Majorana Centre: the
friendly atmosphere of the Erice centre provided a very effective environment to enjoy the company of colleagues and friends during breaks and after
sessions, and to discuss problems of mutual interest. The courtesy,
efficiency and devotion of the secretarial and technical staff was also
appreciated, and greatly contributed to make the Workshop a smoothly run-
ning one.

The Scientific Secretary
Donatella Senatra
Department of Physics
University of Florence (Italy)

INTRODUCTION

The decision to publish, in a more permanent form than heretofore, the Proceedings of the Workshop on "Progress in Microemulsion" of the International School of Quantum Electronics, which was held in Erice (Italy) from October 26 to November 1st, 1985, under the auspices of the "Ettore Majorana" Centre for Scientific Culture, will prove to be a sound one. This workshop has attracted eminent lecturers both from Italy and abroad and those students fortunate enough to be able to attend, although now from many countries, are but a small audience for the countless hours of diligent preparation by such distinguished scientists. Our thanks are gratefully given to all the authors for their willing cooperation in summarizing both the present state of their researches and the background behind them.

In editing this material we did not modify the original manuscripts except to assist in uniformity of style. The papers are presented without reference to the chronology of the Course but in the following topical arrangement:

A. "Fundamental aspects of surfactant association structures", a group of tutorial papers;
B. "Theories and models", where theories and models of microemulsion structures are reviewed;
C. "Experimental studies of microemulsion systems", a discussion of experimental data related to specific systems;
D. "Lyotropic mesophases – Liquid crystals", a very detailed report on the characteristics of this class of microemulsions.

Before concluding, we acknowledge the invaluable help of the Scientific Secretary of the workshop, Prof. Donatella Senatra, for much of the organization of the workshop and the editing of these proceedings. Our thanks are also due to the organizations who sponsored the workshop, especially the E. Majorana Centre for Scientific Culture, whose support made this workshop possible. It has also been a pleasure to work with Mrs. Madeleine Carter of Plenum Press in the preparation of this volume.

A.N. Chester
Hughes Aircraft Company
El Segundo, CA (USA)

S. Martellucci
Engineering Faculty
The Second University
of Rome (Italy)

Directors of the International School of
Quantum Electronics

CONTENTS

FUNDAMENTAL ASPECTS OF SURFACTANT ASSOCIATION STRUCTURES

PHASE BEHAVIOR AND STRUCTURE OF MICROEMULSIONS

E. Ruckenstein

Institut für Physikalische Chemie I der
Universität Bayreuth, Postfach 3008
D-8580 Bayreuth, West-Germany

I. INTRODUCTION

Various pathways can be employed to prepare microemulsions. One of
them starts from micellar solutions. Surfactants dissolved in water form,
above the critical micelle concentration, a large number of micelles.
Hydrocarbon molecules, though sparingly soluble in water, can be solubilized in the hydrophobic core of the micellar aggregates. The solubilized
molecules are usually located among the hydrocarbon tails of the micelles.
For some surfactants or, more generally, when a medium chain length alcohol
(a cosurfactant) is also present, the solubilized molecules can form a core
covered by a layer of surfactant and alcohol molecules. Of course, some
hydrocarbon molecules and part of such molecules remain located among the
hydrocarbon tails of the interfacial layer of surfactant and cosurfactant.
A dispersion containing the latter kind of microstructures is called a
microemulsion. Micellar aggregates as well as solubilized micellar aggregates containing hydrocarbon molecules among the hydrocarbon tails of the
surfactant molecules are thermodynamically stable. This suggests, by
extrapolation, that the microemulsions can also constitute thermodynamically stable dispersions.

The preferred pathway to prepare a microemulsion starts, however, from
an emulsion stabilized by the adsorption of surfactant molecules on the
surface of the globules. The addition of a cosurfactant generates a new
type of emulsion, containing globules of almost uniform size, lying between
10^2 and 10^3 Å. The small size of these globules is responsible for the
name "microemulsion" which is attributed to such a dispersion. In contrast
to the conventional (macro) emulsions, whose stability has a kinetic
origin, microemulsions are thermodynamically stable. For ionic surfactants, the kinetic stability of the macroemulsions is determined by the
potential barrier generated between two globules as a result of the competition between the attractive van der Waals interactions and the repulsive
double layer forces[1]. The thermodynamic stability of microemulsions can
be explained as follows[2]. The dispersion of one phase into the second
one is accompanied by an increase in the entropy of the system and by the
adsorption of surfactant and cosurfactant (surfactants) on the large interfacial area thus created. Because of this adsorption, the interfacial
tension of the water-oil interface is decreased from about 50 dynes/cm
(value which is characteristic for a water-oil interface devoid of surfactants) to a very low value. In addition, the dilution of the two media of

the microemulsion in surfactants, as a result of adsorption, decreases the
chemical potentials of the surfactants. The corresponding decrease in the
free energy of the dispersion could be called the dilution effect. Thermodynamically stable dispersions can arise when the negative free energy
change due to the entropy of dispersion and to the dilution effect over-
comes the positive free energy contribution provided by the product of the
low interfacial tension and the large interfacial area between the two
media of the microemulsion.

A third pathway which starts with a liquid crystal, allows one an
instructive comparison between random globular and liquid crystal like
microemulsions. Let us consider a lamellar liquid crystal of water, oil,
surfactant and cosurfactant to which alcohol is added. The thermodynamic
stability of this liquid crystal is determined (see later in the paper) by
the interfacial tension (defined for an infinite distance between two
layers of the same kind) and by the interaction forces (double layer,
dispersion, hydration, etc.) between the layers. The entropy of dispersion
does not play a major role in this case. The addition of alcohol may
decrease the interfacial tension. Consequently, as shown later in the
paper, the long and thin lamellar can become unstable to thermal perturbations and rupture into smaller elements. While this may decrease the van
der Waals and double layer interactions, it can increase the entropy of the
system to such a degree that the overall free energy of the system is
diminished.

The spherical shape attributed to the globules of the dispersed phase
is a result of both our image about the macroemulsions as well as of numerous experimental investigations[3-9]. Recent experiments have, however,
indicated that additional structures of the dispersion are also possible.
The first of these observations is related to what happens when the amount
of salt is increased for fixed amounts of oil, water, surfactant and cosurfactant[10-12]. For relatively low amounts of salt an oil in water microemulsion coexists with excess oil phase, at intermediate salinities a
(middle phase) microemulsion coexists with both water and oil excess
phases, while at sufficiently high ionic strength a water in oil microemulsion is in equilibrium with excess water phase. Concerning the structure, it is relevant to note that the microemulsion contains spherical
globules of almost uniform size as long as it coexists in equilibrium with
the excess dispersed phase, but that its structure changes, becoming extremely complex, as soon as it coexists with both excess phases. Similar
phenomena have been observed when, instead of the salt concentration, the
temperature or the number of ethylene oxide groups of the surfactant with
polyoxyethylene head group were changed. Concerning the single phase
microemulsions, the NMR experiments of Lindman, et al.[13] show that the
self-diffusion coefficients of water, hydrocarbon and alcohol are in some
circumstances very large. This suggests that the interfaces that separate
the hydrophilic from the hydrophobic regions are in these cases unstable;
they open up and reform at a short time scale.

The aim of the present paper is to present a thermodynamic theory of
microemulsions[14-16] that can explain the above phenomena. The main
features of the theory have been presented at the Lund meeting on surfactants in 1982. In essence, the thermodynamic considerations that follow
demonstrate the existence of transition points in the vicinity of which the
spherical shape of the globules becomes unstable. Such transition points
exist in some single phase microemulsions as well as at the border between
a microemulsion coexisting with the excess dispersed phase and one coexisting with both excess phases. Qualitative explanations of the above phenomena are presented in the next section. Single phase microemulsions are
treated in the third section, where two equations are derived from the
condition that the free energy of the system should be a minimum with

respect to both the radius r of the globules as well as the volume fraction
ϕ of the dispersed phase. The first equation can be used to calculate the
equilibrium radius, while the second, a generalized Laplace equation, can
be employed to determine the conditions under which the spherical interface
becomes unstable. Equations for the phase equilibrium in the two phase
region are derived in the fourth section. They are used to demonstrate
that the spherical interface is stable in this case to thermal perturbations as well as to derive, in the fifth section, expressions for the
interfacial tensions at the surface of the globules and between microemulsion and excess dispersed phase. The sixth section is reserved to the
middle phase microemulsion and the final, the seventh, to birefringent
microemulsions and to a comparison between random globular and possible
liquid crystal like microemulsions. The final section also contains kinetic considerations regarding the formation of microemulsions from liquid
crystals.

II. THE TWO LENGTH SCALES THAT CHARACTERIZE A MICROEMULSION AND THEIR
     THERMODYNAMIC IMPLICATIONS

Let us consider a single phase microemulsion that contains spherical
globules of a single size. Because the globules are macroscopic bodies,
two macroscopic length scales should characterize a microemulsion, namely
the scale of the globules and the scale of the microemulsion, the latter
being much larger than the former. The thermodynamic pressure should be
defined at the scale of the microemulsion, on the basis of a free energy
which includes the macroscopic behavior reflected in the entropy of dispersion of the globules in the continuous phase. An individual globule
does not sense the thermodynamic pressure and the entropy of dispersion, in
the same way as a single molecule in a gas does not feel the pressure and
entropy of the system, which characterize a large number of molecules.
Being a macroscopic body it has, however, its own pressure p_2 and senses
the pressure p_1 from the space among the globules, in its vicinity. The
latter does not include the effect of the collective behavior of the globules. Only a "macroscopic" part of the microemulsion, which is large
compared to the size of a globule can sense the thermodynamic pressure p.
The pressure p_2 inside the globules and the pressure p_1 near the globule in
the continuous phase will be called micropressures to emphasize that they
are defined on the microscale of the globules and to contrast them to the
thermodynamic (macro)pressure p which is defined on the scale of the entire
microemulsion. Consequently, the micropressures are defined on the basis
of a free energy which does not include the entropy of dispersion of the
globules in the continuous phase (an effect which manifests itself at a
scale of the order of the microemulsion), while the thermodynamic pressure
is defined on the basis of a free energy which includes that effect. It is
important to stress that p_2 and p_1 are real pressures on the scale of the
globules, and that the pressure p is a real pressure on a scale that is
large compared to that of the globules. The thermodynamic pressure is
equal to the pressure p_1 in the space among the globules plus an osmotic
contribution due to the entropy of dispersion of the globules in the continuous phase. Because of mechanical equilibrium, the thermodynamic pressure equals the pressure of the environment (external pressure).

Let us now consider one globule at the interface between microemulsion
and environment. As long as $p_2 < p$, the globule will remain in the microemulsion. For $p_2 > p$, the globule will disappear from the continuous
phase, giving way to separate oil and water phases. An excess dispersed
phase in equilibrium with the microemulsion will therefore form when $p_2 =
p$. Similarly, one can conclude that a third phase, the excess continuous
phase, forms when $p_1 = p$. Hence, a three-phase system composed of both
excess phases and a microemulsion arises when $p_2 = p_1 = p$. The equality p_2

$= p_1$ is not, however, compatible with the existence of spherical globules, because under such conditions, the interface becomes unstable to thermal perturbations. Thus, the change in structure observed experimentally near the transition from a microemulsion in equilibrium with an excess dispersed phase to one coexisting with both excess phases is a result of the instability of the spherical interface of the globules. The micropressures p_2 and p_1 can become equal without any phase separation, i.e., without their common value being equal to p. An instability of the spherical interface can therefore also occur in some single phase microemulsions. The above simple considerations already explain the main features concerning microemulsions, namely the formation of two and three phases and the change in structure that arises near the transition from two to three phases.

III. SINGLE PHASE MICROEMULSIONS

III.1 Thermodynamic Equations

The microemulsion is assumed to contain spherical globules of a single size. Their dispersion in the continuous phase is accompanied by an increase in the entropy of the system and we denote by Δf the corresponding free energy change per unit volume of microemulsion. For reasons suggested in the previous section, the Helmholtz free energy f per unit volume of microemulsion is written as

$$f = f_o + \Delta f, \tag{1}$$

where the free energy f_o is expressed in terms of the micropressures p_2 and p_1.

For curved interfaces, characterized by the principal curvatures c_1 and c_2, the Gibbs phenomenological theory leads to the following expression for f_o [17-19]:

$$df_o = \gamma dA + C_1 dc_1 + C_2 dc_2 + \Sigma \, \mu_i dn_i - p_2 d\phi - p_1 d(1-\phi), \tag{2}$$

where γ is a generalized interfacial tension, C_1 and C_2 are bending stresses associated with the curvatures c_1 and c_2, respectively, A is the interfacial area between the two media of the microemulsion per unit volume, μ and n are the chemical potentials per molecule and the number of molecules of species i, respectively, and ϕ is the volume fraction of the dispersed phase. The actual physical surface of the globules (to the extent it can be defined) of radius r is selected as the Gibbs dividing surface.

Some comments concerning γ are necessary. The interfacial tension γ, which is defined as the variation of the free energy f_o with area at constant ϕ, n and T, also includes those interactions between globules, such as the van der Waals, double layer and hydration forces, which change with surface area A. However, the virtual change in A at constant ϕ, which is used to define γ, changes also the shortest distance, 2h, between the surfaces of two neighboring globules and hence the spatial distribution of the globules. Therefore, γ includes not only the effect of the above interactions at a given distance 2h, but also the effect of the work done against these forces, as h changes. Denoting by τ the force per unit area and considering this force positive for repulsion, the change 2dh produces, per unit volume of microemulsion, the work $A\tau dh$. A more familiar interfacial tension γ_o, from which the effect of the above variation is eliminated, can be defined as

$$\gamma dA = \gamma_h \, dA - A\tau \left(\frac{\partial h}{\partial A} \right)_{\phi, n_i, T} dA. \tag{3}$$

It is shown later in the paper that

$$\tau = - \left(\frac{\partial \gamma_h}{\partial h} \right)_{\mu_i, T} \tag{4}$$

hence that

$$\gamma_h = \gamma_\infty + {}_h\!\int^\infty \tau dh, \tag{5}$$

where γ_∞ represents the interfacial tension when the distance between the globules is very large.

For spherical globules, $c_1 = c_2$ and $C_1 = C_2 = C/2$, and Eq. (2) becomes:

$$df_o = \gamma dA + Cd(1/r) + \Sigma \, \mu_i dn_i - p_2 d\phi - p_1 d(1-\phi), \tag{6}$$

which, combined with Eq. (1), leads to

$$df = \gamma dA + Cd(1/r) + \Sigma \mu_i dn_i - p_2 d\phi - p_1 d(1-\phi) + d\Delta f. \tag{7}$$

The equilibrium state of a microemulsion is completely determined by n_i, the temperature T and the external pressure p. The values of r and ϕ will therefore emerge from the condition that the microemulsion be in internal equilibrium, i.e., f be a minimum with respect to r and ϕ (or A and ϕ). One thus obtains

$$\gamma = - \left(\frac{\partial \Delta f}{\partial A} \right)_\phi - C \left(\frac{d(1/r)}{dA} \right)_\phi \qquad \text{(constant } n_i \text{ and } T) \tag{8}$$

and

$$p_2 - p_1 = \left(\frac{\partial \Delta f}{\partial \phi} \right)_A + C \left(\frac{d(1/r)}{d\phi} \right)_A \qquad \text{(constant } n_i \text{ and } T). \tag{9}$$

Because for spherical globules, A and r are related via the equation

$$A = 3\phi/r, \tag{10}$$

Eqs. (8) and (9) become

$$\gamma = \frac{r^2}{3\phi} \left(\frac{\partial \Delta f}{\partial r} \right)_\phi - \frac{C}{3\phi} \tag{11}$$

and (since $\left(\frac{\partial \Delta f}{\partial \phi} \right)_A = \left(\frac{\partial \Delta f}{\partial r} \right)_\phi \left(\frac{dr}{d\phi} \right)_A + \left(\frac{\partial \Delta f}{\partial \phi} \right)_r$)

$$p_2 - p_1 = \left(\frac{\partial \Delta f}{\partial \phi} \right)_r + \frac{r}{\phi} \left(\frac{\partial \Delta f}{\partial r} \right)_\phi - \frac{C}{r\phi} . \tag{12}$$

A second relationship between p_2 and p_1 is provided by the mechanical equilibrium condition between the microemulsion and environment. The variation of the total free energy F of the microemulsion of volume V can be written in the form

$$dF = \gamma d(AV) + CVd(1/r) + \Sigma \, \mu_i dN_i$$

$$- p_2 d(V\phi) - p_1 d(V(1-\phi)) + d(V\Delta f), \tag{13}$$

where N_i is the number of molecules of species i in the entire micro-emulsion. Considering a virtual variation dV of the volume V at constant T, p and N , the mechanical condition of equilibrium with the environment yields

$$\gamma d(AV) + VCd(1/r) - p_2 d(V\phi) - p_1 d(V(1-\phi)) + d(V\Delta f)$$

$$- pdV_e = 0, \tag{14}$$

where $dV_e = -dV$ is the variation of the volume of the environment. Combining Eqs. (11, (12), and (14), one obtains

$$\frac{3\phi}{r}\,\gamma - (p_2 - p_1)\phi + (p - p_1) + \Delta f = 0. \tag{15}$$

The system of Eqs. (12) and (15) leads (after the interfacial tension γ is eliminated with the help of Eq. (11)) to the following expressions for p_2 and p_1 :

$$p_2 = p + \Delta f + (1 - \phi)\left(\frac{\partial \Delta f}{\partial \phi}\right)_r - \frac{C}{\phi r} + \frac{r}{\phi}\left(\frac{\partial \Delta f}{\partial r}\right)_\phi \tag{16}$$

and

$$p_1 = p + \Delta f - \phi\left(\frac{\partial \Delta f}{\partial \phi}\right)_r. \tag{17}$$

Eqs. (16) and (17) relate the micropressures p_2 and p_1 to the thermodynamic pressure p, the bending stress C, and the free energy Δf due to the dispersion of the globules in the continuous phase and its derivatives.

Eq. (11) provides an expression for the calculation of the radius. Of course, expressions for γ, C and Δf in terms of r and ϕ are needed in order to perform such a calculation. Expressions for Δf are available[20,21] and will be used in the following section. One may note that γ (and also C which is related to γ) depends on r because:

(1)   its value is determined by the adsorption of the surfactant and co-surfactant upon the internal interface of the microemulsion and hence by their concentrations in the bulk, which, in turn, depend upon the adsorption area $3\phi/r$, and

(2)   there is also a curvature effect on γ (which is discussed later in the paper).

Eq. (12) constitutes a generalized Laplace equation. It can be re-written in the more revealing form

$$p_2 - p_1 = \frac{2\gamma}{r} - \frac{C}{3\phi r} + \frac{r}{3\phi}\left(\frac{\partial \Delta f}{\partial r}\right)_m, \tag{18}$$

where $m = \phi / \frac{4}{3}\pi r^3$ is the number of globules per unit volume of microemulsion and

$$\left(\frac{\partial \Delta f}{\partial r}\right)_m = \left(\frac{\partial \Delta f}{\partial r}\right)_\phi + \frac{3\phi}{r}\left(\frac{\partial \Delta f}{\partial \phi}\right)_r.$$

It is instructive to compare Eq. (18) with the Laplace equation which was derived for a single liquid droplet surrounded by its vapors[18]. In the latter case, the right hand side contains only the first two terms, and, since the mechanical equilibrium condition demands $p_2 > p_1$, their difference must be positive. The existence of the third term $\frac{r}{3\phi}(\frac{\partial \Delta f}{\partial r})_m$ in Eq. (18), which is always positive, permits the mechanical equilibrium condition to be satisfied even when the difference of the first two terms is negative. The term $\frac{r}{3\phi}(\frac{\partial \Delta f}{\partial r})_m$ is positive because the increased volume exclusion, which arises when the radius r is increased at constant number m of globules per unit volume, decreases the disorder. This diminishes the entropy of dispersion of the globules in the continuous phase and thus increases Δf. It will be shown later in the paper, that while in micro-emulsions the difference $(2\gamma/r) - (C/3\phi r)$ is, in general, negative, the positive value of $\frac{r}{3\phi}(\frac{\partial \Delta f}{\partial r})_m$ more than compensates for this negative difference.

III.2 **Instability of the Spherical Interface**

As already noted, a particular globule, which has inside the pressure p_2, senses outside the micropressure p_1. The spherical shape of the globules is stable to thermal perturbations if $p_2 > p_1$ and becomes unstable in the vicinity of the point where

$$p_2 - p_1 = 0. \tag{19}$$

Employing Eq. (12), condition (19) becomes

$$\left(\frac{\partial \Delta f}{\partial \phi} \right)_r + \frac{r}{\phi} \left(\frac{\partial \Delta f}{\partial r} \right)_\phi - \frac{C}{r\phi} = 0. \tag{20}$$

Eqs. (11) and (20) permit one to determine the values of r and ϕ in the neighborhood of which the spherical interface of the globules is no longer stable to thermal perturbations. Of course, expressions for γ, C and Δf are necessary to perform such calculations. Here, we will use only an expression for Δf and calculate the values of γ and C (as a function of r and ϕ) for which such an instability arises.

A lattice model was employed to derive upper and lower bounds for Δf[20]; now, an alternative expression, which is based on the Carnahan-Starling approximation for hard spheres, will be used[21]:

$$\Delta f = - \frac{3\phi kT}{4\pi r^3} \left[1 - \ln\phi - \phi \frac{4 - 3\phi}{(1 - \phi)^2} + \ln\left(\frac{4\pi r^3}{3v_c} \right) \right], \tag{21}$$

where k is the Boltzmann constant, T is the absolute temperature and v_c is the molecular volume of the solvent. From Eqs. (11) and (20) one obtains

$$\gamma = - \frac{r}{3} \left(\frac{\partial \Delta f}{\partial \phi} \right)_r \tag{22}$$

and

$$C = r^2 \left(\frac{\partial \Delta f}{\partial r} \right)_\phi + \phi r \left(\frac{\partial \Delta f}{\partial \phi} \right)_r, \tag{23}$$

which, when combined with Eq. (21), lead to

$$\gamma = \frac{kT}{4\pi r^2} \left[-\ln\phi + \ln\left(\frac{4\pi r^3}{3v_c}\right) - \frac{\phi(8 - 9\phi + 3\phi^2)}{(1 - \phi)^3} \right] \tag{24}$$

and

$$C = \frac{3kT\phi}{4\pi r^2} \left[2\ln\left(\frac{4\pi r^3}{3v_c}\right) - 2\ln\phi + \frac{12\phi^2 - 6\phi^3 - 4\phi}{(1 - \phi)^3} \right]. \tag{25}$$

Table 1 provides some values at 25°C for γ and C as a function of ϕ and r and shows that the spherical shape becomes unstable if the bending stress C is sufficiently large compared to the interfacial tension γ. A similar conclusion can be drawn more directly from Eq. (18), which, when combined with condition (19), becomes

$$\frac{2\gamma}{r} - \frac{C}{3\phi r} + \frac{r}{3\phi} \left(\frac{\partial \Delta f}{\partial r}\right)_m = 0. \tag{26}$$

Because, as already noted, $\left(\frac{\partial \Delta f}{\partial r}\right)_m$ is a positive quantity, a necessary condition for the spherical interface to become unstable is

$$\frac{C}{3\phi} > 2\gamma. \tag{27}$$

As shown later, C is related to γ via the expression

$$\frac{\partial \gamma}{\partial r} = -\frac{C}{3\phi r} \quad \text{(constant } \mu_i \text{ and T)}.$$

As a result, the necessary condition for instability becomes

$$2\gamma + r\left(\frac{\partial \gamma}{\partial r}\right)_{\mu_i} < 0. \tag{27a}$$

Since $\gamma > 0$, the inequality (27a) is satisfied only if $\left(\frac{\partial \gamma}{\partial r}\right)_{\mu_i}$ is sufficiently negative (C sufficiently positive).

An equation for γ in the presence of surfactant, cosurfactant and salt was recently established[22], though for a planar interface. While its extension to a curved interface can be carried out (see section V.2) by using the expression for the bending energy proposed by Helfrich[23], the latter expression is not entirely satisfactory. For this reason, the calculations concerning the values of ϕ and r at the instability point are not included here. However, the above considerations do explain the phenomena that have been detected by means of the NMR experiments[13]. In the vicinity of the point where $p_2 = p_1$, the spherical interface becomes unstable to thermal perturbations. This means that the interface does not maintain its identity, the dispersed units break up and coalesce and the system acquires the chaotic characteristics of the turbulence. As a result, the apparent self diffusion coefficients of the various components are tremendously increased.

Table 1. The Occurrence of the Instability of the Spherical Shape in Single Phase Microemulsions

ϕ	$r \cdot 10^7$ (cm)	γ (dyne/cm)	C (dyne/cm)
0.2		0.27	0.417
	3		
0.3		0.167	0.595
0.2		0.087	0.127
	6		
0.3		0.061	0.183
0.2		0.043	0.062
	9		
0.3		0.032	0.09
0.2		0.018	0.025
	15		
0.3		0.014	0.037
0.2		0.008	0.011
	24		
0.3		0.006	0.016
0.2		0.003	0.004
	42		
0.3		0.0023	0.006
0.2		0.0015	0.002
	60		
0.3		0.0012	0.003
0.2		0.0009	0.0013
	78		
0.3		0.0008	0.002

IV. TWO PHASE SYSTEMS

IV.1 Thermodynamic Equations for the Phase Equilibrium

The chemical potential in the microemulsion phase is defined as $\frac{\partial f}{\partial n_i}$ at constant T, A, ϕ and n_j (with $j \neq i$). Because Δf depends only on ϕ and r, it follows that $\frac{\partial f}{\partial n_i}$ is equal to the chemical potential μ_i, which was defined in Eq. (2) on the basis of f_o. Assuming the concentrations of various components to be the same in the globules and the excess dispersed phase, the equality of the chemical potentials is equivalent to the equality of the pressures. The chemical potentials μ_i in the dispersed phase are expressed in Eq. (2) at the micropressure p_2, while in the excess dispersed phase the pressure is equal to the external pressure p. Therefore, a microemulsion is in equilibrium with an excess dispersed phase when

$$p_2 = p, \tag{28}$$

which together with Eq. (16) leads to

$$\Delta f + (1 - \phi)\left(\frac{\partial \Delta f}{\partial \phi} \right)_r - \frac{C}{\phi r} + \frac{r}{\phi}\left(\frac{\partial \Delta f}{\partial r} \right)_\phi = 0. \tag{29}$$

Eqs. (11) and (29) can provide the values of ϕ and r at which an excess dispersed phase begins to form. Any additional excess dispersed phase, having the same composition as the dispersed phase, will not change r and ϕ. Eqs. (11) and (29) will now be used to express γ and C in terms of r and ϕ. In this respect, it is convenient to rewrite them in the form:

$$\gamma = - \frac{1}{3} r\Delta f - \frac{1}{3} r(1 - \phi)\left(\frac{\partial \Delta f}{\partial \phi} \right)_r \tag{30}$$

and

$$C = r\phi\Delta f + r\phi(1 - \phi)\left(\frac{\partial \Delta f}{\partial \phi} \right)_r + r^2\left(\frac{\partial \Delta f}{\partial r} \right)_\phi . \tag{31}$$

Employing Eq. (21) for Δf, one obtains

$$\gamma = \frac{kT}{4\pi r^2} \left[\ln\left(\frac{4\pi r^3}{3v_c \phi} \right) - \frac{8\phi - 5\phi^2}{(1 - \phi)^2} + \phi \right] \tag{32}$$

and

$$C = \frac{3\phi kT}{4\pi r^2} \left[-2\ln\phi + 2\ln\left(\frac{4\pi r^3}{3v_c} \right) + \frac{6\phi^2 - 5\phi - \phi^3}{(1 - \phi)^2} \right]. \tag{33}$$

Some numerical results are listed in Table 2. They show that the interfacial tension γ at the "actual" surface of the globules is indeed very small, between about 10^{-1} and 10^{-2} dynes/cm, and that its value decreases with increasing values of r.

The use of appropriate expressions for γ and C in Eqs. (32) and (33) can provide the values of r and ϕ at the transition from a single microemulsion phase to one coexisting with an excess dispersed phase. As already noted, the expression available for the bending energy of the interface is not yet satisfactory. For this reason, we delay the presentation of the results obtained regarding the values of r and ϕ at the point of transition.

IV.2 The Stability of the Spherical Interface

Eq. (28) shows that when a microemulsion is in equilibrium with an excess dispersed phase $p_2 = p$. Let us prove that $p_2 > p_1$ under such circumstances and hence that the spherical interface can be stable to thermal perturbations. Eq. (12) can be rewritten in the form

$$p_2 - p_1 = \frac{2\gamma'}{r}$$

where, using Eq. (21) for Δf and Eq. (33) for C,

$$\gamma' = \frac{3\phi kT}{8\pi r^2} (1 + \phi + \phi^2 - \phi^3)/(1 - \phi)^3 . \tag{34}$$

Eq. (34) shows that $\gamma' > 0$ for all positive, finite values of r. The interfacial tension γ' becomes, however, extremely small in the neighborhood of the transition from two to three phases (see section VI.1). As a result, large fluctuations are expected to occur in that region of the two phase system. It is also instructive to note that the numerical calculations listed in Table 2 show that the difference $\frac{2\gamma}{r} - \frac{C}{3\phi r}$ is negative, but that the third term $\frac{r}{3\phi}\left(\frac{\partial \Delta f}{\partial r} \right)_m$ in Eq. (18) is, indeed, a positive quantity which

more than compensates for the negative difference of the previous two
terms.

V. THE INTERFACIAL TENSION BETWEEN MICROEMULSION AND EXCESS DISPERSED PHASE

V.I. <u>A Generalized Gibbs Adsorption Equation</u>

Let us integrate Eq. (7) at constant $1/r$, γ, μ_i, p_2 and p_1. One
obtains

$$f = A\gamma + \Sigma\, n_i\mu_i - p_2\phi - p_1(1-\phi) + \Delta f, \tag{35}$$

which when differentiated leads to

$$df = Ad\gamma + \gamma dA + \Sigma\, n_i d\mu_i + \Sigma\, \mu_i dn_i - p_2 d\phi - p_1 d(1-\phi) - \phi dp_2$$
$$- (1-\phi)dp_1 + d\Delta f. \tag{36}$$

The Gibbs–Duhem equations for the continuous and dispersed phases have the
forms

$$\Sigma\, n_i'' d\mu_i - \phi dp_2 = 0 \tag{37a}$$

and

$$\Sigma\, n_i' d\mu_i - (1-\phi)dp_1 = 0, \tag{37b}$$

where the superscripts prime and double prime refer to the continuous and
dispersed phases, respectively. Subtracting (37a) and (37b) from Eq. (36)
and using Eq. (7), one finally obtains:

$$d\gamma + \Sigma\Gamma_i d\mu_i = \frac{rC}{3\phi}\, d(1/r), \tag{38}$$

where $\Gamma_i = [n_i - (n_i' + n_i'')]/A$ is the surface excess of component i.

The chemical potentials of the surfactants are the same on the surface
of the globules and at the interface between the two phases. Therefore,
the interfacial tension γ at the "actual" surface of the globules can be
related to the interfacial tension σ between microemulsion and excess
dispersed phase by integrating Eq. (38) at constant chemical potentials, to
obtain:

$$\gamma(r \to \infty) - \gamma = {}_r\!\int^\infty \frac{rC}{3\phi}\, d(1/r) \equiv - {}_r\!\int^\infty \frac{C}{3\phi r}\, dr. \tag{39}$$

$\gamma(r \to \infty)$ is identified as σ, by neglecting the effect of the presence of
the globules of finite size r in the neighborhood of the planar interface.
This consideration is justified, because experiment shows that the interfacial tension between excess phase and microemulsion is almost equal to
that between the continuous medium of microemulsion and the excess phase
[27].

If one considers $C > 0$, as suggested by Table 2, Eq. (39) shows that

$$\sigma < \gamma. \tag{40}$$

Table 2A. Microemulsions in Equilibrium with Excess Dispersed Phase

$$\phi = 0.2$$

	3	6	9	15	24
$r \cdot 10^7$ cm	3	6	9	15	24
C dyne/cm	0.406	0.124	0.061	0.0250	0.011
γ dyne/cm	0.287	0.091	0.045	0.019	0.008
γ' dyne/cm	0.026	0.007	0.003	0.001	0.0004
$\dfrac{2\gamma}{r} - \dfrac{C}{3\phi r}$ dyne/cm^2	-339,000	-42,300	-12,500	-2,710	-661
$\dfrac{r}{3\phi}\left(\dfrac{\partial \Delta f}{\partial r}\right)_m$ dyne/cm^2	515,000	64,300	19,100	4,120	1,010
$\dfrac{\dfrac{2\gamma}{r} - \dfrac{C}{3\phi r}}{\dfrac{r}{3\phi}\left(\dfrac{\partial \Delta f}{\partial r}\right)_m}$	-0.66	-0.66	-0.66	-0.66	-0.66
$\dfrac{\dfrac{2\gamma}{r} + \dfrac{r}{3\phi}\left(\dfrac{\partial \Delta f}{\partial r}\right)_m}{\dfrac{C}{3\phi r}}$	1.07	1.064	1.06	1.05	1.046
$\dfrac{C}{3\phi\gamma}$	2.35	2.28	2.25	2.22	2.19
$\dfrac{r^2}{3\phi}\left(\dfrac{\partial \Delta f}{\partial r}\right)_\phi$ dyne/cm	0.964	0.298	0.15	0.06	0.026

V.2 A Relation for the Interfacial Tension σ

An equation relating σ to γ can also be established on the basis of the following intuitive considerations. The interfacial tension γ between the continuous and dispersed media of a microemulsion is determined by the adsorption of surfactant and cosurfactant from the bulk upon the curved interface separating the two phases. The interfacial tension σ between a microemulsion and the excess dispersed phase is similarly determined by the adsorption of the surfactant and cosurfactant upon the "planar interface" separating the two. Since the concentrations of surfactant and cosurfactant are the same in the globules and in the excess dispersed phase, γ and σ should differ only because the adsorbed layer of surfactant and cosurfactant is bent to a relatively large curvature in the former case. Denoting by ε the bending energy per unit area, one can therefore write that

$$\sigma = \gamma - \varepsilon. \tag{41}$$

Obviously, from Eqs. (39) and (41),

$$\varepsilon = \int_r^\infty \frac{C}{3\phi r}\, dr. \tag{42}$$

Table 2B. Microemulsions in Equilibrium with Excess Dispersed Phase

	$\phi = 0.3$				
$r.10^7$ cm	3	6	9	15	24
C dyne/cm	0.556	0.173	0.086	0.035	0.015
γ dyne/cm	0.211	0.072	0.037	0.016	0.007
γ' dyne/cm	0.065	0.016	0.007	0.003	0.001
$\dfrac{2\gamma}{r} - \dfrac{C}{3\phi r}$ dyne/cm^2	$-650,000$	$-81,500$	$-24,200$	$-5,200$	$-1,270$
$\dfrac{r}{3\phi}\left(\dfrac{\partial \Delta f}{\partial r}\right)_m$ dyne/cm^2	1090,00	136,000	40,300	8,710	2,130
$\dfrac{\dfrac{2\gamma}{r} - \dfrac{C}{3\phi r}}{\dfrac{r}{3\phi}\left(\dfrac{\partial \Delta f}{\partial r}\right)_m}$	-0.6	-0.6	-0.6	-0.6	-0.6
$\dfrac{\dfrac{2\gamma}{r} + \dfrac{r}{3\phi}\left(\dfrac{\partial \Delta f}{\partial r}\right)_m}{\dfrac{C}{3\phi r}}$	1.21	1.17	1.15	1.14	1.12
$\dfrac{C}{3\phi\gamma}$	2.92	2.68	2.60	2.50	2.45
$\dfrac{r^2}{3\phi}\left(\dfrac{\partial \Delta f}{\partial r}\right)_\phi$ dyne/cm	0.83	0.264	0.132	0.054	0.024

The bending energy ε can be evaluated for a spherical globule, in terms of the elastic energy of the bent layer, from the expression[23-25].

$$\varepsilon = 2K\left(\frac{1}{r} - \frac{1}{R}\right)^2, \tag{43}$$

where the constant K is of the order of 0.1 ev[24,25] and R is the so called natural radius of curvature of the surfactant layer. The natural radius was considered in the previous evaluations of ε as an empirical parameter. One may note that the concept of natural radius was introduced to explain the natural tendency of some surfactants to lead to either oil in water (o/w) or water in oil (w/o) emulsions, hence to bend either towards one or the other of the two immiscible liquids. Because microemulsions are thermodynamically stable dispersions, it is clear that one kind of dispersion is preferred to the other when its free energy is smaller. The radius of the globules as well as the kind of dispersion

should therefore be predicted by the thermodynamic equilibrium condition.
One can still think of the natural radius as a quantity which is useful in
evaluating the bending energy, but a less hazy physical meaning should be
attached to it. The following considerations regarding single droplets of
liquid or radius r, immersed in their vapors, are suggestive from this
point of view. In the latter case, one can define a surface of tension,
characterized by a radius r', for which the bending stress is zero and
therefore the Laplace equation $p_2 - p_1 = 2\gamma'/r'$ is valid. The elastic
energy per unit area due to bending is expected to be greater when $|(1/r) -
(1/r')|$ is larger. In the case of a microemulsion, the effect of the
conventional interfacial bending $-C/3\phi r$ is decreased by the additional term
$(r/3\phi)(\frac{\partial \Delta f}{\partial r})_m$ (see Eq. (18)), which as already explained is always a positive quantity. One can again define a surface of tension, characterized by
the radius R_o and volume fraction ϕ_o, for which the Laplace equation holds

$$P_2 - P_1 = \frac{2\gamma}{r} - \frac{C}{3\phi r} + \frac{r}{3\phi} \left(\frac{\partial \Delta f}{\partial r} \right)_m \equiv \frac{2\gamma'''}{R_o} \; , \tag{44}$$

where R_o and ϕ_o are related via the expression

$$\frac{C(R_o,\phi_o)}{3\phi_o R_o} = \frac{R_o}{3\phi_o} \left(\frac{\partial \Delta f}{\partial r} \right)_m'. \tag{45}$$

The superscript prime indicates that r and ϕ should be replaced in the
derivative by R_o and ϕ_o. The energy of bending per unit area is expected
to be larger when $|(1/r) - (1/R_o)|$ is larger. Comparing with Eq. (43), it
appears reasonable to equate the natural radius R to the radius of the
surface of tension. Employing Eq. (21) for Δf, Eq. (45) becomes

$$2\ln\left(\frac{4\pi R^3}{3v_c\phi_o} \right) = \frac{17\phi_o - 24\phi_o^2 + \phi_o^3}{(1 - \phi_o)^2} + \frac{6\phi_o^2(4 - 3\phi_o)}{(1 - \phi_o)^3} \; . \tag{46}$$

Since the number of globules should be the same,

$$(\phi_o/R^3) = (\phi/r^3). \tag{47}$$

Consequently, the natural radius R is a function of r and ϕ, as it should
be. Eq. (43) for ε is not completely satisfactory. The presence of alcohol
in the interfacial layer makes the latter more fluid and its description as
an elastic body less adequate. In addition, Eq. (43) disregards a number
of effects which might be significant, particularly because of the low
interfacial tensions involved. There is, for instance, a curvature contribution due to the double layer, since the double layer affects the interfacial tension of a planar and curved interface differently. A similar
comment can be made concerning the effect of hydration forces. The above
effects are probably relevant at sufficiently low ionic strengths. At high
ionic strengths, situations which are frequently encountered in microemulsions, the electric field is completely shielded and the ions no longer
form "independent" units of hydrated ions, but compete for water. The
double layer and the conventional hydration forces are thus no longer
relevant in these cases. Instead, the presence of an oil-water interface,
the competition of ions for water and the interactions between them, very
likely generate a liquid crystal like structure in the vicinity of the
interface, which probably increases the rigidity of the latter. This
provides some support to the elastic assumption involved in Eq. (43).
While this equation is not satisfactory, we still use it, because a better
one is not yet available.

16

Combining Eqs. (41), (43) and (32), the following equation is obtained for σ

$$\sigma = \frac{kT}{4\pi r^2} \left[\ln\left(\frac{4\pi r^3}{3v_c \phi} \right) - \frac{8\phi - 5\phi^2}{(1 - \phi)^2} + \phi \right] - 2K\left(\frac{1}{r} - \frac{1}{R} \right)^2, \tag{48}$$

where R is given by Eq. (46).

It is important to note that γ accounts for the effect of both the entropy of dispersion and curvature. Considering $K = \alpha kT$, where α is a constant, Eq. (48) shows that $(\sigma r^2)/(kT)$ is not a constant, but has a weak dependence on the radius r. An equation of a similar form, relating σ to the entropy of dispersion of the globules in the continuous phase, was first established in Ref. 26.

Numerical calculations have been carried out to estimate γ and σ as a function of r and ϕ for an oil in water microemulsion in equilibrium with excess oil and the results are summarized in Table 3. The values obtained are of the right order of magnitude, since experiment[27] provides for σ values between 10^{-1} and 10^{-2} dynes/cm. A detailed comparison with experiment is not yet possible because r and ϕ have not been simultaneously determined. Because Eq. (43) is approximate, and, in addition, K is very likely dependent upon the ratio of surfactant to cosurfactant, one cannot expect quantitative agreement between theory and experiment. We do, however, expect the interfacial tension γ given by Eq. (32) to provide an upper bound for σ, the ratio between the two being not larger than about 2.

V.3 A Relation for the Bending Stress C

Eqs. (42) and (43) provide the following expression for C

$$2K\left(\frac{1}{r} - \frac{1}{R} \right)^2 = {}_r\!\int^\infty (C/3\phi r)\,dr, \tag{49}$$

from where one obtains

$$\frac{C}{3\phi r} = 4K\left(\frac{1}{r} - \frac{1}{R} \right)\left(\frac{1}{r^2} - \frac{1}{R^2} \frac{\partial R}{\partial r} \right). \tag{50}$$

Combining Eqs. (33) and (50), yields

$$\frac{kT}{4\pi r^3} \left[2\ln\left(\frac{4\pi r^3}{3v_c \phi} \right) + \frac{6\phi^2 - 5\phi - \phi^3}{(1 - \phi)^2} \right] = 4K\left(\frac{1}{r} - \frac{1}{R} \right)\left(\frac{1}{r^2} - \frac{1}{R^2} \frac{\partial R}{\partial r} \right). \tag{51}$$

The values calculated for r by using Eq. (51) and taking $K = 0.16$ eV are 12 Å for $\phi = 0.1$ and 9 Å for $\phi = 0.15$, which are too small. This indicates again that Eq. (43) for ε is not entirely satisfactory.

V.4 Experimental Method for the Determination of the Bending Energy ε

Equation (41) suggests a simple experimental method to evaluate ε. Indeed, σ can be measured experimentally, while the values of γ can be calculated by means of Eq. (32) for the experimentally measured values of r and ϕ. The results could be correlated by observing that dimensional analysis suggests that $\frac{\varepsilon r^2}{\kappa T}$ is a function of r/R and ϕ.

Table 3. The Interfacial Tensions and the Natural Radius R Against r and ϕ for a Microemulsion in Equilibrium with Excess Dispersed Phase

ϕ	$10^7 r$(cm)	$10^7 R$(cm)	γ(dyne/cm)	σ(dyne/cm)
0.1	3	3.84	0.353	0.097
	6	7.76	0.107	0.039
	9	11.68	0.053	0.021
	12	15.61	0.032	0.014
	15	19.54	0.021	0.010
	18	23.48	0.015	0.007
	21	27.41	0.012	0.006
0.15	3	3.74	0.319	0.109
	6	7.59	0.099	0.040
	9	11.44	0.049	0.022
	12	15.31	0.029	0.014
	15	19.18	0.020	0.010
	18	23.05	0.014	0.007
	21	26.93	0.011	0.006
0.20	3	3.61	0.286	0.134
	6	7.35	0.090	0.045
	9	11.12	0.045	0.024
	12	14.89	0.027	0.015
	15	18.68	0.018	0.010
	18	22.47	0.013	0.008
	21	26.27	0.010	0.006
0.25	3	3.45	0.250	0.160
	6	7.06	0.081	0.052
	9	10.70	0.041	0.026
	12	14.35	0.025	0.016
	15	18.03	0.017	0.011
	18	21.70	0.012	0.008
	21	25.39	0.009	0.006

VI. THE MIDDLE PHASE MICROEMULSION

VI.1 The Origin of the Instability of the Spherical Interface

If in addition to $p_2 = p$,

$$p_1 = p,$$

a middle phase microemulsion will coexist with both excess phases. However, the equality $p_2 = p_1$ can be satisfied only if $r \to \infty$. Indeed, Eq. (34) shows that the radius r has to diverge in order to have $p_2 = p_1 = p$. The interfacial tension γ and the bending stress C tend to zero proportionally to the second power of the curvature. In other words, the equalities $p_2 = p_1 = p$ are not compatible with a finite curvature and a change in the structure of the microemulsion must occur. A planar structure of alternate oil and water layers will have a zero curvature, but will be also unstable. Indeed, the entropy of dispersion of such an ordered structure is zero, because the number of distinguishable configurations is equal to one. In this case,

$$df = \gamma dA + \Sigma \; \mu_i dn_i - p_2 d\phi - p_1 d(1-\phi). \tag{52}$$

The equilibrium condition leads to

$$\gamma = 0$$

and

$$p_2 = p_1$$

Because $\gamma = 0$, the planar interface is likely to be unstable to thermal perturbation (this point is discussed again later). Consequently, both the spherical and the planar interfaces are unstable in the middle phase microemulsion. This suggests that the stable interface has unsteady, oscillatory motions. It is, therefore, appropriate to define the average micropressures $\bar{p}_2$ and $\bar{p}_1$, where the bar indicates a temporal or spatial averaging. Intuition suggests to rewrite the thermodynamic equilibrium condition in the form

$$\bar{p}_2 = \bar{p}_1 = p. \tag{53}$$

The oscillating interfaces are compatible with this condition of equilibrium since the spatial (or temporal) average of the positive and negative deviations of $p_2 - p_1$ (caused by the oscillations) can cancel out. Eqs. (53) suggest to extrapolate the condition of zero curvature, which was obtained from the condition of thermodynamic equilibrium $p_2 = p_1 = p$ for spherical globules, to the entire middle phase domain, as the condition of zero average curvature. An oscillating interface can satisfy such a condition. Indeed, a multitude of globules whose interfaces oscillate can have positive curvatures in some regions of their surface and negative curvatures in other regions and can thus satisfy the condition of zero average curvature.

VI.2 The Structure of the Middle Phase Microemulsion

There ia an analogy between the state of the middle phase microemulsion and that of a liquid film which flows along a vertical wall. In the latter case, the planar interface is unstable to perturbations and an unsteady motion with chaotic characteristics develops[28,29]. The free interface does not disappear, but acquires complicated spatial and temporal nonperiodical oscillations. Similarly, the interface between the two media of the microemulsion does not disappear but oscillates in a disorderly way. A possible scenario is as follows. Near the transition point from two to three phases, the dispersion in the size distribution of the globules is likely to increase, but more importantly, their interface is in a state of unsteady motion of large amplitude. As a result, the globules can break up and coalesce, the layer of surfactants covering them as well as the molecules of the dispersed medium being continuously exchanged between the globules. As one advances in the three phase region, by increasing for instance the ionic strength, the volume fraction of the globules increases and a percolation threshold occurs as a result of transient interconnections between a large number of globules. These transient interconnections are facilitated by the oscillations and by the van der Waals interactions between globules and provide an additional pathway for rapid communication between them. While the van der Waals interactions between the various elements of the dispersed phase destabilize the internal interface, repulsive interactions which have a stabilizing effect are also probably involved, particularly when ϕ is sufficiently large. At relatively low salt concentrations, the ions are hydrated and each hydrated ion maintains its own identity. Because of the hydrated ions, repulsive hydration forces are generated between two approaching surfaces, in addition to the more conventional double layer forces. Various mechanisms have been suggested

for their origin [30-32]. However, in the middle phase microemulsions
which are of practical interest, the ionic strength is large and, therefore, the electric field is completely shielded. Both the free ions and
the ion pairs complete for the water molecules and one can no longer
identify hydrated ions as individual entities. Instead, the presence of
interactions among the species involved probably facilitate the organization of water and ions in a quasi-rigid structure (i.e., liquid crystal
like) structure. The presence of an oil-water interface modifies somewhat
this structure because of the negative surface excess of salt on such an
interface. When two such surfaces approach one another, the increase of
the free energy caused by the overlap of the liquid-crystal like layers
formed in their vicinity results in a repulsive, stabilizing force. (The
occurrence of aggregates of polyhedral droplets in concentrated emulsions
at high salt concentrations, with the drplets being separated by thin
planar films[33], involves a repulsive force which can probably be
explained in this manner). The possible existence of this repulsive force
was emphasized for two reasons:

(1)   the interface may become less unstable with increasing ionic strength,
      and increasing values of ϕ, and
(2)   near the phase inversion point ($\phi \approx 0.5$), the middle phase micro-
      emulsion may be similar to a kind of bicontinuous sponge (correspond-
      ing to a zero average curvature of the kind examined mathematically by
      Schwartz[34] and Neovius[35] and suggested by Scriven[36] for the
      description of the middle phase microemulsion).

Near the transition point from two to three phases, where ϕ is relatively
small, the oscillations of the interface of the globules ensure that the
condition of zero average curvature is satisfied. At relatively large
values of ϕ, in the neighborhood of the inversion point ($\phi \approx 0.5$), the
system may organize in a bicontinuous structure which can satisfy this
condition by itself. Such a possibility should not be excluded and the
repulsive force just identified may have the appropriate stabilizing effect
to ensure the formation of this "rigid" ordered sponge. It is, however,
more likely that a more random structure of the type suggested by Talmon
and Prager[37,38] is generated which exhibits oscillations and whose average curvature is zero. For volume fractions of oil greater than about 0.5,
the middle phase microemulsion has the tendency to transform into a water
in oil microemulsion. Near the transition point from three to two phases,
the volume fraction of water can be sufficiently small and the zero average
curvature can be again achieved via the oscillations of the internal interface alone.

VI.3   <u>The Phase Inversion</u>

     A middle phase microemulsion is a state between an oil in water micro-
emulsion in equilibrium with excess oil and a water in oil microemulsion in
equilibrium with excess water. The phase inversion process which takes
place in the middle phase microemulsion as the ionic strength increases is
caused by the salting out effect. The addition of salt has the following
effects: At low concentrations, in systems containing ionic surfactants,
it shields the electric field produced by the adsorption of the charged
surfactant molecules on the oil-water interface and thus facilitates an
even greater adsorption of surfactant. This reduces the interfacial
tension. At the large ionic strengths which exist in the middle phase
microemulsion, the electrical field is completely shielded and the double
layer effects are negligible. However, at large concentrations, salt has
another effect on both ionic and nonionic surfactants, because of the
extensive organization of water molecules by ions with relatively small
ionic radius, such as Na+. As a consequence, the interactions between
water and the hydrocarbon tails of the surfactant become even less favor-

able and the favorable interactions between the polar head groups and water
are diminished. The result is that the activity coefficient of the surfactant increases with the addition of salt, altering the equilibrium
distribution between water and oil phases. While the surfactant was in-
itially concentrated in water, it gradually transfers to the oil as water
becomes less favorable as a solvent. Throughout this salting out process,
the surfactant chemical potential rises as follows from the fact that its
concentration in oil (where the salt has no effect) increases. The chemical potential in the water also rises because the increase in the activity
coefficient with salt outweighs the effect of the fall in concentration.

The gradual transfer to oil of the surfactant affects the value of the
interfacial tension between the two media and changes the trend of the
dispersion from an oil in water to a water in oil microemulsion, because
the free energy becomes thus smaller. (It must be emphasized that the
middle phase microemulsion is not just an oil in water or a water in oil
dispersion, but has only some trend in a direction or the other). Indeed,
the free energy of the interface is smaller when the adsorption of the
surfactants on the internal interface takes place from a favorable environment for the surfactants. Consequently, the interfacial layer will be bent
so as to expose its larger of the two areas, towards the phase which is
more favorable to the surfactant (as much as this is compatible with the
other constraints acting on the system, such as oscillations and percolation).

The change in trend from an oil in water to a water in oil microemulsion occurs int he middle phase at about ϕ = 0.5 because at this point
the free energies of the two types of dispersions are equal. Indeed, one
can assume that the free energy is given by the expression

$$\overline{f} = \overline{A\gamma} + \Sigma\, n_i\overline{\mu}_i - \overline{p}_2\phi - \overline{p}_1(1-\phi) + \overline{\Delta f}, \tag{54}$$

where the bar indicates a spatial or temporal average. Assuming a random
geometry of interspersed oil and water domains generated by a Voronoi
tesselation, one obtains[37]

$$\overline{\Delta f} \sim kT\left(\frac{\sigma^*}{\phi(1-\phi)}\right)^3\left\{\ln\left(\frac{\sigma^*}{\phi(1-\phi)}\right)^3 + \phi\ln\phi + (1-\phi)\ln(1-\phi)\right\}, \tag{55}$$

where σ^* is proportional to the amount of surfactant involved. Because $\overline{p}_2$
= $\overline{p}_1$ = $\overline{p}$, the two entropically dominated free energies will be equal (or
nearly equal) when ϕ = 1-ϕ, hence when ϕ = 1/2. For this value of ϕ, $\overline{A\gamma}$
(which is very small) and $\Sigma\, n_i\overline{\mu}_i$ (which has a weak dependence on ϕ) are
also expected to be nearly the same in both types of dispersions.

The temperature or the number of oxyethylene groups of the nonionic
surfactant with polyoxyethylene head group have the same effects as the
salt because they affect the distribution of the surfactant between oil and
water in a similar way.

VI.4 The Interfacial Tension

As shown in a previous section, the interfacial tension σ between a
microemulsion and the excess dispersed phase is given by an expression of
the form

$$\frac{\sigma r^2}{kT} = \varphi(r/R, r^3/v_c, \phi), \tag{56}$$

where φ is a weak function of the radius r of the globules. Eq. (56) can be
established on the basis of dimensional analysis. The form of the function φ
can be obtained by comparison with Eq. (48). In the neighborhood of the

transition from two to three phases, Eq. (56) (which involves the assumption of spherical globules of a single size) is no longer valid, because the spherical interface becomes unstable to thermal perturbations. In the three phase region, there are two interfaces. In the vicinity of and after the transition point from two to three phases the continuous water phase is in contact with excess water and their interfacial tension is very low. The interfacial tension between microemulsion and excess dispersed phase is also small, but greater than the previous one. As soon as the instability of the internal interface sets in, the size of the globules is no longer the relevant length to be included in the expression of σ. The maximum extension L of a globule, which arises because of its oscillations, is now the relevant length, since it determines the excluded volume associated with an oscillating globule. In the case of spherical droplets, the excluded volume was determined by the radius r, and the interfacial tension γ was related to r via an expression of the form $\gamma \approx kT/r^2$. In the present case, one can write, by analogy, for the interfacial tension between microemulsion and excess oil, the expression

$$\sigma_{mo} = \frac{kT}{L^2} (1 - \phi).$$
(57)

The factor $1-\phi$ is included in Eq. (57) because the value of the interfacial tension σ_{mo} is determined by the contact between water and excess oil and $1-\phi$ represents the volume fraction of water. The fraction of the interface where the globules of oil contact the excess oil phase is likely to be smaller than ϕ for relatively small values of ϕ, and to approach the value of ϕ for larger values. Similarly, one can write, for the interfacial tension σ_{mw} between microemulsion and excess water, the expression

$$\sigma_{mw} = \frac{kT}{L^2} \phi,$$
(58)

which, as Eq. (57), is probably valid for sufficiently large values of ϕ. One may note that for $\phi = 1/2$, σ_{mo} becomes equal to σ_{mw}, as observed experimentally[27].

VII. ISOTROPIC AND BIREFRINGENT MICROEMULSIONS

VII.1 <u>Various Isotropic Microemulsions</u>

In many single phase microemulsions as well as in microemulsions coexisting with the excess dispersed phase, the globules are spherical and of almost uniform size. There are, however, single phase microemulsions in which the spherical interface is not stable to thermal perturbations. In the three phase systems, where a microemulsion coexists with both excess phases, the spherical interface is always unstable. The characteristics of the instability are different in the single phase microemulsions and in the middle phase microemulsions. While in the former case the spherical globules become unstable at a finite radius, in the latter case the instability is a result of the thermodynamic equilibrium condition between the three phases, which demands the average curvature of the internal interface to be zero. This explains the ill defined, fluctuating interfaces which are observed in the first case, and the more complex behavior similar (but not the same) to that which occurs near a critical point observed in the latter case.

One can also have two globular microemulsion phases in equilibrium. Assuming that the dispersed phase is the same in both microemulsions, considerations similar to those already employed in previous sections provide the conditions of equilibrium in the form

$$p_1' = p_1'' \tag{59}$$

and

$$p_2' = p_2'', \tag{60}$$

where the prime and double prime indicate the two microemulsion phases. In this case, there is a critical point at which the two microemulsions become identical and near the critical point strong fluctuations occur.

The two microemulsions can have different dispersed and continuous media. In such cases, the conditions of equilibrium are

$$p_1' = p_2 \tag{61a}$$

and

$$p_2' = p_1''. \tag{61b}$$

In addition, the two microemulsions could also coexist with one or both excess phases.

VII.2 Birefringent, Lamellar Microemulsions

In isotropic, random, globular microemulsions (with stable or unstable spherical shapes), the entropy of dispersion and the interfacial contributions to the free energy (such as the interfacial tension and the bending stress) play an important role in the thermodynamic stability of the system. Of course, the van der Waals interactions and other types of interactions can also have an effect. In fact, the occurrence of two microemulsion phases in equilibrium is possible only if the latter interactions do play a role. The ordered structure of lamellar microemulsions makes their entropy of dispersion extremely small. Therefore, the interaction forces and the interfacial tension may be in these cases primarily responsible for the thermodynamic stability. The scope of this section is to derive the thermodynamic equilibrium conditions for such cases.

The variation of the Helmholtz free energy F of the system has in this case the form

$$dF = \gamma d(VA) + \Sigma \mu_i dN_i - p_2 d(V\phi) - p_1 d(V(1-\phi)), \tag{62}$$

where, for reasons already noted in section III.1,

$$\gamma = \gamma_h - \tau A \left(\frac{dh}{dA} \right)_{N_i,V,T,\phi}, \tag{63}$$

where γ_h is the interfacial tension when the distance between two successive lamellae is equal to 2h.

At equilibrium the distance between the lamellae (or the thickness of the lamellae), for given values of N , V, ϕ and T is given by

$$\gamma = \left(\frac{\partial F}{\partial (VA)} \right)_{N_i,V,T,\phi} = 0. \tag{64}$$

One may note that the minimum of F with respect to ϕ leads to

$$p_2 = p_1 \tag{65}$$

and that, finally, the mechanical equilibrium condition with the environment yields

$$p_2 = p_1 = p. \tag{66}$$

However, as already noted (see Eq. (3)), γ includes the effect of the interaction forces between the lamellae as well as the effect of the reversible work done against these forces when because of the change in area, the distance 2h between two lamellae changes. Let us, therefore establish an expression for γ in which the various contributions are better identified. In order to achieve this, Eq. (62) is integrated at constant γ, μ_i, h and T, to obtain:

$$F = \gamma_h VA + \Sigma \mu_i N_i - pV \tag{67}$$

which, differentiated at constant p, becomes

$$dF = \gamma_h d(VA) + (VA)d\gamma_h + \Sigma \ \mu_i dN_i + \Sigma N_i d\mu_i - pdV. \tag{68}$$

The Gibbs–Duhem equations for the two media of the microemulsion have the form

$$\Sigma \ N_i' d\mu_i = 0 \tag{69}$$

and

$$\Sigma \ N_i'' d\mu_i = 0, \tag{70}$$

where the superscripts prime and double prime refer to the two media of the microemulsion. Equating Eqs. (62) and (68) and subtracting Eqs. (69) and (70), one finally obtains

$$d\gamma_h + \Sigma \ \Gamma_i d\mu_i + \tau dh = 0, \tag{71}$$

where

$$\Gamma_i = \left[N_i - (N_i' + N_i'') \right]/VA \tag{72}$$

is the surface excess of species i. Consequently[39,40],

$$\tau = - \left(\frac{\partial \gamma_h}{\partial h} \right)_{\mu_i,T} \tag{73}$$

from where it results that

$$\gamma_h - \gamma_\infty = {}_h\!\int^\infty \tau dh. \tag{74}$$

The interfacial tension γ can therefore be decomposed in the sum of three contributions

$$\gamma = \gamma_\infty + {}_h\!\int^\infty \tau dh - \tau A \frac{dh}{dA} . \tag{75}$$

The first term represents the interfacial tension when the distance between lamellae is very large, the second represents the work due to the interaction forces between the lamellae and the third term represents the work done against these forces because of the virtual infinitesimal change in area used to define γ.

In addition,

$$hA = (1-\phi) = \text{const.}, \tag{76}$$

which yields

$$\frac{dh}{dA} = -\frac{h}{A}. \tag{77}$$

Consequently, Eq. (75) becomes

$$\gamma = \gamma_\infty + {}_h\!\int^\infty \tau\, dh + \tau h. \tag{78}$$

Because

$${}_h\!\int^\infty \tau\, dh = {}_h\!\int^\infty d(\tau h) - {}_h\!\int^\infty h\, \frac{d\tau}{dh}\, dh,$$

Eq. (78) yields, assuming that $(\tau h) \xrightarrow[h\to\infty]{} 0$,

$$\gamma = \gamma_\infty - {}_h\!\int^\infty h\, \frac{d\tau}{dh}\, dh. \tag{79}$$

The basic equation (64) for equilibrium can therefore be written in the form

$$\gamma_\infty = {}_h\!\int^\infty h\, \frac{d\tau}{dh}\, dh. \tag{80}$$

The force τ per unit area was considered positive for repulsion and negative for attraction. It is physically reasonable to assume that the repulsive (τ_r) and attractive (τ_a) forces have a smooth dependence on h and tend to zero for large values of h. One can therefore conclude that

$$\frac{d\tau_r}{dh} < 0 \quad \text{and} \quad \frac{d\tau_a}{dh} > 0. \tag{81}$$

As a result,

$$\gamma_\infty = I_a - I_r, \tag{82}$$

where

$$I_a = {}_h\!\int^\infty h\, \frac{d\tau_a}{dh}\, dh \tag{83a}$$

and

$$I_r = - {}_h\!\int^\infty h\, \frac{d\tau_r}{dh}\, dh, \tag{83b}$$

I_a and I_r being positive quantities.

Considering γ_∞ a small positive quantity, Eq. (82) shows that thermodynamically stable lamellar microemulsions can arise only if the attractive interactions can compensate for the repulsive ones and for the interfacial

tension γ_∞. Repulsive interactions alone cannot lead in this case to
thermodynamically stable lamellar structures. If the attractive interactions are too weak, one can still have a stable microemulsion, but as a
random, globular dispersion, whose thermodynamic stability is facilitated
by the entropy of dispersion. If the attractive interactions are too
strong in comparison with the other contributions, they can enhance the
thermal perturbations of the internal interfaces of the lamellar structure
and the globular dispersions can become the thermodynamically preferred
state. While the favorable effect of the attractive interactions may be
thus decreased, the entropy of dispersion can more than compensate for this
decrease. The interfacial tension is also increased and the interface
becomes more stable. Indeed, the increased area per unit volume when the
dispersion is globular instead of lamellar leads to the adsorption of a
greater amount of surfactants. This decreases the concentrations of the
surfactants in the bulk and results in a decrease of the amount adsorbed
per unit area of the internal interface of the microemulsion. Therefore,
the interfacial tension increases.

Strong van der Waals interactions associated with sufficiently strong
repulsive interactions can lead to lamellar (or other types of liquid
crystalline structures) of microemulsions. The destabilizing effect of the
attractive interactions on the interfacial stability is compensated in this
case by the stabilizing effect of the repulsive interactions.

VII.3 Ordered Globular and Cylindrical Microemulsions

The entropy of dispersion of the globules in the continuous phase is
extremely small for possible microemulsions containing ordered arrangements
of spherical globules. If the entropy of dispersion is negligible, Eqs.
(11) and (12) show that such microemulsions could exist only if the bending
stress C (which depends upon the proportion of surfactant and alcohol)
becomes negative. Cylindrical structures with ordered arrangements also
have very small entropies of dispersion. The equations of section III.1
can easily be adapted to this case, by observing that now $A = 2\phi/r$ and that
one of the curvatures is zero. The conclusion is the same. If the entropy
of dispersion is negligible, such an ordered structure can exist only if
the bending stress is negative. A random distribution of flexible cylinders is also, at least in principle, possible. While the entropy of a
random globular dispersion of smaller units is greater, the long cylinders
may have stronger van der Waals interactions among them than the globules.
The behavior of flexible, cylindrical aggregates is similar to that of the
polymeric molecules, with one major difference. While the linear dimension
of the polymeric molecule is fixed such a constraint does not apply to
microemulsion aggregates. If the overlap of the aggregates is so large
that their free motion is impeded and the gain in attractive interaction
cannot compensate for the loss in entropy, the size of the aggregates will
decrease in order to increase the entropy and thus satisfy the condition of
minimum free energy.

VII.4 Globular vs. Lamellar Structures

Let us first consider a single thin film of oil in water which has a
layer of surfactants at its free surface. Such a film is generally unstable and ruptures into smaller units. In order to explain this, let us
consider the thin film to be planar and apply small periodical perturbations to its interfaces. Although the total surface free energy
increases because of the increased surface area associated with these
deformations, the free energy can decrease because the deformations alter
the overall van der Waals interactions of the molecules. Indeed, because

26

the film is thin, the range of the interaction forces between one of its
molecules and all the other molecules of the system is greater than the
thickness of the film. The interaction potential at a planar interface of
a thin film of uniform thickness, h, has for dispersion interactions the
form $\phi = \phi_o + \dfrac{A}{6\pi h^3}$, where ϕ_o is negative and A is a positive quantity (the
Hamaker constant). By applying a periodical perturbation to the surface of
the film, some molecules will be displaced from smaller to larger distances
from the central plane of the film, thus making their interaction potential
more negative. If this negative energy change can overcome the positive
free energy change due to the interfacial free energy, the perturbation
will grow and the film will rupture. Among the perturbations, the so
called dominant perturbation, whose wave length causes the maximum growth,
is of particular interest, since it dominates the rupture process. The
dominant wave length is given by the expression[41-43].

$$\lambda_D \sim \left(\frac{\gamma}{A} \right)^{1/2} h_o^2, \tag{84}$$

where h_o is the thickness of the planar unperturbed film. The film of
initial thickness h_O will rupture in units having a size of the order of
λ_D. For small values of γ and h_o, λ_D is small and therefore the small
perturbations caused by the thermal fluctuations can lead to the rupture of
the film. For large values of h_o and/or γ, λ_D is much larger and the
thermal perturbations are no longer as dangerous. One may also note that
since A is of the order of kT and considering $\lambda_D \approx h_o$, Eq. (84) leads to
$\lambda_D{}^2 \sim \dfrac{kT}{\gamma}$, an expression which has the same form as Eq. (32) that relates
the radius r of the globules in a microemulsion to the interfacial tension
γ.

If, instead of a single lamella, one considers a large number of
parallel lamellae, the mechanical stability of their interfaces can be
increased by the existence of repulsive interactions between them while
their thermodynamic stability is facilitated by the van der Waals attractive interactions. The latter interactions have, however, a destabilizing
effect on the interfaces. If they are too strong, the lamellae may break
into globules. The gain in entropy can be greater than the loss in van der
Waals interactions and thus the random globular dispersion is preferred
thermodynamically. If the repulsion is too strong, the system may also
prefer the random globular dispersion whose gain in entropy will minimize
the free energy of the system: Moderate van der Waals interactions as well
as moderate repulsive interactions may lead to lamellar or other liquid
crystalline structures of the microemulsions.

SUMMARY

A thermodynamic approach is developed to explain the phase behavior of
microemulsions as well as the relation between the phase behavior and their
structure. The treatment is extended to lamellar microemulsions, for which
the conditions of equilibrium are established. A comparison between random
globular and liquid crystalline structure of microemulsions is also
included in the paper.

REFERENCES

1. E. J. W. Verwey and J. Th. G. Overbeek, "Theory of the Stability of Lyophobic Colloids," Elsevier, Amsterdam (1948).
2. E. Ruckenstein, Chem.Phys.Lett. 56:518 (1978).
3. T. P. Hoar and J. H. Schulman, Nature 152:102 (1943).
4. J. H. Schulman and D. R. Riley, J.Colloid Sci. 3:383 (1948).
5. W. Stoeckenius, J. H. Schulman, and L. M. Prince, Kolloid Z. 169:170 (1960).
6. J. E. L. Bowcott and J. H. Schulman, Z. Electrochem. 59:283 (1955).
7. J. W. Falco, R. D. Walker Jr., and D. O. Shah, AIChEJ. 20:510 (1974).
8. R. Hwan, C. A. Miller, and T. Fort, J.Colloid Interface Sci. 68:221 (1979).
9. A. M. Cazabat, D. Langevin, J. Meunier, and A. Pouchelon, Adv.Colloid Interface Sci. 16:175 (1982).
10. P. A. Winsor, "Solvent Properties of Amphiphilic Compounds," Butterworth, London (1954).
11. R. N. Healy, R. L. Reed, and D. G. Stenmark, Soc.Pet.Eng.J.Trans.AIME 261:147 (1976).
12. M. Bourel, C. Koukounis, R. Schechter, and W. Wade, J.Dispersion Sci.Tech. 1:13 (1980).
13. B. Lindman, P. Stilbs, and E. Moseley, J.Colloid Interface Sci. 83:569 (1981).
14. E. Ruckenstein, Chem.Phys.Lett. 98:573 (1983).
15. E. Ruckenstein, Fluid Phase Equilibria 20:189 (1985).
16. E. Ruckenstein, in: "Macro and Microemulsions," D.O. Shah, ed., ACS Symposium Series No. 272:21 (1985).
17. J. W. Gibbs, "Collected Works," Vol. 1, Yale University Press, New Haven, CT (1948).
18. F. R. Buff, J.Chem.Physics 19:159 (1951).
19. S. Ono and S. Kondo, in: "Handbuch der Physik," Vol. X, p. 134, S. Flugge, ed., Springer, Berlin (1960).
20. E. Ruckenstein and J. C. Chi, J.C.S. Faraday II 71:1690 (1975).
21. J. Th.G. Overbeek, Faraday Discussions of the Chem.Soc. 65:7 (1978).
22. E. Ruckenstein and I. V. Rao (to be published).
23. W. Helfrich, Z.Naturforsch. 28c:693 (1973).
24. S. A. Safran, L. A. Turkevick, and P. A. Princus, J.Phys.(Paris) 45:L69 (1984).
25. S. A. Safran, J.Chem.Phys. 78:2073 (1983).
26. E. Ruckenstein, Soc.Pet.Eng.J.Trans.AIME 21a:593 (1981).
27. A. Pouchelon, J. Meunier, D. Langevin, D. Chatenay, and A. M. Cazabat, Chem.Phys.Lett. 76:277 (1980).
28. G. I. Sivashinsky and D. M. Michelson, Prog.Theor.Phys. 63:2112 (1980).
29. T. Shlang and G. I. Sivashinsky, J.Physique 43:459 (1982).
30. S. Marcelja and N. Radič, Chem.Phys.Lett. 42:129 (1976).
31. E. Ruckenstein and D. Schiby, Chem.Phys.Lett. 95:439 (1983).
32. E. Ruckenstein and D. Schiby, Langmuir 1:612 (1985).
33. H. M. Princen, M. P. Aronson, and G. C. Moser, J.Colloid Interface Sci. 75:246 (1980).
34. M. A. Schwartz, "Gesammelte Mathematische Abhandlung," Vol. 1, Springer, Berlin (1890).
35. E. R. Neovius, "Minimalflächen", J. C. Frenkel, Helsingfors, (1883).
36. L. E. Scriven, in: "Micellization, Solubilization and Microemulsions," K. L. Mittal, ed., Plenum Press, New York (1977).
37. Y. Talmon and S. Prager, J.Chem.Phys. 69:2984 (1978).
38. J. Jouffroy, P. Levinson, and P. G. de Gennes, J.Physique 43:1241 (1982).

39. E. L. Mackor and J. H. van der Waals, _J.Colloid Sci._ 7:535 (1952).

40. S. G. Ash, D. H. Everett, and C. Radke, _J.C.S. Faraday II_ 69:1256 (1973).

41. A. Vrij, F. Hesselink, J. Lucassen, and M. van den Tempel, _Proc.Kon.Ned.Akad.Wet._ B13:124 (1970).

42. E. Ruckenstein and R. K. Jain, _J.C.S. Faraday II._ 70:132 (1974).

43. C. Maldarelli, R. K. Jain, I. B. Ivanov, and E. Ruckenstein, _J.Colloid Interface Sci._, 18:118 (1980).

A SIMPLE INTERPRETATION OF HYDROPHOBIC INTERACTIONS AND CRITICAL
CONCENTRATIONS IN MICELLAR SOLUTIONS

E. Ruckenstein

Institut für Physikalische Chemie I der
Universität Bayreuth, Postfach 3009
D-8589 Bayreuth, West Germany

INTRODUCTION

Surfactant molecules possess a hydrophobic tail and a hydrophilic head group. Because of this dual property, they form in water (or other polar liquids) a large number of relatively large aggregates, as soon as their concentration becomes greater than the critical micelle concentration[1,2]. In the aggregates, the hydrocarbon tails are shielded from water by the polar head groups. Two factors are mainly responsible for the formation of micelles. One of them, the hydrophobic bonding, is primarily due to the incompatibility of the hydrocarbon tails of the surfactant with the polar solvent. Indeed, the tails interfere with the strong polar interactions between the water molecules and, therefore, the free energy of the system is decreased by diminishing their contact with water. Should this factor act alone, the surfactant molecules will form "infinite size" aggregates, thus leading to a separate surfactant phase. The repulsion that arises between the head groups competes with the hydrophobic bonding and ensures the formation of a large number of finite (but relatively large) size aggregates. The present paper is concerned with the mechanistic interpretation of the hydrophobic bonding, as well as with the critical micelle concentration in both polar and nonpolar solvents. The latter quantity provides information about the conditions under which a large number of aggregates can form.

Information about hydrophobic bonding has been obtained by Frank and Evans[3] from measurements of the standard free energy of transfer of hydrocarbon molecules, from a hydrocarbon to a water environment. The temperature dependence of the solubility of the hydrocarbon molecules in water has enabled Frank and Evans to determine the separate enthalpic and entropic contributions to the hydrophobic bonding. The large negative entropy change thus obtained was interpreted as being due to a higher ordering of the water molecules near the hydrocarbon chains. This "Iceberg" formation was considered responsible for the large positive standard free energy of transfer of the hydrocarbon molecules from a hydrocarbon to a water environment. More recently, Shinoda[4], using the same experimental data, concluded that the formation of a cavity in the solvent (to accommodate the hydrocarbon molecule), is mainly responsible for the hydrophobic bonding. One of the goals of the present paper is to summarize the

above two points of view, and, in addition, to prove by using simple thermodynamic arguments, that, indeed, the free energy of formation of the
cavity constitutes the primary contribution to the hydrophobic bonding.
The "Iceberg" formation plays only a secondary role. Secondly, the meaning
of the critical micelle concentration and its relation to hydrophobic
bonding as well as repulsive interactions is examined, on the basis of the
size distribution of aggregates, for polar solvents. In addition, the
possibility of the existence of a critical micelle concentration in nonpolar solvents is also discussed.

THE FRANK AND EVANS INTERPRETATION OF HYDROPHOBIC BONDING

McBain[5] and Debye[6] considered that the association of the surfactant molecules in micelles arises because of the favorable interactions
between the hydrocarbon chains. In contrast, Frank and Evans have demonstrated that micellization is primarily a result of the strong, attractive
interactions between the water molecules which are perturbed by the presence of the hydrocarbon tails. This conclusion was reached on the basis of
thermodynamic measurements of the dissolution of hydrocarbon molecules in
water. Their solubility in water as a function of temperature was used to
determine the separate enthalpic and entropic contributions to the standard
free energy change of transfer, by employing the following well known
thermodynamic procedure. The solubility X_W of the hydrocarbon molecules in
water being very small, one can write

$$RT\ln X_W = \mu_{HC} - \mu_W^0 = h_{HC} - h_W^0 - T(s_{HC} - s_W^0)$$
$$\equiv - \Delta h^0 + T\Delta s^0, \tag{1}$$

where X_W is the molar fraction of hydrocarbon in water, μ_j (j = HC or W)
are the chemical potentials of the hydrocarbon molecules, h and s are their
partial molar enthalpies and entropies, T is the absolute temperature, R is
the gas constant, the subscripts HC and W indicate the hydrocarbon and
water phases, respectively, and the superscript ° refers to the standard
state of infinite dilution. Combining Eq. (1) with the equation

$$R \frac{d\ln X_W}{d(1/T)} = - \Delta h^0, \tag{2}$$

one can calculate both $\Delta s°$ and $\Delta h°$ as a function of temperature.

For the interpretation of the thermodynamic results, it is useful to
identify the processes which arise when a hydrocarbon molecule is introduced in water. First, a cavity is formed by the breaking of a number of
hydrogen bonds. This process is associated with an increase both in enthalpy and entropy. Second, the hydrocarbon molecule introduced into the
cavity organizes the surrounding water molecules. The "Iceberg" formation
decreases the entropy, because of the increased order it introduces, and
diminshes the enthalpy, because of the dispersion interactions between the
hydrocarbon molecule and the surrounding (organized) water molecules.

The ratio between the values obtained for $T\Delta s°$ and $\Delta h°$ is large. For
instance, for butane, at 25°C, $T\Delta s° = -6850$ cal/mol and $\Delta h° = -800$ cal/mol.
On this basis, Frank and Evans concluded that the decrease in entropy,
which is associated with the "Iceberg" formation, is primarily responsible
for the large positive values of $\mu_W^° - \mu_{HC}$.

SHINODA'S INTERPRETATION

Because two processes are involved in the dissolution of the hydrocarbon molecules, Shinoda decomposes the enthalpic and entropic changes
into two terms:

$$\Delta h^0 = \Delta h^0_{H-bonds} + n\Delta h^0_i \qquad (3a)$$

and

$$\Delta s^0 = \Delta s^0_{H-bonds} + n\Delta s^0_i, \qquad (3b)$$

where n is the number of molecules of water in the icebergs and the subscripts H-bonds and i refer to hydrogen bonds break up and icebergs, respectively.

The organization of water as icebergs is expected to diminish as the
temperature is increased, and at large temperatures, $n \to 0$. Assuming, in
addition, that $\Delta h^o_{H-bonds}$ and $\Delta s^o_{H-bonds}$ are weakly temperature dependent,
their values can be approximated by the high temperature values of Δh° and
Δs°. (In fact, experiment shows that the slope of $\ln X_W$ against $1/T$ becomes
independent of temperature at large temperatures. Therefore, Δh° does,
indeed, become independent of temperature, at high temperatures). This
permits the calculation of $n\Delta h^o_i$ and $n\Delta s^o_i$ as a function of T. For butane,
at 25°C, such calculations provide the values:

$$\Delta h^0 \quad = \quad -800 \text{ cal/mol}; \quad T\Delta s^0 \quad = -6850 \text{ cal/mol};$$

$$\Delta h^0_{H-bonds} = \quad 11200 \text{ cal/mol}; \quad T\Delta s^0_{H-bonds} = \quad 2411 \text{ cal/mol};$$

$$n\Delta h^0_i \quad = -12000 \text{ cal/mol}; \quad nT\Delta s^0_i \quad = -9261 \text{ cal/mol}.$$

These simple considerations provide an alternate mechanistic picture of the
hydrophobic interactions. "Iceberg" formation is, indeed, accompanied by a
large decrease in entropy. This is, however, more than compensated by an
even greater enthalpic effect due to the interactions between the hydrocarbon molecule and the neighboring water molecules. The formation of the
cavity via the breaking of the hydrogen bonds is, therefore, primarily
responsible for the low solubility of the hydrocarbons in water and for the
hydrophobic bonding.

THERMODYNAMICS OF "ICEBERG" FORMATION

A slightly different thermodynamic treatment is as follows: The free
energy change $\mu^o_W - \mu_{HC}$ contains two contributions: (1) a free energy
change $\Delta\mu_c$ associated with the formation of the cavities and (2) a free
energy change $\Delta\mu_i$ due to the "Iceberg" formation:

$$\mu^0_W - \mu_{HC} = \Delta\mu_c + \Delta\mu_i. \qquad (4)$$

The "Iceberg" formation is associated with the enthalpic change $n\Delta h^\circ$ which
occurs because n water molecules change their state from that in the bulk
that in the neighborhood of the hydrocarbon molecules; $n\Delta s^o_i$ represents the
corresponding entropic change. There is a third contribution as well. The
transition from water to iceberg is accompanied by a change in "the interfacial free energy" from the value σ_c in the cavity to the value σ between

the hydrocarbon molecule and water. Denoting by $\Delta\sigma = \sigma - \sigma_c$ the change in
"the interfacial free energy" and by A_h the "area" between the hydrocarbon
molecules and water per mole of hydrocarbon molecules, the third contri-
bution to $\Delta\mu_i$ is given by $A_h\Delta\sigma$. The difference $\Delta\sigma$ is expected to be negative because "the interfacial tension" between the hydrocarbon molecule and
(organized) water is expected to be smaller than that of the cavity.

Consequently,

$$\Delta\mu_i = n\Delta h_i^0 - Tn\Delta s_i^0 + A_h\Delta\sigma. \qquad (5)$$

However, the water molecules in the icebergs are in thermodynamic equilibrium with those in the bulk. As a result,

$$n\Delta h_i^0 = Tn\Delta s_i^0 \qquad (6)$$

and

$$\Delta\mu_i = A_h\Delta\sigma < 0. \qquad (7)$$

In other words, the entropic contribution $Tn\Delta s_i^0$ is compensated by the
enthalpic contribution $n\Delta h_i^0$, and the free energy change associated with the
"Iceberg" formation is only due to $A_h\Delta\sigma$. Of course, $A_h\Delta\sigma$ represents the
free energy of interaction between the hydrocarbon molecules and the surrounding water molecules.

Neglecting the high temperature value of $A_h\Delta\sigma$ in comparison with $\Delta\mu_c$,
calculations similar to those of the previous section lead for butane, at
25°C, to the values

$$\Delta\mu_c = 8789 \text{ cal/mol and } A_h\Delta\sigma = -2739 \text{ cal/mol.}$$

In conclusion, the contribution of "Iceberg" formation to the free
energy change $\mu_W^0 - \mu_{HC}$ is negative, and the primary contribution to the
standard free energy of transfer comes from the free energy $\Delta\mu_c$ of formation of the cavity. It is also important to note that hydrogen bonding is
not necessary for micellization. Micellization will occur in any polar
liquid in which the dipole-dipole interactions are sufficiently strong.
Because the hydrocarbon tails of the surfactants interfere with these
interactions they will be forced to aggregate as micelles.

It may also be noted in passing that various procedures have been
developed to calculate the free energy of dissolution of simple gases in
water[7a,7b]. A recent paper[7c] provides some of the important references. However, they are outside the scope of the present paper.

THE CRITICAL CONCENTRATION IN POLAR MEDIA

While hydrophobic bonding constitutes the driving force for aggregation, the decrease in the number of individual particles (because of
aggregation) diminishes the entropy of the system. Obviously, the latter
factor opposes aggregation. At sufficiently low concentrations, the latter
factor dominates and the size distribution spectrum of the aggregates is a
monotone decreasing function of the aggregation number. At sufficiently
large surfactant concentrations, the free energy of aggregation, determined
by the hydrophobic bonding and additional repulsive interactions, dominates
and a large number of relatively large aggregates form. Let us examine the
above behavior in some detail.

The surfactant molecules form all kinds of aggregates. Considering the aggregates of different sizes as distinct chemical species, one derives that the molar fraction X_i of the aggregates containing i molecules is related to the molar fraction X_1 of the single surfactant molecules, via the expression[8,9]:

$$X_i = X_1^i \exp(-i\Delta\mu_i^0/kT), \tag{8}$$

where $\Delta\mu_i^0 = \mu_i^0/i - \mu_1^0$, μ_i^0 is the standard chemical potential of an aggregate containing i molecules, k is the Boltzmann constant and T is the absolute temperature. The quantity $i\Delta\mu_i^0$ represents the standard free energy change of formation of an aggregate containing i molecules from i single surfactant molecules. Equation (8) involves the assumption that the system is dilute.

The factor X_1^i arises because of the decrease in entropy associated with the reduction of the number of independent species when i single molecules aggregate. For sufficiently small surfactant concentrations, X_1 is small and the size distribution of aggregates is a rapidly decreasing function of i (even though the second factor in Eq. (8) increases with i). For sufficiently large values of X_1, X_i decreases with increasing i for relatively small values of i, but its behavior for larger values of i is determined by the manner in which $i\Delta\mu_i^0$ depends on i. A large number of relatively large micelles will form if there is a surfactant concentration above which the size distribution has a maximum for relatively large values of i. Because for small values of i the size distribution curve should still be a monotone decreasing function, the above mentioned curve should also have a minimum before the maximum. It is clear that the transition from a monotone decreasing size spectrum to one which has a minimum and a maximum should be a decreasing size spectrum with a horizontal inflexion point (Fig. 1). Consequently, a large number of finite size micelles of relatively large size will form if there is a critical concentration X_{1c} and a value i_c (with i_c sufficiently large) for which[9]

$$\frac{dX_i}{di} = \frac{d^2X_i}{di^2} = 0. \tag{9}$$

It is convenient to replace Eq. (9) with the equivalent expressions

$$\frac{d\ln X_i}{di} = \frac{d^2\ln X_i}{di^2} = 0. \tag{10}$$

In Fig. 1, we plot $\ln X_i$ as a function of i for all the situations mentioned above. It is worthwhile to note the analogy with the van der Waals equation of state; the critical point and the critical concentration are analogous quantities.

THE CRITICAL SIZE AND THE CRITICAL CONCENTRATION

In order to better identify the roles of hydrophobic bonding, repulsive interactions and entropy in the aggregation process, a simple expression for $i\Delta\mu_i^0$ will be employed.

The hydrophobic bonding is expected to contribute to $i\Delta\mu_i^0$ with a term of the form

$$(i\Delta\mu_i^0)_{HC} = -\alpha'i, \tag{11}$$

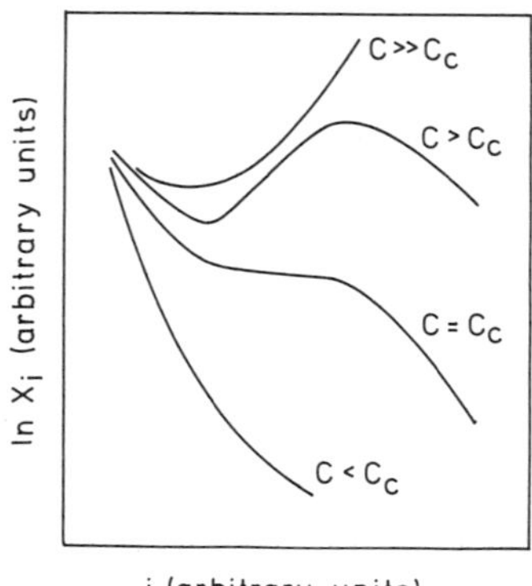

Fig. 1. Qualitative micellar size distributions. The curves represent size distributions when the total surfactant concentration c is less than the critical concentration c_c, equal to c_c, larger than c_c and much larger than c_c.

where $-\alpha'$ represents the standard free energy change per surfactant molecule due to the hydrophobic interactions.

For steric reasons, there is an incomplete shielding of the micellar core from water. This incomplete shielding leads to a term of the form

$$(i\Delta\mu_i^0)_{IS} = \beta'(A - a)i, \tag{12}$$

where A is the area of the micellar core per surfactant molecule, a is the cross-sectional area of the hydrocarbon chain and β' is a kind of interfacial free energy.

Finally, there are repulsive contributions which arise because of the electrostatic repulsions between the charged head groups and of the surface exclusion caused by their finite size. For nonionic surfactants, only the latter repulsive contribution remains. Let us write the surface exclusion repulsion in the power form

$$(i\Delta\mu_i^0)_{SE} = \gamma i^P. \tag{13}$$

In the vicinity of the critical concentration the micelles are, in general, spherical. As a result,

$$A = (36\pi)^{1/3}v^{2/3}i^{-1/3} \equiv A_o i^{-1/3}, \tag{14}$$

where v is the volume of the hydrocarbon tail of the surfactant. Therefore, for nonionic surfactants,

$$\begin{aligned}i\Delta\mu_i^0 &= -(\alpha' + \beta'a)i + \beta'A_o i^{2/3} + \gamma i^P \\ &\equiv -\alpha i + \beta i^{2/3} + \gamma i^P.\end{aligned} \tag{15}$$

Starting with Eq. (8) and employing Eq. (15) for $i\Delta\mu_i$, Eqs. (10) lead to

$$\ln X_{1c} + \frac{1}{kT}\left[\alpha - \frac{2}{3}\beta i_c^{-1/3} - p\gamma i_c^{P-1}\right] = 0 \tag{16}$$

36

and

$$\frac{2}{9} \beta i_c^{-4/3} - p(p-1)\gamma i_c^{p-2} = 0. \tag{17}$$

Eq. (17) clearly shows that i_c has finite values only if $p > 1$. In other words, micellization can occur when the hydrophobic interactions are coupled with repulsive interactions which have a sufficiently strong dependence on i. Tanford has chosen the value $4/3$ for the exponent p, in which case Eqs. (16) and (17) become

$$i_c = (\frac{\beta}{2\gamma})^{3/2} \tag{18}$$

and

$$kT\ln X_{1c} = 1.886(\beta\gamma)^{1/2} - \alpha. \tag{19}$$

One may note that the critical size is independent of hydrophobic bonding. It is greater when the interfacial tension β' as well as the volume v of the hydrocarbon tail of the surfactant are larger and the surface exclusion repulsion smaller (hence when the size of the head group is smaller). Instead, the critical concentration X_{1c} depends upon the hydrophobic bonding, being smaller when the latter is greater. This is as expected. The "repulsive" effects due to the interfacial free energy and to the surface exclusion increase the value of X_{1c}, again as expected.

Obviously, the existence of finite values of i_c does not ensure that a large number of relatively large aggregates form. The critical size i_c should be sufficiently large for this to happen. When water is the solvent, the values of β and γ are such that i_c is large. Indeed, typical values for nonionic surfactants are [1] $\beta/kT = 9$ and $\gamma/kT = 0.6$; they lead to $i_c = 21$.

In practice, the critical micelle concentration (CMC) above which micellization occurs is determined experimentally as the concentration at which a sharp change occurs in any of a wide variety of properties of the surfactant solution (such as electrical conductance, transference number, dye absorption, surface tension, etc.). It was demonstrated by Nagarajan [10] that the critical concentration, which is defined on the basis of the size distribution spectrum (Eqs. (10)), provides a very close lower bound of the CMC. The critical concentration thus defined has a simple and clear physical meaning.

AGGREGATION IN NONPOLAR MEDIA – DOES A CRITICAL MICELLE CONCENTRATION EXIST?

In contrast to polar solvents, the structure of nonpolar solvents is not so profoundly affected by the presence of the surfactant molecules. In fact, the interactions of the amphiphilic tails with the solvent molecules can be just as favorable as those with the other amphiphiles. Therefore, in this case, there is no strong driving force comparable to hydrophobic bonding to lead to the formation of large aggregates. The dipole-dipole interactions[11-15] as well as intermolecular bonding of a quasichemical nature[11,13,16] facilitate, however, the formation of relatively small aggregates. Experiment confirms that aggregation occurs even at low surfactant concentrations. The average aggregation numbers are, however, in general, much smaller than in aqueous systems. In addition, in aqueous systems, as the surfactant concentration is increased, the properties of

the system undergo an abrupt change over a narrow range of concentrations
(CMC). Concerning the behavior of nonpolar solutions the opinions are
divided. Kertes, [11,12] who has discussed this problem in detail,
expresses the opinion that in nonpolar solvents there is no CMC. Even
though values for the CMC have been reported for some solvents, a critical
examination of these results by Kertes led him to the conclusion that a CMC
is absent in these systems as well, the aggregation being triggered by the
presence of traces of water. This opinion is, however, challenged by other
researchers [11,17]. Let us, therefore, dwell for a moment on this
problem.

The forces involved in the aggregation of surfactant molecules are
very different in polar and nonpolar solvents, being dominated by hydrophobic bonding in the former and by dipole-dipole interactions in the
latter case. The aggregation process can be described in both cases by the
same thermodynamic approach, based on a size spectrum. Such calculations
have been carried out for nonpolar solvents[18] and show that the size
distribution is either a monotone decreasing function of size, or possesses
a weak minimum and a weak maximum at relatively small sizes. The number
average aggregation number

$$\langle g \rangle = \frac{\sum g X_g}{\sum X_g}$$

and the weight average aggregation number

$$\langle g \rangle_W = \frac{\sum g^2 X_W}{\sum g X_g}$$

have been calculated as a function of the total amphiphilic concentration.
Most of the physical, colligative or spectral properties are proportional
to one or the other of the above quantities. Consequently, one can consider that these plots represent these physical properties against the
concentration. Many of the curves show a gradual change as the total
amphiphilic concentration is increased. However, some of the curves do
show a less gradual change (see Figs. 5 and 6 of Ref. 18) and it is possible to discriminate two regions separated by a concentration which has
much in common with the CMC.

In aqueous solutions, the CMC appears because the size distribution
curve has a horizontal inflexion point and the corresponding critical size
i_c is relatively large. In nonpolar media, the size distribution is, in
general, a monotone decreasing function (and even if it has an inflexion
point for a given concentration, the value of i_c is very small i $\approx$ 2). The
CMC arises in this case due to a different reason. A possible explanation
is suggested by Fig. 2. For low concentrations, the size distribution,
while monotone decreasing, is very steep (curve A). For higher concentrations, the curve is much less steep (curve B). If the transition from
steep to less steep occurs in a small range of concentrations, the plot of
a property against the concentration can have nearly an abrupt change.
Hence, in some cases a CMC can exist even in nonpolar solvents. The two
types of CMC, the CMC in polar and the CMC in nonpolar media occur, however, due to different causes.

SUMMARY

The points of view of Frank and Evans and of Shinoda regarding hydrophobic bonding are presented and thermodynamic arguments are brought to
conclude that the free energy of formation of a cavity in the polar solvent

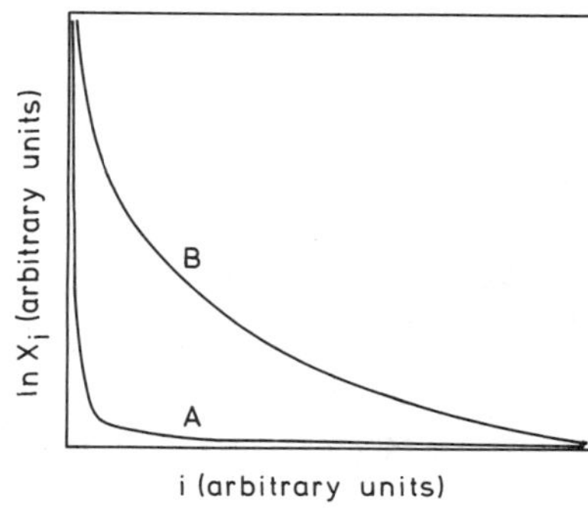

Fig. 2. Possible size distributions in nonpolar solvents. A: size
distribution at low concentrations; B: size distribution at large
concentrations.

is mainly responsible for hydrophobic bonding. A critical concentration is
defined on the basis of the size distribution spectrum of the micellar
aggregates and is further used to emphasize the important role played by
the repulsive interactions between the head groups in the micellization
process. If the surface exclusion is assumed in the form of i^P, where i is
the number of molecules in an aggregate, the exponent p must be greater
than unity for a critical concentration to exist and hence for micellization to take place. The question of the existence of a critical micelle
concentration in nonpolar solvents is also examined to conclude that it can
exist under particular conditions. Its origin is, however, shown to be
different from that in polar solvents.

REFERENCES

1. C. Tanford, "The Hydrophobic Effect: Formation of Micelles and Bio-
    logical Membranes," 2nd ed. Wiley-Interscience, New York (1980).
2. K. Shinoda, "Principles of Solution and Solubility," M. Decker, New
    York (1977).
3. H. S. Frank and M. W. Evans, J.Chem.Phys. 13:507 (1945).
4. K. Shinoda, J.Phys.Chem. 81:1300 (1977).
5. J. W. McBain, "Colloid Science," D. C. Heath and Co., Boston (1950).
6. P. Debye, Ann.N.Y.Acad.Sci. 51:575 (1949).
7. (a) R. Pierotti, J.Phys.Chem. 67:1840 (1963); 69:281 (1965).
    (b) L. R. Pratt and D. Chandler, J.Chem.Phys. 67:3683 (1977).
    (c) R. Fernandez-Prim, R. Czovetto, M. L. Japas, and D. Laria,
        Acc.Chem.Res. 18:207 (1985).
8. C. Tanford, J.Phys.Chem. 78:2469 (1974).
9. E. Ruckenstein and R. Nagarajan, J.Phys.Chem. 79:2622 (1975).
10. R. Nagarajan and E. Ruckenstein, J.Colloid Interface Sci., 91:500
    (1983).
11. A. S. Kertes and H. Gutmann, in: "Surface and Colloid Science," Vol.
    8, E. Matijevic, ed., Interscience, New York (1975).
12. A. S. Kertes, in: "Micellization, Solubilization and Microemulsions,"
    K. L. Mittal, ed., Plenum Press, New York (1977).
13. A. S. Kertes, H. Gutmann, O. Levy, and G. Y. Markovits, Isr.J.Chem.,
    6:421 (1968).
14. N. Muller, J.Phys.Chem. 79:287 (1975).
15. N. Muller, J.Colloid Interface Sci. 63:383 (1978).
16. R. Debye and h. Coll, J.Colloid Sci. 17:220 (1962).
17. K. Kon-No, A. Kitahara, and O. A. El Seoud, in: "Nonionic Surfactants:
    Physical Chemistry," M. J. Schick, ed., (in press).
18. E. Ruckenstein and R. Nagarajan, J.Phys.Chem. 84:1349 (1980).

FROM MICELLIZATION TO MICROEMULSION FORMATION

E. Ruckenstein

Institut für Physikalische Chemi I der
Universität Bayreuth, Postfach 3008, D-8580 Bayreuth
West Germany

INTRODUCTION

Surfactant molecules aggregate in dilute aqueous solutions if their
concentration is sufficiently large, generating a variety of microstructures which minimize the contact between their hydrophobic tails and water
[1-5]. The hydrophobic interactions constitute the main driving force of
this process. They arise mainly because the strong interactions between
the water molecules are impeded by the presence of the hydrocarbon tails of
the surfactant molecules. The resulting microstructures can be either in
the form of compact aggregates (micelles) or as spherical bilayers (vesicles). The micellar core as well as the spherical shell of the vesicles
constitute nonpolar micro-environments which can accommodate nonpolar or
slightly polar solutes. The low solubility of the latter molecules in
water is thus enhanced by the presence of these microstructures in solution. The disolution of solubilizates in the micellar aggregates can
occur in two different ways. The solubilizate molecules can either be
located among the hydrocarbon tails of the surfactant molecules or form, in
addition, a core inside the aggregate. The first case is usually referred
to as solubilization and the second as microemulsification. Whereas the
nonionic surfactants as well as the double chain surfactants can lead by
themselves to the formation of microemulsions, the ionic surfactants need
the cooperation of a medium length alcohol - a cosurfactant. A simple
molecular thermodynamic approach of micellization was suggested earlier[6].
That treatment was found to predict satisfactorily the observed critical
micelle concentration and average aggregation numbers for a variety of
nonionic, ionic and zwitterionic surfactants. The formation of vesicles
and of mixtures of micelles and vesicles have also been predicted. The
treatment was later extended to the case of solubilization[7,8]. The scope
of the present paper is to demonstrate again that the above procedure has
predictive power and can be applied to all the types of aggregation behavior from micelles to microemulsions. Various contributions to the free
energy of formation of an aggregate are identified and simple expressions
are proposed for their calculation. While the number of examples treated
by us is large, the procedure will be illustrated here by considering the
effect of alcohols on the surfactant aggregation as well as their effect on
solubilization and microemulsion formation, particularly on the transition
from solubilization to microemulsions. A schematic representation of
various types of aggregates is given in Fig. 1.

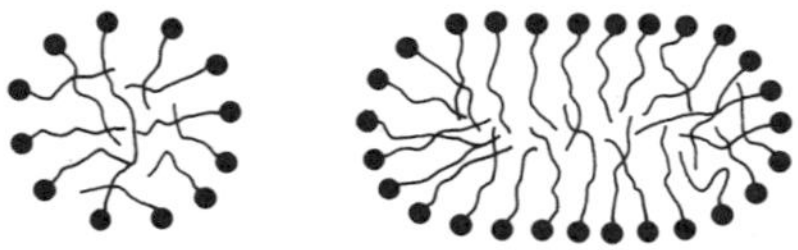

MICELLES

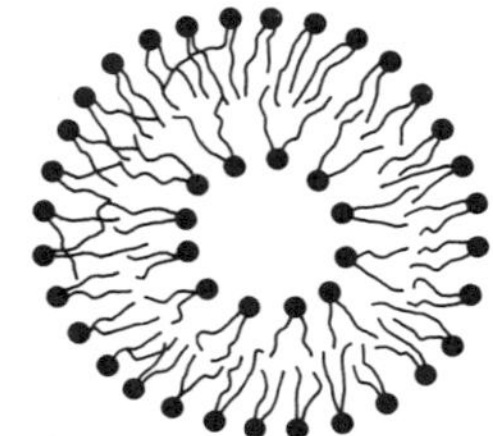

VESICLES

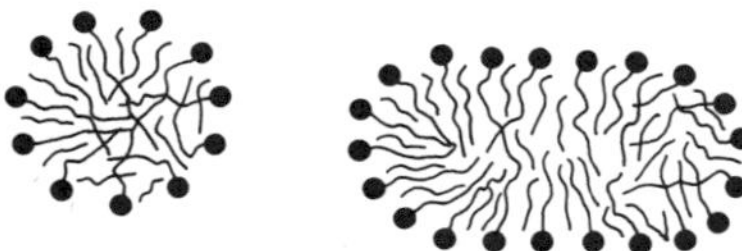

SOLUBILIZATION

MICROEMULSION
OIL/WATER

Fig. 1.   Schematic representation of various kinds of aggregates.

THE SIZE SPECTRUM AND THE FREE ENERGY OF SURFACTANT AGGREGATION

To better emphasize the main physical features involved in the present calculations, let us first discuss the more "simple" case of dilute surfactant solutions. By considering the aggregates of different sizes as distinct species and neglecting the interactions between them (because the system is dilute), the following expression is obtained for the size spectrum of aggregates[4,5]:

$$X_n = X_1^n \exp(-(\mu_n^0 - n\mu_1^0)/kT),\tag{1}$$

where X_m is the mole fraction of the micelles which contain n surfactant molecules, X_1 is the molar fraction of the singly dispersed surfactant molecules, k is the Boltzmann constant, T is the absolute temperature and μ_n^0 and μ_1^0 are the standard chemical potentials of an aggregate containing n molecules and of the singly dispersed molecules, respectively, defined for infinite dilution.

The standard free energy change of aggregation $\mu_n^\circ - n\mu_1^\circ$ can be written as the sum of a number of contributions:

$$\Delta\mu_n^0 = \mu_n^0 - n\mu_1^0 = n\Delta\mu_{HC/W}^0 + n\Delta\mu_c^0 + n\sigma(a - a_s)$$
$$- nkT \ln(1-a_p/a) + \Delta\mu_{electr}. \qquad (2)$$

The first term on the right hand side represents the standard free energy change when the tail of the surfactant is transferred from the solvent to a liquid hydrocarbon phase. However, the aggregation behavior is not entirely equivalent to such a process. In micellization, the hydrocarbon tails are constrained because of the location of the polar head groups on the micelle–water interface, while in a hydrocarbon phase they are free to possess any orientation. The second term in Eq. (2), $n\Delta\mu_c^\circ$, accounts for this constraint. The head groups of the surfactant molecules do not shield completely the hydrocarbon core of the micelle from water. The free energy of formation of the interface between the two is accounted for by the third term, where σ is the interfacial tension between hydrocarbon and water phases, a is the micellar core surface area per amphiphile and a_s is the area, again per amphiphile, that is shielded from water by the polar head group. The fourth term is due to the surface exclusion caused by the finite size of the head group, a_p being the cross–sectional area of the head group. The last, fifth term, represents the free energy due to the electrostatic interactions. There are also dispersion and dipole–dipole interactions between the head groups, which are, however, neglected here.

The core surface area per amphiphile is given for spherical micelles by $a = 4\pi r_O^2/n$, where $r_O = (3nv_S/4\pi)^{1/3}$ and v_S is the molecular volume of the hydrocarbon tail. At a particular value of n, the radius r_O becomes equal to the extended chain length l_S ($l_S = 1.50 + 1.26n_c$ (Å), where n_c is the number of carbon atoms in the hydrocarbon tail of the surfactant). Starting with this value of n, the micelle can no longer maintain a spherical shape and it will be assumed that it acquires the shape of a cylinder with hemispherical ends. The latter micelles grow at constant radius l_S by increasing their length L, where $L = (nv_S - 4\pi l_S^3/3)/\pi l_S^2$.

EVALUATION OF THE TERMS OF EQ. 2

Experimental investigations concerning the solubility of the hydrocarbon molecules in water[3,9] indicate that the free energy change for the transfer of an aliphatic hydrocarbon chain from water to a hydrocarbon phase at 25°C is given by

$$\Delta\mu_{HC/W} = -(2.05+1.49n_c)kT. \qquad (3)$$

Because of the constraint imposed on the hydrocarbon tails by the location of the head groups at the interface, the above expression for the free energy change should be corrected. A simple evaluation of this correction can be made by observing that the hydrocarbon tail can rotate in a bulk hydrocarbon phase in all three directions, whereas in a micelle it can rotate only around the long axis. Such a model provides a correction of about +0.25kT for each CH_2 group. An alternate way is to assume that

$$\Delta\mu_c = (\alpha' + \alpha''n_c)kT$$

and to determine the coefficients α' and α'' which provide good agreement
between the measured and calculated values of the CMC for a homologous
series of surfactants.

One thus obtains

$$\Delta\mu_c^{\circ} = -(0.50 - 0.24n_c)kT. \tag{4}$$

The interface between the micellar core and water is similar to a
hydrocarbon-water interface. For this reason the value of σ is chosen to
be 50 dynes/cm. The area a_s, which is shielded from contact with water, is
taken to be equal to the cross-sectional area a_h of the hydrocarbon chain
(21 Å^2) when a_p, the cross-sectional area of the head group, is greater
than a_h (because the molecules of water can still penetrate in this case to
the region around the hydrocarbon chain) and equal to a_p when $a_p < a_h$. In
the calculation of the molecular volumes of the hydrocarbon tails, values
of 27 Å^3 are used for each CH_2 group and 54 Å^3 for each CH_3 group.

The calculation of the electrostatic free energy is a difficult task,
primarily because:

(1)   the interface is extremely complex, and
(2)   only some of the head groups are dissociated, the remaining forming,
      probably, ionic head group-bound counterion pairs.

In addition, the degree of dissociation is a function of the size and shape
of the micelle[10]. This dependence arises because of two opposite tend-
encies: the dissociation increases the entropy and allows hydration of
ions; it, however, also increases the electrostatic energy. The minimum of
the free energy determines the value of the dissociation constant. Therefore, when the distance between two surfactant molecules is larger, the
electrostatic energy will be smaller and a greater dissociation is expected
to occur. The distance between two surfactant molecules is larger on a
spherical than on a cylindrical surface of equal radius. As a result, the
degree of dissociation is expected to be greater for the spherical than for
the cylindrical micelle of the same radius. The detailed calculations
carried out by Beunen[10] provide the value of the dissociation constant in
terms of size, shape and electrolyte concentration. They are, however, too
complex to be included in the extensive calculations involving a size
distribution. In the present calculations, a simple expression, based on
the linearized Poisson-Boltzmann equation, is employed for spherical
micelles[4]

$$\Delta\mu^0_{electr} = (n_{ion}^2 e^2 \beta / 2\varepsilon r)[(1 + \kappa a_i)/(1 + \kappa a_i + \kappa r)], \tag{5a}$$

where r is the radius of the surface where the charges are located ($r = r_o +$
δ, r_o being the radius of the micellar core and δ the separation distance
between the core and the charges), κ is the reciprocal Debye length
($\kappa = (c_1 + c_{Add})^{1/2}/(3.08 \cdot 10^{-8})\text{cm}^{-1}$ at 25°C for an 1-1 electrolyte, c_1 being
the molar concentration of the singly dispersed surfactant and c_{Add} the
molar concentration of the added salt), n_{ion} is the number of charges
(considered equal to the number of surfactant molecules for a 1-1 ionic
surfactant), a_i is the radius of the counterion, e is the electronic
charge, ε is the dielectric constant. A universal value of 0.46 for β was
found to give good agreement between calculated and experimental values of
the CMC for a wide variety of surfactants and the same value was used in
the present calculations.

For cylindrical micelles whose length to diameter ratio is less than 3 (i.e. for $L/2l_s \leqslant 3$), the expressions valid for spherical micelles are still used with the equivalent radius $r_{eq} = [4\pi(l_s+\delta)^2 + 2\pi(1+\delta)L]/4\pi^{1/2}$ replacing $r = l_s + \delta$ in Eq. (5a). For sufficiently long cylindrical micelles (i.e. for $L/2l_s > 3$), the following expression is employed for the cylindrical part of length L:

$$\Delta\mu^0_{electr} = \frac{n^2_{ion}e^2\beta}{\varepsilon L} \left\{ \frac{K_o(\kappa r)}{\kappa r K_1(\kappa r)} + \ln(1 + a_i/r) \right\}, \tag{5b}$$

where K_o and K_1 are modified Bessel functions of order 0 and 1, respectively. In Eq. (5b), $r = l_s + \delta$. For the hemispherical ends, Eq. (5a) is employed. The area a is calculated as for a cylinder, for the cylindrical part, and as for a sphere, for the spherical ends.

THE SIZE SPECTRUM AND THE FREE ENERGY OF AGGREGATION IN THE
PRESENCE OF ALCOHOLS

The equations of the previous sections can be easily extended to micellization in the presence of alcohols. The size distribution of aggregates containing g molecules of surfactant and m molecules of alcohol has, in dilute systems, the form:

$$X_{gm} = X^g_{s1} X^m_{a1} \exp - \{(\mu^0_{gm} - g\mu^0_{s1} - m\mu^0_{a1})/kT\}, \tag{6}$$

where X_j are mole fractions and μ^o_j are standard chemical potentials (j = gm for aggregates, j = sl for the singly dispersed surfactant molecules, j = al for the singly dispersed alcohol molecules).

The standard free energy change can be written, by an obvious generalization of Eq. (2), as

$$\begin{aligned}
\Delta\mu^0_{gm} = \mu^0_{gm} - g\mu^0_{s1} - m\mu^0_{a1} &= g[\Delta\mu^0_{HC/W} + \Delta\mu^0_c]_s \\
&+ m[\Delta\mu^0_{HC/W} + \Delta\mu^0_c]_a + \sigma[na - (ga_s + ma_a)] \\
&- nkT\cdot\ln[(1 - (ga_{ps} + ma_{pa})/na)] + \Delta\mu^0_{electr} \\
&+ kTg\cdot\ln\{gv_s/(gv_s + mv_a)\} + kTm\cdot\ln\{mv_a/(gv_s + mv_a)\}.
\end{aligned} \tag{7}$$

In Eq. (7), the subscripts s and a refer to the surfactant and alcohol, respectively, n = g + m and all the other symbols have the same meaning as in Eq. (2). The micelles are assumed spherical as long as $r_0 \leqslant (gl_s + ml_a)/(g + m) \equiv r_c$, where $r_0 = \{3(gv_s + mv_a)/4\pi\}^{1/3}$. The surface area per molecule is given for spherical micelles by $an = 4\pi r_0^2$. For cylindrical micelles of radius r_c, with $L/2r_c \leqslant 3$, the area per amphiphile is given by $an = (4\pi r_c^2 + 2\pi r_c L)$, where the length L is given by $L = (gv_s + mv_a - \frac{4}{3}\pi r_c^3)/\pi r_c^2$. For cylindrical micelles with $L/2r_c > 3$, the area per amphiphile is taken as for a cylinder for the cylindrical part and as for a sphere for the hemispherical ends. The last term in Eq. (7) accounts for the entropy of mixing of the surfactant and alcohol tails. Equation (7) should also include free energy contributions due to the van der Waals interactions between the head groups, but are neglected here.

It is important to emphasize that Eqs. (6) and (7) are valid only for sufficiently long chain alcohols. The short chain alcohols are not incor-

porated into the micelles because of their strong favorable interactions
with water and weak interactions with the hydrocarbon tails of the surfactant. In addition, their presence affects the structure of the solvent,
making the latter more compatible with the hydrocarbon tails of the surfactant molecules. Therefore, they increase the CMC. In contrast, the
higher alcohols have more unfavorable interactions with water and relatively strong interactions with the hydrocarbon tails of the micelles. As
a result, they are incorporated into the micelles and decrease the CMC.

COMPARISON BETWEEN CALCULATED AND MEASURED CMC AND AVERAGE AGGREGATION
NUMBERS

A large number of experimental determinations regarding the critical
micelle concentration and aggregation numbers are available in the literature regarding both single surfactants as well as surfactants and alcohols.
The effect of alcohols has recently been the subject of extensive experimental investigations[11–19], in which, in addition to the CMC, the aggregation numbers were also determined.

The expressions of the previous section can be used to calculate the
average aggregation numbers for the different species and the critical
micelle concentration. The average aggregation number is defined as

$$\langle n \rangle = \langle g \rangle + \langle m \rangle = \sum_{g,m} nX_{gm} / \sum_{g,m} X_{gm} \qquad (n \geqslant 2) \tag{8}$$

and the average mole fraction of the surfactant in the aggregates may be
calculated from

$$\langle X_s \rangle = \sum_{g,m} gX_{gm} / \sum_{g,m} nX_{gm}. \tag{9}$$

The average aggregation numbers of the surfactant and alcohol are related
to $\langle n \rangle$ via expressions of the form

$$\langle g \rangle = \langle X_s \rangle \langle n \rangle. \tag{10}$$

The critical micelle concentration of a single surfactant in water is
calculated as the value of the monomer concentration X_{s1} at which

$$X_{s1} = \sum_{n=2}^{\infty} nX_n, \tag{11}$$

i.e. as the point at which half the amount of the total surfactant is
aggregated. Similarly, for alcohol containing micelles, the CMC is defined
as the monomer surfactant concentration X_{s1} at which

$$X_{s1} + X_{a1} = \sum_{g,m} gX_{gm} + \sum_{g,m} mX_{gm}. \qquad (g > 1, \ m > 1) \tag{12}$$

At the experimentally determined CMC, the fraction of aggregated
surfactant is usually small so that although the measured quantity is the
total concentration, this quantity is not very different from that of the
monomer. In addition, changing the fraction of aggregates from 0.5 to,
say, 0.05 in the definition of the CMC alters the calculated monomer concentration by less than 5%. Therefore one can equate the theoretical CMC
defined by Eq. (12) to the experimental value without serious error.

One can see from Eq. (6) that X_{gm} depends on X_{s1} and X_{a1}, i.e. the
mole fractions of the singly dispersed surfactant and alcohol molecules,

46

respectively. The average aggregation numbers have been therefore calculated, for fixed values of the ratio $\alpha_{sl} = X_{sl}/(X_{sl} + X_{al})$, as a function of the total concentration c_T of the surfactant and alcohol mixtures. The computations show that the average aggregation number and the average molar fractions in the micelles have a very weak dependence on the total concentration c_T (between the CMC and 100 CMC), for fixed values of α_{sl}. This weak dependence of the average aggregation number on c_T can be explained intuitively on the basis of the plausible assumption that the aggregate area per surfactant molecule is mainly dependent on the interfacial interactions. The molar fraction in the micelles and therefore the interfacial interactions are expected to depend on α_{sl} only. As a result, the aggregate area per surfactant molecule should also depend only on α_{sl}. For spherical micelles,

$$a_s = \frac{4\pi\{3(gv_s + mv_a)/4\pi\}^{2/3}}{g} = f(\alpha_{sl}).$$

Since, in addition.

$$\frac{\langle g \rangle}{\langle m \rangle} = f_1(\alpha_{sl}),$$

one can conclude that $\langle g \rangle = f_2(\alpha_{sl})$.

Similar considerations can be extended to cylindrical micelles.

An important quantity in the calculations is the total mole fraction of the surfactant on a solvent free basis given by the expression:

$$\alpha_{st} = \frac{X_{sl} + \Sigma g X_{gm}}{X_{sl} + \Sigma g X_{gm} + X_{al} + \Sigma m X_{gm}}, \tag{13}$$

where the singly dispersed molecules are excluded from the summations. The mole fraction α_{st} was calculated as a function of the total molar concentration (surfactant + alcohol), c_T, for various fixed values of α_{sl}. Fixing the values of α_{st} and c_T one can therefore obtain the value of α_{sl}, and hence one can calculate the average aggregation numbers.

Tables 1 to 5 contain some of the computed results and the comparison with experiment. A more complete presentation of the calculations is available elsewhere[20].

It can be seen that for ionic surfactants the CMC decreases with the addition of the alcohol. This is not surprising. The CMC-decreasing effect is due to the solubilization of the alcohol molecules into the micelle with their polar groups in the palisade layer. This increases the average distance between the charges, thus reducing the repulsive interactions between the charged head groups. In addition, the entropy of mixing of the two molecular species also generates a favorable free energy change which enhances the tendency for micellization. The greater hydrophobic interactions that occur for longer chains favor the aggregation process. Therefore, it is expected that the higher alcohols will bring about a greater CMC-decrease. Of course, for a given alcohol, the greater its amount, the smaller the CMC. The CMC-decreasing effect of alcohols on nonionic surfactants is expected to be small compared to ionic surfactants, because the electrostatic interactions are absent in the latter case. All these trends are observed both experimentally and in the calculations. It

Table 1. CMC Values for SDS in Water + Alcohol Mixtures at 25°C

Alcohol Concentration (mMol/litre)	Critical Micelle Concentration (mMol/litre)		
	Experimental		Calculated
	Ref.(13)	Ref.(17)	
0	8.01	8.20	7.88
1-Buthanol			
43.74	6.96	7.20	5.30
116.90	5.32	5.85	3.90
204.40	4.18	4.95	2.90
1-Pentanol			
15.41	6.86	6.00	5.50
31.15	5.70	4.85	4.50
64.45	4.17	3.65	3.10
1-Hexanol			
2.683	7.32	6.90	6.25
5.584	6.61	5.70	5.55
9.227	5.89	4.50	4.85
12.918	–	3.85	4.31
26.200	–	2.00	2.92
1-Heptanol			
0.735	7.54	–	6.45
1.918	6.89	–	5.60
3.178	6.45	–	4.90

Table 2. Comparison of the Aggregation Numbers for SDS $\langle g\rangle$ + Alcohol $\langle m\rangle$ Systems at 25°C

SDS (Mol/litre)	Alcohol (Mol/litre)	Experimental*		Calculated	
		$\langle g\rangle$	$\langle m\rangle$	$\langle g\rangle$	$\langle m\rangle$
0.0081	0.0	62	–	60	–
1-Butanol					
0.0337	0.109	63.2	21.5	50	24
0.0330	0.326	40.7	41.4	44	40
0.132	0.326	48.7	37.5	44	37
1-Pentanol					
0.0339	0.0923	49.0	46.0	49	38
0.136	0.018	68.6	6.1	59	6
0.134	0.129	61.2	39.0	49	34
0.135	0.073	62.0	22.5	54	20
1-Hexanol					
0.0341	0.0156	62.4	19.4	57	19
0.0340	0.0320	49.4	31.4	53	34
0.0338	0.0800	38.5	64.4	48	66
1-Heptanol					
0.0341	0.0141	60.7	25.1	60	22
0.0340	0.0282	52.7	43.7	56	43
0.0339	0.0423	42.6	53.1	56	58

* Experimental values from Reference (12)

Table 3. CMC Values for Alkyltrimethyl Ammonium Bromides as a Function of the Number of Carbon Atoms n_c in the Alkyl Chain, in Water at 25°C.

n_c	Critical Micelle Concentration (Mol/litre)	
	Experimental*	Calculated
8	2.9×10^{-1}	2.81×10^{-1}
10	6.4×10^{-2}	5.96×10^{-2}
12	1.5×10^{-2}	1.36×10^{-2}
14	3.4×10^{-3}	3.13×10^{-3}
16	8.5×10^{-4}	6.42×10^{-4}

* Experimental data from Reference (11)

Table 4. CMC Values for Potassium Dodecanoate in the Presence of Alcohols at 10°C

Additive Concentration (Mol/litre)	Critical Micelle Concentration (Mol/litre)	
	Experimental*	Calculated
0.0	2.90×10^{-2}	3.34×10^{-2}
1-Butanol		
2.722×10^{-2}	2.80×10^{-2}	2.72×10^{-2}
7.08×10^{-2}	2.63×10^{-2}	2.36×10^{-2}
1.65×10^{-1}	2.27×10^{-2}	1.83×10^{-2}
2.74×10^{-1}	1.86×10^{-2}	1.44×10^{-2}
4.22×10^{-1}	1.30×10^{-2}	1.08×10^{-2}
1-Pentanol		
9.259×10^{-3}	2.80×10^{-2}	2.78×10^{-2}
2.389×10^{-2}	2.64×10^{-2}	2.39×10^{-2}
5.583×10^{-2}	2.29×10^{-2}	1.86×10^{-2}
1.175×10^{-1}	1.61×10^{-2}	1.31×10^{-2}
1.780×10^{-1}	9.42×10^{-3}	9.39×10^{-3}
1-Hexanol		
8.148×10^{-3}	2.60×10^{-2}	2.44×10^{-2}
1.889×10^{-2}	2.20×10^{-2}	1.89×10^{-2}
3.917×10^{-2}	1.45×10^{-2}	1.31×10^{-2}
1-Heptanol		
1.418×10^{-3}	2.76×10^{-2}	2.69×10^{-2}
3.921×10^{-3}	2.50×10^{-2}	2.22×10^{-2}
6.441×10^{-3}	2.26×10^{-2}	1.93×10^{-2}

*Values from Reference (18)

Table 5. CMC Values for Polyoxyethylene Lauryl Ether ($C_{12}E_{31}$) in the Presence of Alcohols at 20°C

Additive Concentration (Mol/litre)	Critical Micelle Concentration (Mol/litre)	
	Experimental*	Calculated
0.0	3.10×10^{-4}	2.95×10^{-4}
1-Butanol		
2.525×10^{-3}	3.09×10^{-4}	2.80×10^{-4}
5.278×10^{-3}	3.06×10^{-4}	2.78×10^{-4}
1.062×10^{-2}	3.02×10^{-4}	2.72×10^{-4}
2.530×10^{-2}	2.91×10^{-4}	2.56×10^{-4}
4.699×10^{-2}	2.75×10^{-4}	2.36×10^{-4}
1-Pentanol		
2.450×10^{-3}	3.04×10^{-4}	2.72×10^{-4}
5.014×10^{-3}	2.98×10^{-4}	2.64×10^{-4}
9.750×10^{-3}	2.86×10^{-4}	2.50×10^{-4}
2.172×10^{-2}	2.56×10^{-4}	2.19×10^{-4}

*Values from Reference (14)

is surprising, that the extremely simplified model employed provides in many cases such good agreement with experiment.

SOLUBILIZATION AND MICROEMULSIONS

The approach employed in the previous sections is now extended to the aggregation in systems containing surfactant, alcohol, solubilizate (hydrocarbons), water and electrolyte. A transition from solubilized aggregates to microemulsion aggregates is thus identified when the amount of alcohol is increased. This transition is examined in terms of the chain length of the alcohol, the nature of the solubilizate, chain length of the surfactant and electrolyte concentration. As already mentioned in the Introduction, the micellar environment enhances the solubility of the hydrocarbon molecules in water. The latter molecules can penetrate either among the hydrocarbon tails of the amphiphiles or can form, in addition, a core of molecules. The inclusion of the hydrocarbon molecules among the tails of the surfactant molecules is associated with an increase of the water-hydrocarbon contact but also with an increase in entropy because of the mixing of the two species. The large micelles can no longer possess a spherical shape but instead acquire the shape of rods, which in the present calculations are assumed to be cylinders with hemispherical ends. A microemulsion forms because the spherical globules which contain a core of hydrocarbon molecules expose a smaller surface area to the water phase than the long rods containing solubilizate molecules among the tails of the surfactant. The alcohol molecules, whose cooperation is, in general, necessary for the formation of a microemulsion, decrease the hydrocarbon-water contact by their adsorption in the interfacial layer among the surfactant molecules. For ionic surfactants, they also increase the distance between the charged groups thus decreasing the electrostatic interactions between them. Because the chains located in the core no longer mix with the tails of the surfactant and cosurfactant, there is an entropy loss compared to solubilization. This entropy loss is, however, more than compensated by a more advantageous interfacial free energy.

Let us consider a multicomponent system consisting of surfactant, cosurfactant (alcohol), solubilizate (a nonpolar or slightly polar liquid), a polar liquid (here water) and electrolyte. Aggregates of types I and II (type I = solubilizate microstructure and type II = microemulsion microstructure) of different sizes and compositions are considered as distinct species. The aggregates of type II are considered always spherical, whereas the aggregates of type I are considered spherical when they are sufficiently small and cylindrical with hemispherical ends when the equivalent radius of their volume exceeds the limiting value r_c determined by the length of the surfactant and alcohols molecules. The system is dilute and therefore the interactions between the different species can be neglected.

The minimization of the Gibbs free energy of the entire system leads to the following generalized form of Eq. (2) for the size and composition spectrum of the aggregates:

$$X_{gjm} = X_{sl}^g X_{sol}^j X_{al}^m \exp - \{(\mu_{gjm}^0 - g\mu_{sl}^0 - j\mu_{sol}^0 - m\mu_{al}^0)/kT\}, \tag{14}$$

where X_{gjm} is the mole fraction of an aggregate which contains g surfactant molecules, j solubilizate molecules and m molecules of alcohol, X_{sl}, X_{sol} and X_{al} are the mole fractions of the singly dispersed surfactant, solubilizate and alcohol molecules, respectively, and the μ^0 are the standard chemical potentials. The difference $\Delta\mu_{gjm}^o = \mu_{gjm}^o - g\mu_{sl}^o - j\mu_{sol}^o - m\mu_{al}^o$ represents the standard free energy change when g single surfactant molecules, j single solubilizate molecules and m single alcohol molecules form an aggregate.

The various contributions to $\Delta\mu_{gjm}^o$ are very similar to those already considered in the previous sections. In the present case, there are, however, two kinds of aggregates. In type I aggregates, the solubilized molecules are present among the tails of the surfactant and alcohol molecules and contribute to the entropy of mixing. In type II aggregates, a part of the solubilizate and alcohol molecules constitute a core of their own and do not mix with the surfactant and alcohol tails of the interfacial layer, while another part is present in the latter layer. The standard free energy change can be therefore written in the form:

$$\Delta\mu_{gjm}^0 =$$

$$= \left[g(\Delta\mu_{HC/W}^0 + \Delta\mu_c^0)_s + j(\Delta\mu_{HC/W}^0 + \Delta\mu_c^0)_{so} + m_I(\Delta\mu_{HC/W}^0 + \Delta\mu_c^0)_{aI}\right.$$

$$\left.+ m_c(\Delta\mu_{HC/W}^0 + \Delta\mu_c^0)_{ac}\right] + \left[\sigma(A - ga_s - m_I a_a)\right]$$

$$- kT(g + m_I)\ln\left\{(A - ga_{ps} - m_I a_{pa})/A\right\} + \Delta\mu_{electr}^0$$

$$+ kT\left[\left\{g\ln\frac{gv_s}{gv_s + j_I v_{so} + m_I v_a} + j_I\ln\frac{j_I v_{so}}{gv_s + j_I v_{so} + m_I v_a}\right.\right. \tag{15}$$

$$\left.+ m_I\ln\frac{m_I v_a}{gv_s + j_I v_{so} + m_I v_a}\right\} + \left\{j_c\ln\frac{j_c v_{so}}{j_c v_{so} + m_c v_a'}\right.$$

$$\left.\left.+ m_c\ln\frac{m_c v_a'}{j_c v_{so} + m_c v_a'}\right\}\right].$$

In Eq. (15) the subscripts s, so and a refer to the surfactant, solubilizate and alcohol, respectively, the additional subscripts I and c to the interfacial region and to the core, a is the area of the core which a particular molecule (identified by the subscript) shields from contact with water, A is the surface area of the aggregate, v is the volume of the hydrocarbon tail of the species identified by the subscript and v_a' is the molecular volume of the cosurfactant. For type I aggregates $j_C = m_C = 0$, whereas for type II aggregates the alcohol and hydrocarbon molecules are distributed between the interfacial and core regions. Hence,

$$m = m_I + m_c \tag{16a}$$

and

$$j = j_I + j_c. \tag{16b}$$

The first term in Eq. (15) represents the standard free energy change of transfer of the hydrocarbon tail of the surfactant, solubilizate, hydrocarbon tail of the alcohol and of the entire alcohol molecule from water into a hydrocarbon environment. However, as already noted in previous sections, the interior of the interfacial layer of an aggregate differs from that of a bulk hydrocarbon phase because of the ordering imposed by the location of the head groups of the surfactant and cosurfactant at the aggregate-water interface. The terms $\Delta\mu_c^\circ$ account for this correction. Eq. (4) is used to evaluate the values of $\Delta\mu_c^\circ$ for the surfactant and cosurfactant molecules present in the interfacial layer. For the aliphatic solubilizate, $\Delta\mu_c^\circ$ is taken to be zero, because its ends are not constrained to be located at the interface. Similarly, for the alcohol present in the core $\Delta\mu_c^\circ$ is also taken to be zero. For the aromatic solubilizates, their presence in type I aggregates probably gives rise (because of the rigidity of the aromatic ring) to a slightly increased local packing density which generates an enhancement in the van der Waals interactions. If this effect is taken to be $-$ 1kT for each aromatic molecule, good agreement with experiment is obtained[8]. For type II aggregates, this effect is taken to be zero even in the interfacial layer, because the latter layer is not as rigid as in the type I aggregates. A different, perhaps more plausible, interpretation of the above aromaticity effect is suggested below.

The second term in Eq. (15) represents the free energy of formation of the water-core interface. The interfacial tension σ is taken to be equal to the interfacial tension between the aliphatic hydrocarbon and water, i.e. $\sigma = 50$ dynes/cm. However, the interfacial tension between the aromatic hydrocarbons (which are slightly polar) and water is smaller, $\sigma \simeq 35$ dynes/cm. When the latter molecules constitute the solubilized molecules, the interfacial tension will be smaller than 50 dynes/cm. Because all our calculations involve the value of 50 for the interfacial tension, the negative contribution needed to obtain agreement with experiment, noted in the previous paragraph, might arise as a result of the lower interfacial tension between water and core.

The third term is a result of the surface exclusion, the fourth is due to the electrostatic interactions and the fifth represents the entropy of mixing of the species present in the interfacial region plus the entropy of mixing of the species present in the core region.

Eqs. (14) and (15) are now employed to examine the transition from solubilization to microemulsions. An anionic surfactant, sodium dodecylsulfate (SDS) is considered as the surfactant, n-butanol, n-pentanol or n-hexanol as cosurfactants and toluene, n-hexane or n-decane as the solubilizates. The solubilized or microemulsion phase is considered in equilibrium with an excess hydrocarbon phase.

Using Eqs. (14) and (15), one can calculate the average aggregation numbers $\langle g \rangle$, $\langle j \rangle$, $\langle m \rangle$ of the surfactant, solubilizate and alcohol molecules in the aggregates from the expressions

$$\langle g \rangle = \Sigma g X_{gjm}/\Sigma X_{gjm}, \tag{17a}$$

$$\langle j \rangle = \Sigma j X_{gjm}/\Sigma X_{gjm} \tag{17b}$$

and

$$\langle m \rangle = \Sigma m X_{gjm}/\Sigma X_{gjm}. \tag{17c}$$

The fractions of alcohol and solubilizate in the interfacial region of the type II aggregates are given by

$$\langle j_I \rangle / \langle j \rangle = \Sigma j_I X_{gjm}/\Sigma j X_{gjm} \tag{18a}$$

and

$$\langle m_I \rangle / \langle m \rangle = \Sigma m_I X_{gjm}/\Sigma m X_{gjm}. \tag{18b}$$

TRANSITION FROM SOLUBILIZATION TO MICROEMULSION FORMATION

The results of the calculations for SDS + n-pentanol + toluene are plotted in Figs. 2 to 4. The molar ratio of total toluene solubilized to total SDS is plotted against the molar ratio of total alcohol/total surfactant in Fig. 2. For small values of the latter ratio only type I aggregates form. When the ratio is between 0.17 and 0.6, both type I and type II aggregates coexist, while for values larger than 0.6, only type II aggregates exist. Additional information is provided by Fig. 3. While the ratio of toluene to SDS is almost the same for the two types of aggregates in their coexisting region, the pentanol/SDS ratio for type II is significantly larger than for type I. Also the number of SDS molecules per aggregate is greater for type II. The core area per surfactant molecule is plotted in Fig. 4. The values are comparable in the coexistence region for the two types but increase appreciably in the type II region. The fraction of toluene in the interfacial region decreases significantly in the type II region. The calculations also indicate that almost all the alcohol molecules are located in the interfacial region.

In order to evaluate the effect of the alcohol chain length, computations concerning the aggregation of SDS + toluene in the presence of various alcohols (n-butanol to n-hexanol) have also been carried out[21]. In each case, there is a small region in which aggregates of types I and II coexist. The area per surfactant molecule as well as the fraction of solubilizate in the interfacial region are almost the same for all three alcohols. The total amount of toluene which is solubilized increases, however, with increasing chain length of the alcohol at a given molar ratio of total alcohol to total SDS.

When toluene was replaced with n-hexane or n-decane as solubilizates, the region of type I aggregation has virtually disappeared being present only for very small amounts of alcohol[21]. The occurrence of type I aggregates for toluene and their absence for the aliphatics is a result of the aromaticity effect discussed in a previous section. For the rigid planar aromatic molecules to be accommodated in the structure of the type I aggregates, which is relatively rigid but has still some flexibility, a small variation in the packing density of the aggregate is necessary, which enhances the van der Waals interactions. This does not happen for the flexible aliphatic molecules in type I aggregates and for aromatics in the

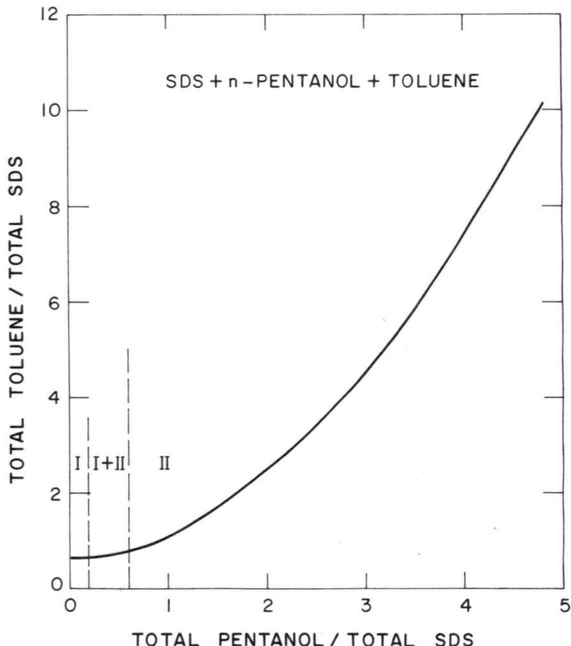

Fig. 2. Molar ratio of toluene solubilized to total SDS against the molar ratio of total n-pentanol to total SDS. This dependence is independent of the concentration of SDS if the latter is greater than about 0.1 M.

more flexible interfacial region of type II aggregates. The favorable contribution of this effect delays the occurrence of type II aggregates for aromatics. An alternate interpretation, along the lines mentioned in the previous section, is also possible.

Concerning the amount of aliphatics solubilized, the calculations show that, for the same ratio of alcohol/surfactant, the molar ratio of aliphatic molecules/surfactant molecules is greater for the smaller size solubilizates.

Some of these computational results can be explained intuitively on the basis of the plausible assumption that the aggregate area per surfactant molecule is primarily determined by the interfacial interactions, namely by the aggregate core-water interfacial free energy, by the surface exclusion and by the electrostatic interactions among the head groups. Therefore, the area per surfactant molecule should depend on the molar ratio of alcohol to surfactant and be largely independent of the size of the solubilizate. As a result, at a given ratio of alcohol to surfactant, the radius of the aggregate should be the same and the number of the smaller solubilizates larger. The fraction of solubilizate in the interfacial region decreases with increasing molecular volume of the solubilizate. This happens probably because the volume available in the interfacial region is the same and can therefore accommodate a larger number of small solubilizates.

The addition of electrolyte causes a decrease in the electrostatic repulsion between the head groups of the surfactant molecules and has the same effect as the alcohol, favoring the formation of type II aggregates. The area of the aggregate per SDS molecule is substantially decreased and

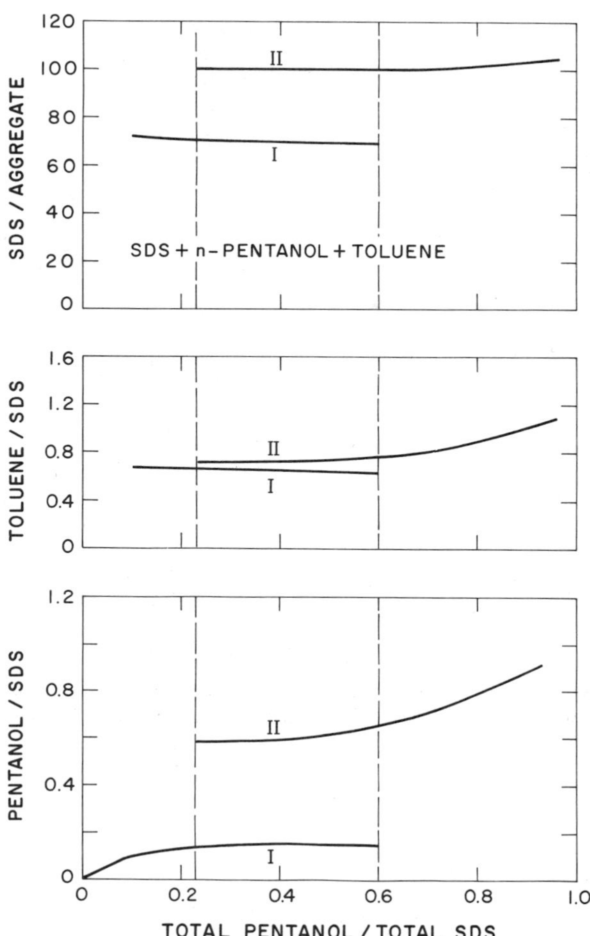

Fig. 3.   The number of molecules of the various components per aggregate or
per number of surfactant molecules in the corresponding type of
aggregate against the molar ratio of total pentanol/total SDS for
types I and II aggregates in the region of coexistence and in its
neighborhood.

the amount of toluene solubilized for a given ratio of alcohol to SDS is
greatly increased (see Figs. 5,6). It is interesting to note that the area
per surfactant molecule does not vary appreciably with the molar ratio
alcohol/SDS, whereas in the systems without electrolyte the variation was
important (compare Figs. 4 and 7) for type II aggregates. In the latter
case, the electrostatic interactions decrease slowly with increasing
alcohol concentrations. But, in the presence of 0.05 M NaCl the shielding
of the electrostatic interactions is sufficiently strong and the effect of
the alcohol/surfactant ratio is no longer important.

In order to examine the effect of the surfactant chain length, SDS was
replaced by sodium decylsulfate (SDeS). In this case there is a region of
coexistence of types I and II aggregates of smaller extent than for SDS.
The aggregate surface areas per surfactant molecule are almost the same for
SDeS and SDS. However, in the present case, the amount of toluene solubilized is greater. In addition, the number of SDeS molecules per aggregate

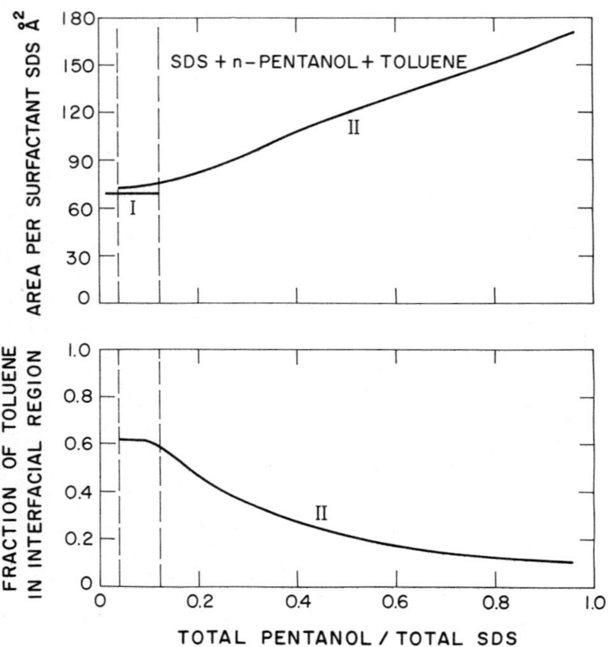

Fig. 4.   Area per surfactant molecule and fraction of toluene in the
interfacial region against the molar ratio of total pentanol/total
SDS.

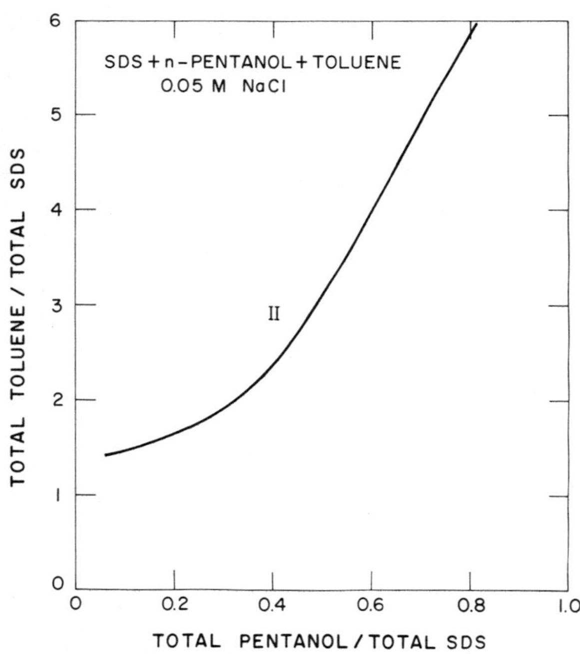

Fig. 5.   Molar ratio total toluene/total SDS against molar ratio total
pentanol/total SDS for 0.05 M NaCl.

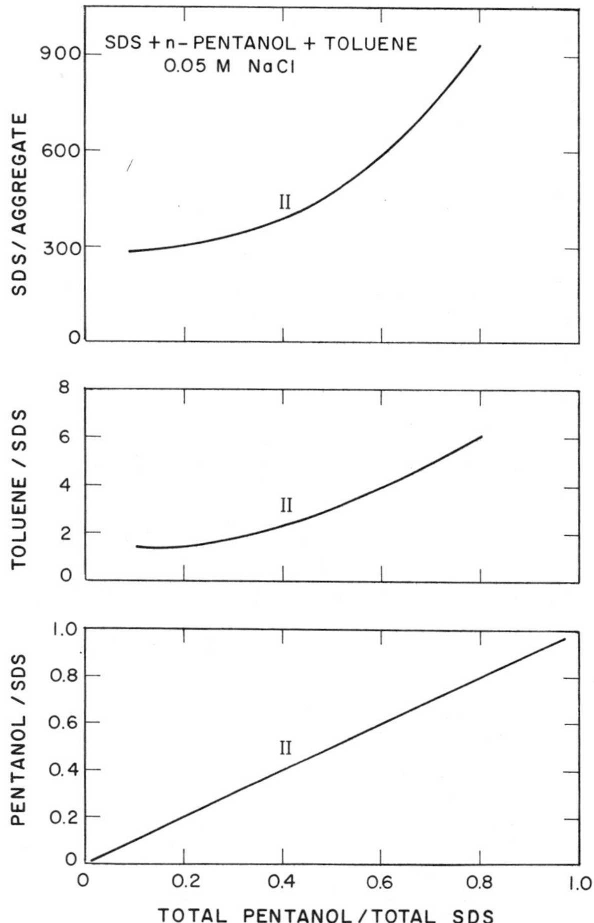

Fig. 6. Molecules SDS/aggregate, molar ratio toluene/SDS, molar ratio pentanol/SDS against molar ratio total pentanol/total SDS for 0.05 M NaCl.

is larger and the fraction of toluene in the interfacial region is somewhat decreased.

SUMMARY

A molecular treatment is developed to predict the aggregation behavior of single surfactants in solution, of mixtures of surfactants, of mixtures of surfactants and alcohols, as well as the aggregation behavior when solubilization and/or formation of microemulsions occur. Expressions are proposed for the standard free energy of aggregation in which the free energy of transfer of a hydrocarbon tail from water to the hydrocarbon environment of a micelle accounts for the ordering of the tails imposed by the constraint that the head groups should be located at the interface. In addition, the free energies due to the core-water interface and to the surface exclusion caused by the finite size of the head groups, as well as various entropies of mixing, are included in the standard free energy change. On this basis, the effect of the presence of alcohols on the critical micelle concentration and on the average aggregation number of

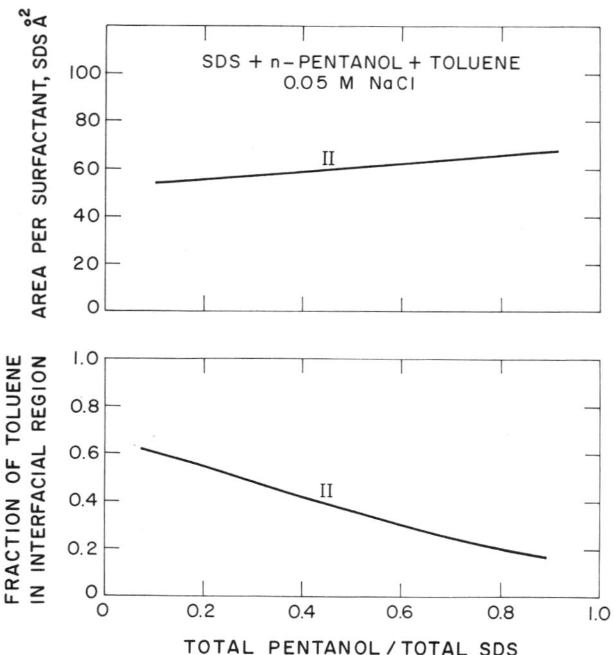

Fig. 7. Area per surfactant and molar fraction of toluene in the interfacial region against molar ratio of total pentanol/total SDS for 0.05 M NaCl.

micellar aggregates as well as the effect of alcohol on the solubilization of hydrocarbons and on the transition from solubilization to microemulsion formation are examined. General conclusions concerning the effect of the chain length of alcohol, surfactant and solubilizate are drawn from the present calculations. It is shown that the alcohol, ionic strength and shorter tails of the surfactant act in the same direction, namely, to increase the number of molecules of hydrocarbon which are solubilized, the latter number being greater for shorter chains of hydrocarbon molecules. It is also shown that for aromatic solubilizates there is a coexistence region of type I and type II aggregates (in type I the molecules are solubilized among the tails of the surfactant molecules, whereas in type II they form, in addition, a core of solubilized molecules) for sufficiently small values of the molar ratio of alcohol to surfactant molecules. On the other hand, for aliphatic molecules only the type II aggregates exist starting from very small values of the above ratio.

REFERENCES

1. P. Mukerjee, Adv. Colloid Interface Sci. 1:241 (1967).
2. P. Mukerjee, J.Pharm.Sci. 68:972 (1974).
3. C. Tanford, "The Hydrophobic Effect," Wiley, New York, (1973).
4. C. Tanford, J.Phys.Chem. 78:2469 (1974).
5. E. Ruckenstein and R. Nagarajan, J.Phys.Chem. 79:2622 (1975).
6. R. Nagarajan and E. Ruckenstein, J.Colloid Interface Sci., 60:221 (1977); 71:580 (1979).
7. R. Nagarajan and E. Ruckenstein, Sep.Sci.Technol. 16:1429 (1981).
8. R. Nagarajan, M. A. Chaiko and E. Ruckenstein, J.Phys.Chem. 88:2916 (1984).
9. M. H. Abraham, J.Chem.Soc.Faraday Trans.I. 80:153 (1984).

10. J. A. Beunen, and E. Ruckenstein, J.Colloid Interface Sci. 96:469 (1983).

11. R. Zana, S. Yio, C. Strazielle, and P. Lianos, J.Colloid Interface Sci. 80:208 (1981).

12. M. Almgren, S. Swarup, J.Colloid Interface Sci. 91:256 (1983).

13. K. Hayase, S. Hayano, Bull.Chem.Soc.,Japan 50:83 (1977).

14. N. Nishikido, Y. Moroi, H. Uehara, R. Matuura, Bull.Chem.Soc.,Japan 47:2634 (1974).

15. K. Hayase, S. Hayano, J.Colloid Interface Sci. 63:446 (1978).

16. K. Shirahama, T. Kashiwabara, J.Colloid Interface Sci 36:65 (1971).

17. A. K. Jain and R. P. B. Snigh, J.Colloid Interface Sci., 81:536 (1981).

18. K. Shinoda, J.Phys.Chem. 58:1136 (1954).

19. P. Lianos and J. Lang, J.Colloid Interface Sci. 96:222 (1983).

20. I. V. Rao and E. Ruckenstein, J.Colloid Interface Sci., (in press).

21. R. Nagarajan and E. Ruckenstein, in preparation.

PHASE BEHAVIOR OF TERNARY SYSTEMS H_2O - OIL - AMPHIPHILE

AS DETERMINED BY THE INTERPLAY OF THE OIL - AMPHIPHILE GAP

AND THE H_2O - AMPHIPHILE LOOP

M. Kahlweit, R. Strey and J. Jen

Max-Planck-Institut fuer biophysikalische Chemi
Postfach 2841, D-3400 Goettingen, FRG

1. INTRODUCTION

When studying multicomponent mixtures of the type H_2O - oil - nonionic
amphiphile one should distinguish between their macroscopic equilibrium
behavior and their microstructure. The detailed knowledge of their macroscopic properties, i.e. of their phase behavior as function of temperature
(and pressure, if necessary), is a prerequisite for performing meaningful
experiments in order to clarify their microstructure. The situation may be
compared with that of sailing a ship across the unknown waters. It may
then happen that one observes a high surf in some limited area which may
incline one to develop some exotic theory about the origin of that surf,
whereas a detailed map may reveal the surf to be caused by a reef below the
ocean surface. A similar situation may occur when performing, e.g., scattering experiments in homogeneous multicomponent liquid solutions without
knowing the location of the critical lines and their end points. It is for
this reason that we have emphasized the importance of studies of the phase
behavior of such systems. In a recently published review article[1] we
have shown that the phase behavior of ternary and quaternary systems with
either a lyotropic salt (NaCl) or a ionic amphiphile (SDS) as fourth component can be readily understood on the basis of the theory of phase diagrams as developed at the beginning of this century without assuming any
particular microstructure of the solutions.

This statement has met with some objections, namely that it may hold
for the 'simple' systems studied by us, whereas it may not hold for the
'real' systems encountered in application. Although we consider the simple
system as real as the 'real' systems, we have, in a more recently published
paper[2] shown that there is no qualitative difference between systems with
short chain and such with long chain nonionic amphiphiles. The quantitative difference lies in the fact that the long chain amphiphiles show a
higher solubilization capacity, i.e., that less amphiphile is needed to
prepare homogeneous solutions of equal masses of H_2O and oil. On the other
hand, with increasing chain length the amphiphiles show an increasing
tendency to form highly viscous lyotropic mesophases which makes the experiments more time consuming.

In order to study the phase behavior of such systems it is convenient
to vary the oils as well as the nonionic amphiphiles within homologous
series. The oils are then characterized by their carbon number k. As

nonionic amphiphiles we have chosen n-alkyl polyglycol-ethers C_iE_j, where i denotes the carbon number of the hydrophobic part of the molecule, whereas j denotes the number of oxyethylene groups of the hydrophilic part. In this case one may change two parameters: either i at constant j, or vice versa. Accordingly, we have studied the dependence of the phase behavior on the three parameters i, j, and k.

H_2O and oil are practically insoluble in each other. In order to increase their mutual solubility one has to add a third component which is equally soluble in both of them. In the case of H_2O and oil this is, evidently, an amphiphile whether nonionic or ionic. In this paper we shall concentrate on the effect of nonionic amphiphiles. Phase diagrams of multicomponent systems are nowadays mainly taught in metal physics but to a much lesser extend in physical chemistry. For those readers who are, therefore, not too familiar with reading phase diagrams, we recommend the introduction into phase equilibria by Prince[3].

A mixture of m components has m+1 independent variables, namely temperature T, pressure p, and m-1 concentration variables. For representing the phase behavior of a ternary system in three-dimensional space, one thus has to dispense with one of these variables. In view of the rather weak pressure dependence of the phase behavior it is appropriate to dispense with p by keeping it constant, e.g. at 1 bar. The phase behavior may then be represented in an upright prism with the Gibbs triangle H_2O-oil-C_iE_j as base T as ordinate.

II. PHASE BEHAVIOR - GENERAL FEATURES

A prerequisite for the interpretation of the phase behavior of a ternary system is the knowledge of the three corresponding binary phase diagrams H_2O-oil, oil-C_iE_j and H_2O-C_iE_j. The general features of these diagrams are demonstrated on top of Fig. 3 which shows the unfolded phase prism with schematical phase diagrams of the three binary systems. All three systems show for thermodynamic reasons a lower miscibility gap with an upper critical point.

For the system H_2O - oil, this gap extends far beyond the boiling point so that the very low mutual solubility between these two components increases only little between melting and boiling point of the mixture.

The upper critical point of the system oil-C_iE_j lies, in general, below its boiling point, its upper critical solution temperature (UCST) T_α depending strongly on both the hydrophobicity of the oil and the amphiphilicity of the amphiphile, i.e., on all three parameters i, j, and k. Unfortunately, T_α can only be determined with rather hydrophilic amphiphiles, since with the less hydrophilic ones, T_α lies below the melting point of the mixture. For this reason, Fig. 1 shows few data only. On top of this figure we have plotted the dependence of T_α on the carbon number k of some n-alkanes with C_iE_j as parameter. One finds a rise of T_α with increasing k of about 13 degrees per $\Delta k = 2$. In the center we have plotted the dependence of T_α on the carbon number i of C_iE_5, with k as parameter. One finds a drop of T_α with increasing i of about 22 decrees per $\Delta i = 1$. On the bottom of Fig. 1, finally, we have plotted the dependence of T_α on the number j of oxyethylene groups of C_6E_j, with k as parameter. One finds a rise of T_α with increasing j of about 32 degrees per $\Delta j = 1$.

The binary system H_2O-C_iE_j shows in addition to its lower miscibility gap which lies, in general, below its melting point, an upper miscibility gap which is, for thermodynamic reasons, a closed loop with an upper as well as a lower critical point. Position and shape of this loop, in par-

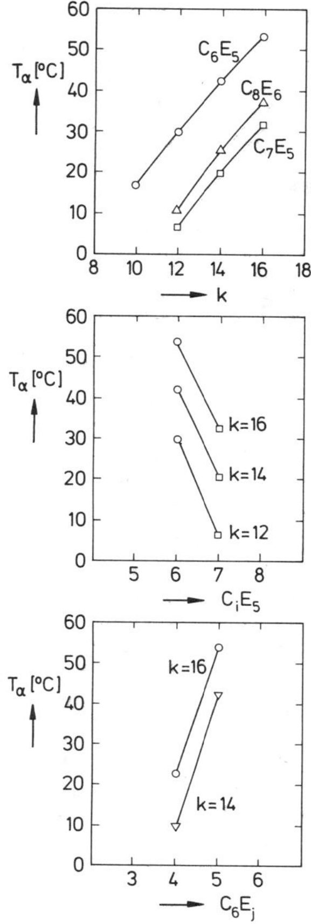

Fig. 1. Dependence of the UCST T_α of the (lower) oil (k) – C_iE_j gap on the parameters i, j, and k. Top: Dependence on k with $C_i^iE_j^j$ as parameter; center: dependence on i at constant j = 5 with k as parameter; bottom: dependence on j at constant i = 6 with k as parameter.

ticular its LCST T_β, again depend strongly on the parameter i and j, but evidently not on k. Fig. 2 shows T_β versus i with j as parameter (left) and versus j with i as parameter (right). The experimentally determined values (points) were taken from various sources including own measurements. As one would expect, T_β drops with increasing i, and rises with increasing j, in both cases, however, flattening off at high i and j, respectively. For the interpretation of these curves one may consider the following crude model: the addition of a hydroxy group to n-alkane leads to an increased mutual solubility with H_2O and, at least for the short chain n-alkanols (C_iE_0), even to a separation of the miscibility gap into a lower and an upper one. Further increase of the hydrophilic part by addition of oxy-ethylene groups leads to a shrinking of this loop and thus to a raise of T_β. For the first oxyethylene groups, this raise appears to be almost linear. With increasing j, however, the effect of each added oxyethylene group becomes increasingly weaker – possibly because the oxyethylene groups adjacent to the alkyl chain become increasingly less effective – so that, at high j, the curve flattens off.

In qualitative agreement with this model, the intersections of the curves on the right with the ordinate (j = 0) are a measure for the effect

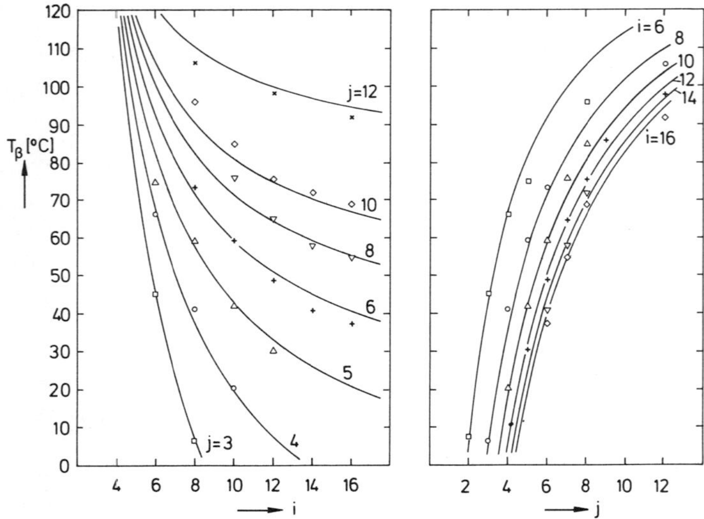

Fig. 2.   Dependence of the LCST T_β of the (upper) H_2O – C_iE_j loop on the parameters i and j. Left: Dependence on i with j as parameter; right: dependence on j with i as parameter. The full lines were calculated by means of equations (1) and (2), respectively.

of the hydroxy group on the mutual solubility between the n-alkanols and H_2O below the melting point: the higher i, the lower this intersection.

In order to describe this model by an empirical equation one may set for the dependence of T_β on j with i as parameter.

$$T_\beta = a(i) + b(i) \cdot j[1 + cj]^{-1}. \tag{1}$$

If one fits this equation to the experimental data on the right of Fig. 2, one finds (full lines)

$$a(i) = -1055 + 4285/i \quad , \quad b(i) = 1448 - 5152/i$$

$$\text{and} \qquad c = 1.2$$

and accordingly, for the dependence of T_β on i with j as parameter (full lines on the left of Fig. 2).

$$T_\beta = -1055 + \frac{1448\,j}{1 + 1.2\,j} + [4285 - \frac{5152\,j}{1 + 1.2\,j}]\,\frac{1}{i}\,. \tag{2}$$

Whether or not this empirical relation can be substantiated by theory[4] has to be seen. We further remark that the shape of the upper H_2O – C_iE_j loop is strongly effected by addition of electrolytes[5]: while the addition of lyotropic salts like NaCl makes the amphiphile effectively more hydrophobic and thus lowers T_β, that of hydrotropic salts and, in particular, of ionic detergents makes the nonionic amphiphile effectively more hydrophilic and thus raises T_β. These effects can be visualized by assuming salts with two strongly hydrated inorganic ions to compete with the oxyethylene groups in binding water, whereas salts with an organic ion tend to associate with the micelles of the nonionic amphiphile, increasing their overall hydration and thus raising the phase separation temperature.

64

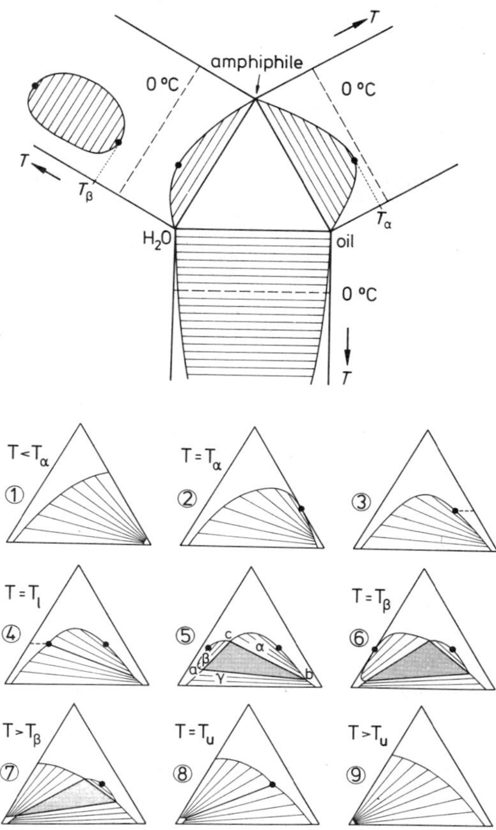

Fig. 3. Top: Unfolded phase prism of the ternary system H_2O oil nonionic amphiphile (schematic). Bottom: Horizontal sections through the phase prism with rising temperature T (for discussion see text).

If one adds oil to the H_2O - $C_i E_j$ mixture (above its cmc), i.e. proceeds into the phase prism, the first oil molecules will be solubilized in the hydrophobic cores of the micelles. With further increasing oil concentration, the solubilization capacity of the micelles will eventually be exhausted. As a consequence, the quasibinary mixture H_2O - ($C_i E_j$ + oil) can no longer 'carry' the oil which then separates as second phase. If one raises the temperature, even the quasibinary mixture separates into two phases at the 'end point' of its critical line cp_β, which starts at T_β and either descends or ascends - depending on the nature of the oil - into the phase prism, at T_l: the system separates into three phases, a H_2O rich, (a) a $C_i E_j$ rich (c) and an oil rich phase (b). The three-phase triangle thus appears as a 'lower critical tie line' which extends from the end point of the critical line cp_β to some point on the surface of the miscibility gap on the oil rich side. With further rising temperature the H_2O-$C_i E_j$ loop widens. Accordingly, phases (a) and (c) move increasingly further apart, the H_2O rich phase towards the H_2O edge of the prism, the $C_i E_j$ rich phase clockwise (if looked at from above) around the surface of the miscibility gap towards the oil edge of the prism.

In order to interpret the phase behavior on the oil rich side we have to consider the effect of H_2O on the lower miscibility gap between oil and $C_i E_j$. Let us first attempt to describe the dependence of T_α on the three parameters i, j and k. For that purpose one may apply Flory's approximation[6] which yields for the UCST of a binary system of components with different molar volumes.

$$T_\alpha [1 + (1 + \rho^{1/2})^{-1}]^2 = \Omega \tag{3}$$

where $\rho \equiv V_2/V_1$, and Ω is a measure for the (temperature independent) interaction energies between the molecules of the components. For a given nonionic amphiphile as component 2, one expects Ω to be some function of k. From Fig. 1 one finds for C_6E_5 a linear dependence, namely

$$\Omega_{C_6E_5} = 369 + 54.7 \; k. \tag{4}$$

Accordingly, one may attempt to set

$$\Omega_{C_iE_j} = K(i,j) + \kappa \cdot k \tag{5}$$

where the first term depends on i and j only, whereas the slope κ can be assumed in first approximation to be independent of i, j, and k. We presume, however, the apparent agreement between T_α and Flory's approximation to be rather accidental, since the corresponding relation for the critical volume fraction, $(\phi_2)_c (1 + \rho^{1/2}) = 1$, is not fulfilled: for a given amphiphile, $(\phi_2)_c$ drops faster with increasing ρ than predicted.

The effect of H_2O on T_α may be then interpreted as follows: dry C_iE_j can be looked at as a very hydrophilic 'oil'. As one adds water, the H_2O molecules are bound to the oxyethylene groups of the amphiphile making it increasingly more hydrophilic. As a consequence, the upper critical point of the quasi-binary system oil – (C_iE_j + H_2O) rises steeply with increasing H_2O concentration. To decrease the interaction between the hydrophilic part of the hydrated amphiphile and the oil, one can visualize the amphiphile to form association colloids with hydrophilic cores. With increasing water concentration, an increasing number of H_2O molecules is solubilized in the cores until the 'inverse' micelles are no longer able to 'carry' the water: the system separates into an oil rich and an amphiphile rich phase.

After the three-phase triangle has been formed at T_1 on the water rich side of the phase prism, this triangle is thus flanked by two-phase regions. Region β on the water rich side is, from this point of view, an extension of the upper H_2O – C_iE_j loop, whereas region α on the oil rich side is the extension of the lower oil – C_iE_j gap. Since with rising temperature the amphiphile becomes increasingly less hydrophilic[7], region β grows whereas region α shrinks until at T_u both the oil rich phase (b) and the amphiphile rich phase (c) of the triangle merge at the upper end point of the critical line cp_α on the oil rich side: the three-phase triangle disappears at its upper critical tie line which extends from the end point of cp_α to the water rich phase (a).

We thus propose to consider the formation of a three-phase body within the phase prism as resulting from the counteracting interplay between the quasibinary system H_2O – (C_iE_j + oil) and oil – (C_iE_j + H_2O). If this was an approach to reality, then one expects the position and extension of the three-phase body on the temperature scale to be mainly determined by the position of the upper H_2O – C_iE_j loop and the lower oil – C_iE_j gap, i.e. by T_β and T_α.

If T_α and T_β are sufficiently far apart, i.e., T_β much higher than T_α, the two quasibinary systems interact in such a way that the two critical

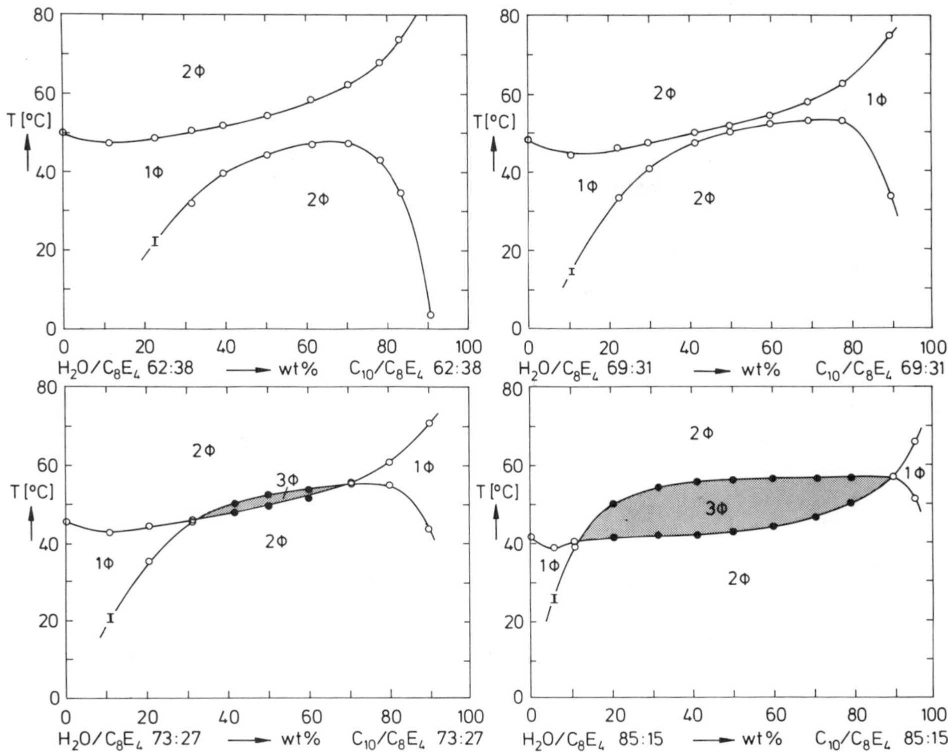

Fig. 4. Vertical sections through the phase prism parallel to the H₂O – oil side of the prism for the system H₂O – n-decane – C₈E₄ with decreasing C₈E₄ concentration (for discussion see text).

lines cp_α and cp_β form a connected critical line which ascends on the oil rich side to change with rising temperature, to the water rich side of the phase prism. If one now increases the hydrophobicity of the oil within a homologous series, i.e. raises T_α at constant T_β(!), the interaction will become increasingly stronger until the connected critical line will break at what is called a tricritical point[8,9]. From then on, one finds a three-phase body the extensions of which grow with further increasing hydrophobicity of the oil.

In order to demonstrate this interplay between the two quasibinary systems we have cut vertical sections through the phase prism parallel to the H₂O – oil – T plane of the prism at decreasing concentrations of the amphiphile. Fig. 4 shows four of such sections for a system with a three-phase body, namely H₂O – n-decane – C₈E₄. The left of each section shows the continuation of the H₂O – C₈E₄ system into the prism, the right that of the decane – C₈E₄ system. Accordingly, on the left the lower boundary of the upper two-phase region terminates at the lower boundary of the upper H₂O – C₈E₄ loop which lowers with decreasing C₈E₄ concentration to approach the LCST T_β. On the right, the upper boundary of the lower two-phase region terminates at the upper boundary of the lower decane – C₈E₄ gap the UCST T_α of which lies below the melting point. As one can see, the lower boundary of the upper two-phase region remains roughly at the temperature of the 'cloud point', whereas the upper boundary of the lower region rises steeply with increasing water concentration. As one decreases the C₈E₄ concentration, the two regions approach each other until they overlap thus shaping the three-phase body within the prism[10]. The lower boundary of

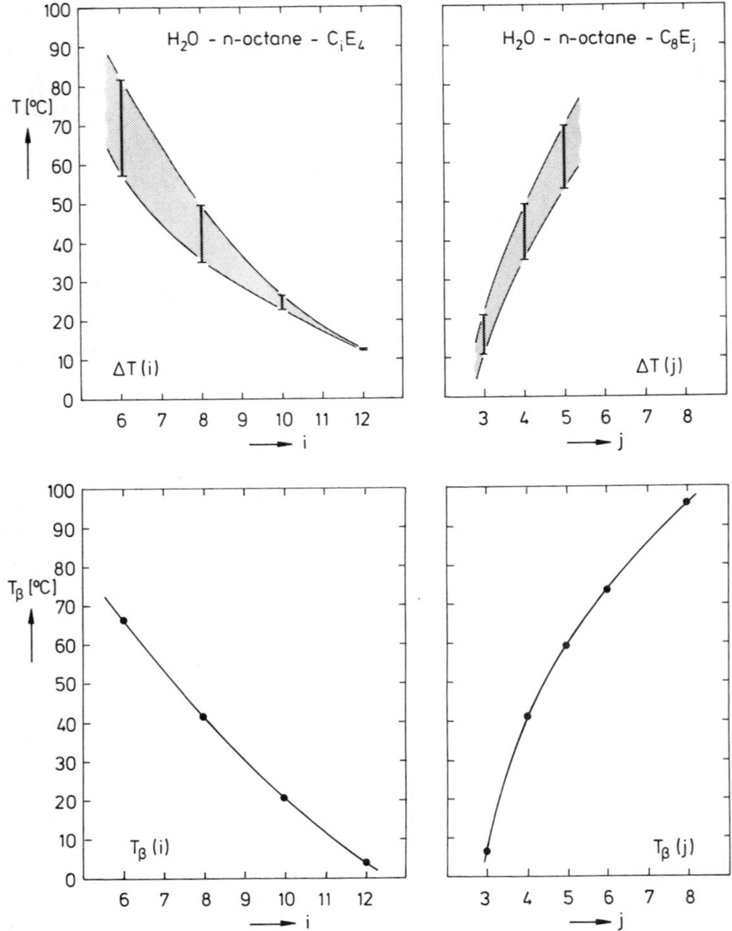

Fig. 5. Top: Three-phase temperature intervals (shaded) of the system H_2O – n-octane – C_iE_j. Left: versus i at constant j = 4; right: versus j at constant i = 8. Bottom: The corresponding LCST T_β of the binary systems H_2O – C_iE_j taken from Figure 2.

this body can be looked at as continuation of the upper H_2O – C_8E_4 loop, its upper boundary as that of the lower decane – C_8E_4 gap. From this one may conclude that the mechanism which leads to the phase separation at the lower boundary of the three-phase body is similar to that which enforces the phase separation in the quasibinary system H_2O – (C_iE_j + oil). The phase separation at the upper boundary of the three-phase body, on the other hand, should be similar to that in the quasibinary system oil – (C_iE_j + H_2O).

From Fig. 4 it further follows that the position of the three-phase body on the temperature scale is mainly determined by the position of the upper H_2O – C_iE_j loop, i.e. by T_β. The effect of the oil appears to be much weaker, expressing itself in a slight rise of the position of the three-phase body with increasing hydrophobicity of the oil, in accord with the dependence of T_α on k (see Fig. 1). This conclusion is supported by Figs. 5 and 6. Fig. 5 shows the dependence of the position of the three-phase interval $\Delta T = T_u - T_l$ on T_β for a given oil (n-octane).

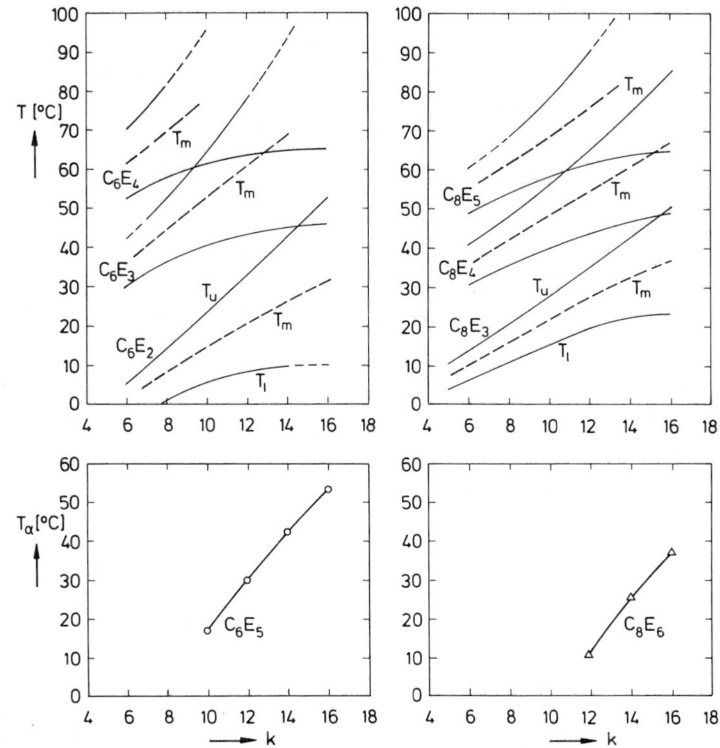

Fig. 6. Top: Three-phase temperature intervals of the systems H_2O – n-alkanes (k) – C_iE_j versus carbon number k of the alkanes. Left: For C_6E_j; right: for C_8E_j, taken from Figure 21 in ref. 1. Bottom: The corresponding UCST T_α of the binary system oil – C_iE_j taken from Figure 1.

On the left we have changed i at constant j = 4, on the right j at constant i = 8. The correlation between the position of the three-phase body and T_β (lower parts of Fig. 5) is evident.

Fig. 6 shows the dependence of the position of the three-phase intervals on the carbon number k of n-alkanes for given C_6E_j and C_8E_j. The dashed lines on the center of each cusp represent the mean temperature $\overline{T}$ = $(T_l + T_u)/2$ of the three-phase intervals. As one can see, the three-phase bodies lie much higher than T_α: while the cusps for C_6E_5 and C_8E_6 lie above the boiling point of the mixtures, their corresponding T_α (below) lie at ambient temperatures. The mean temperatures of the experimentally accessible cusps, however, show the same rise with k as T_α. These conclusions are in good agreement with the results published by Shinoda and Arai[11] who studied the correlation between the 'phase inversion temperature', i.e., between $\overline{T}$ and the cloud points of nonionic amphiphiles.

The fact that position and shape of the three-phase body is mainly determined by the position and shape of the upper H_2O – C_iE_j loop is supported by a comparison between a vertical section through the three-phase body erected on the center line of the base, i.e., on the line H_2O/oil = 1/1 with the shape of the loop. Fig. 7 shows this section through the three-phase body of the system H_2O – decane – C_8E_4. Also shown is the projection of the lower boundary of the H_2O – C_8E_4 loop onto that section, keeping the H_2O/C_8E_4 ratio constant:

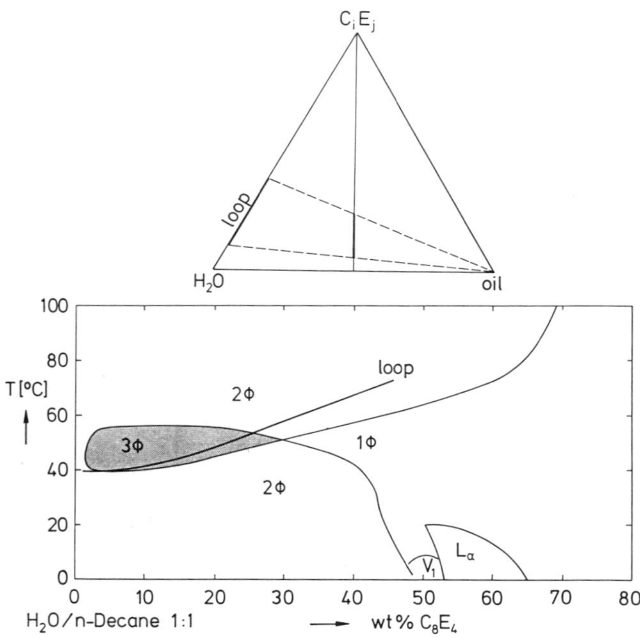

Fig. 7. Vertical section through the phase prism of the system H_2O – n-decane – C_8E_4 at H_2O/oil = 1/1 with section through the three-phase body (shaded). Also shown is the projection of the lower boundary of the upper H_2O – C_8E_4 loop onto that section (constructed as shown on top).

$$C_{(3)} = 100\ C_{(2)}/(200 - C_{(2)}),$$

$C_{(2)}$ and $C_{(3)}$ denoting the wt% of C_8E_4 in the binary and ternary system, respectively. As one can see, the lower boundary of the three-phase body as well as its continuation as lower boundary of the upper two-phase region has almost the same shape as the lower boundary of the H_2O – C_8E_4 loop. The fact that both boundaries coincide even temperaturewise is somewhat accidental: for a less hydrophobic oil like octane the three-phase body lies a little below the projection of the loop, whereas for a more hydrophobic oil like dodecane it lies a little higher.

These results indicate that the interaction between the oxyethylene groups and the H_2O molecules is much stronger than that between the alkyl chains and the oil.

For the detailed discussion of the phase behavior of ternary systems, based on these general features of the two counteracting binary systems, we refer the reader to ref. 1.

REFERENCES

1. M. Kahlwiet, R. Strey, Angew.Chem.(Engl.Ed). 24:654 (1985).
2. M. Kahlweit, R. Strey, P. Firman, J.Phys.Chem., 90:671 (1986).
3. A. Prince, "Alloy Phase Equilibria," Elsevier, Amsterdam, p. 119 ff. (1966).

4. See e.g., R. E. Goldstein, <u>J.Chem.Phys.</u>, 84:3367 (1986); and G. M. Thurston, D. Blankschtein, M R. Fisch, G. B. Benedek, <u>J.Chem.Phys</u>. 84:4558 (1986).

5. P. Firman, D, Hasse, J. Jen., M. Kahlweit, R. Strey, <u>Langmuir</u> 1:718 (1985).

6. E. A. Guggenheim, "Mixtures," Clarendon Press, Oxford 1952, p. 229.

7. For a discussion of the temperature dependence of hydrogen bonding see e.g.: J. C. Lang, <u>in</u>: "Physics of Amphiphiles: Micelles, Vesicles and Microemulsions," V. Degiorgio, and M. Corti, eds., North-Holland, Amsterdam, p. 364 ff. (1985).

8. B. Widom, <u>J.Phys.Chem.</u>, 77:2196 (1973).

9. R. B. Griffiths, B. Widom, <u>Phys.Rev.</u> A8: 2173 (1973).

10. Such sections through three-phase bodies were first published by K. Shinoda, and H. J. Takeda, <u>Colloid Interface Sci.</u>, 32:642 (1970).

11. K. Shinoda, H. Arai, <u>J.Phys.Chem.</u>, 68:3485 (1964).

AQUEOUS AND NON-AQUEOUS MICROEMULSIONS:

A COMPARISON OF SURFACTANT ASSOCIATION STRUCTURES

S. E. Friberg and Y. Liang

Chemistry Department
Clarkson University, Potsdam
New York, 13676 USA

INTRODUCTION

The aqueous microemulsion systems are by now fairly well described[1.2]: they have attracted a considerable attention due to the wave of interest in the tertiary oil recovery[3-5]. The phase behavior of the microemulsion systems have been related to the phases found in the well known Ekwall systems of water, surfactant and amphiphile[6] an approach that gave simple rules for the preparation of microemulsions[7.8]. A later systematics developed by the French school[9] made possible a treatment of multi-component systems using a less number of variables.

Recently, the research in the area of microemulsion has been extended to include systems in which the water is replaced by a polar organic substance such as formamide[10] or glycerol[11.12]. This development was expected extension of earlier contributions in micellar systems[13-15] and liquid crystals[16-19].

With the region of non-aqueous systems having been established, it appears justified to make a first comparison with the phase behavior in aqueous systems[6]. The experimental details are given elsewhere[20] and this tratment will be limited to a comparison of the actual results.

The evaluated microemulsion systems are stabilized by a combination of water soluble ionic surfactant combined and a medium chain length alcohol as cosurfactant.

Aqueous Microemulsion Systems

The basis for these microemulsions is the system of water, surfactant and the medium chain length alcohol: the cosurfactant. The essential features of the phase diagram are shown in Figure 1. with the aqueous micellar solution, the lamellar liquid crystal and the inverse micellar solution being involved in the microemulsion concept.

The W/O microemulsions are formed as a direct continuation of the inverse micellar solution as demonstrated in Figure 2[6]. The addition of hydrocarbon causes only small modifications of the solubility region and the interest may be focussed on the structure of the association colloids found in this region.

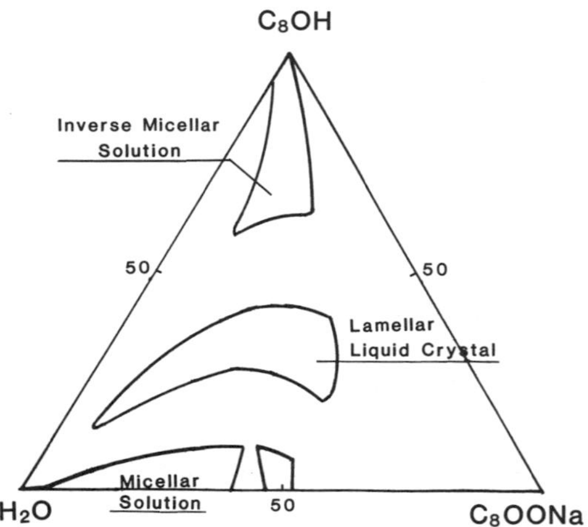

Fig. 1.   In the water (H$_2$O), surfactant (sodium octanoate, NaOOC$_8$), alcohol (octanol, C$_8$OH) system the aqueous micellar solution, the lamellar liquid crystal and the inverse micellar solution form the basis for the microemulsions.

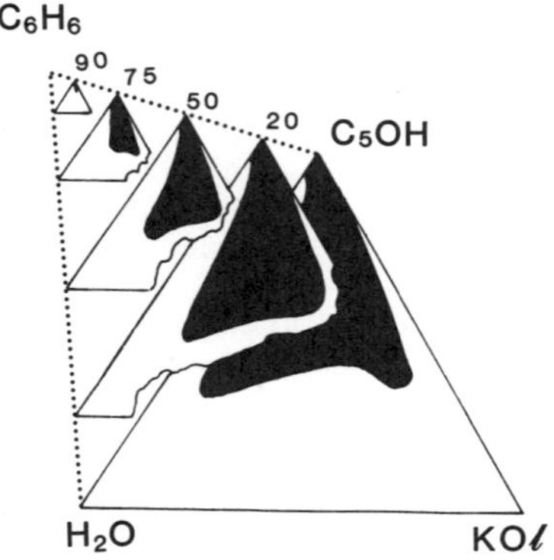

Fig. 2.   The W/O microemulsion regions (black) are a direct extension of the inverse micellar solution in the non-hydrocarbon system (21). H$_2$O = water; KOℓ = potassium oleate; C$_5$OH = pentanol; C$_6$H$_6$ = benzene.

Light scattering[21], NMR[22] and dielectric constant[23] determinations have given results, which agree on the following interpretation. At low water constent, there are no colloidal association structures present.   A few water and alcohol molecules are attached to the polar group of the monomeric surfactant and no aggregates of colloid size exist.   The results from light scattering are the most directly comprehensible results. The scattering intensity at low water content is at a level lower than that obtained from a pure solvent, benzene. Figure 3.

The postulate about monomeric surfactants in this part of the region has received further support from theoretical calculations on the stability of monomeric soap aggregate with attached water molecules[24].   The calculated free energy for such an aggregate versus a combination of crystalline soap and liquid water showed a lower value for a limited range of

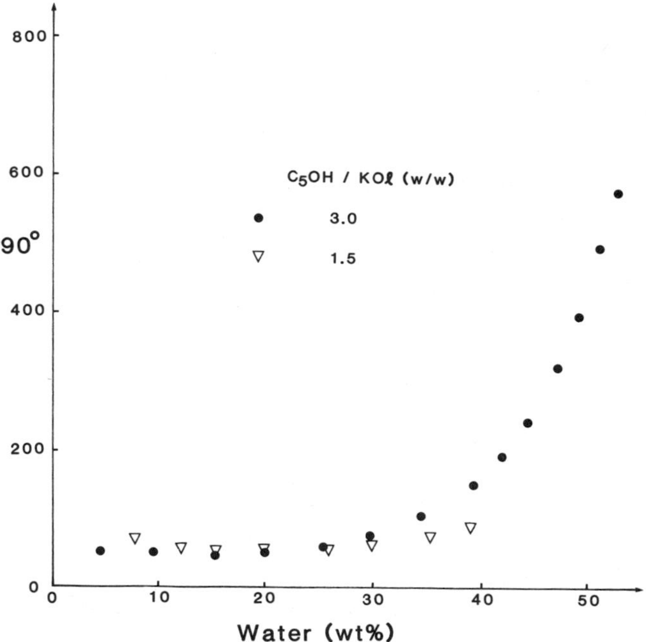

Fig. 3.   The intensity of scattered light from the inverse micellar solution, Figure 2, shows no presence of colloidal aggregates at water concentrations below 30%. The scattered intensity from pure benzene = 100 scale units[21].   ● = C5OH/KOℓ = 3; ▽ = C5OH/KOℓ = 1.5.

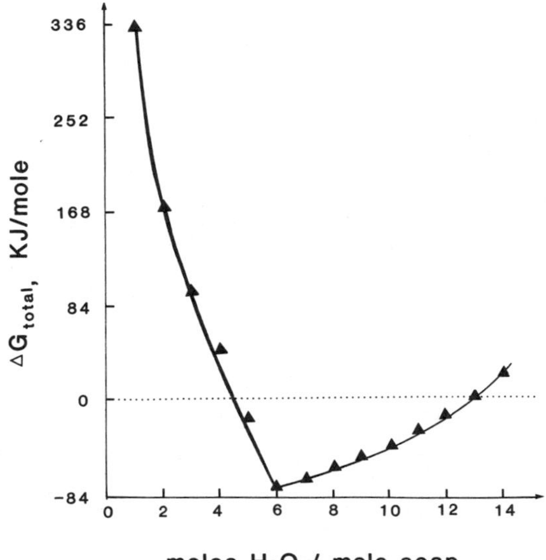

Fig. 4.   The free energy for a surfactant/water aggregate is lower than the level for crystalline soap plus liquid water in a limited water/soap ratio region[24].

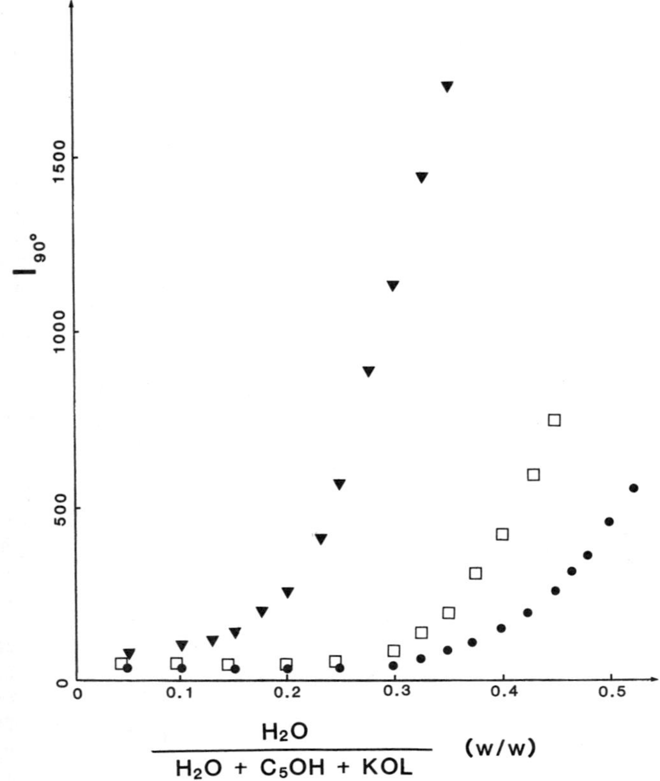

Fig. 5. Addition of hydrocarbon has an influence on the onset of colloidal
association in W/O microemulsion systems[21]. ● = 0% hydrocarbon;
□ = 50% Decane; ▼ = 50% Benzene.

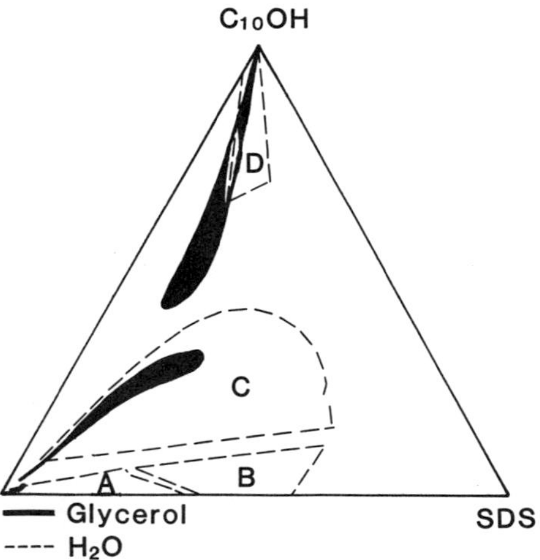

Fig. 6. A comparison of phase regions for water (---) and glycerol (——)
systems with sodium dodecyl sulfate (SDS) and decanol ($C_{10}OH$).
A = aqueous micellar solution; B = liquid crystal of hexagonally
close-packed cylinders; C = lamellar liquid crystal; D = inverse
micellar solution.

the number of water molecules per soap molecule, Figure 4. The lower limit
of this range of lower free energy for the aggregate would correspond to
the onset of solubility; the right hand limit of the solubility regions in
Figure 2. while the upper limit of the negative free energy difference
marks the onset of association of the surfactant molecules to form inverse
micelles.

It is essential to notice that this association to inverse micelles
does not take place for high surfactant/alcohol ratios. In this range the
light scattering value is low in the entire region, the surfactant colloid
association now takes place directly to a lamellar liquid crystal and no
aggregates of colloidal size are formed in the solution, Figure 3.

This onset of surfactant association is influenced only to a moderate
degree by the presence of aliphatic hydrocarbons, Figure 5. Aromatic
hydrocarbons, on the other hand, reduce the water concentration at which
association is initiated in a pronouned manner and it appears probable that
the aromatic hydrocarbons exert a modest molecular interation with the
polar group.

The essential factor for the W/O microemulsion region is the fact that
it is in equilibrium with the lamellar liquid crystalline phase. Any extension into higher water content of the inverse micellar phase on the W/O
microemulsion region must with necessity take place through destabilization
of the lamellar liquid crystal. This destabilization may be followed by a
comparison of the phase regions in the water, surfactant, alcohol system
when the alcohol is varied from hexanol to butanol[6]. The inverse micellar region is enlarged and the lamellar liquid crystalline area is reduced.
The reason for this modification of the solubility regions is fact that the
shorter alcohol induces more disorder into the lamellar phase and it
becomes destabilized. The interdependence of the microemulsion regions and
the lamellar liquid crystalline structure is straight forward for aqueous
systems. With this fact in mind a comparison becomes meaningful with the
conditions in the corresponding systems with glycerol instead of water.

Glycerol Microemulsions

The comparison between aqueous and non-aqueous microemulsions is
initiated by a comparison between the non-hydrocarbon systems, because
these phase equilibria determine the microemulsion regions to a large
extent.

A direct comparison between an aqueous and glycerol combination with
sodium dodecyl sulfate and decanol is shown in Figure 6. The phase diagram
is entirely different. The differences emanate from two factors. A higher
glycerol content is needed to bring the surfactant into solution in the
decanol and the solubility of the surfactant in the glycerol is negligible.
The second factor means that the normal micellar solution and the liquid
crystalline phase of hexagonally packed cylinders in the aqueous system are
absent in the glycerol system. The combination of two factors cause the
region of the lamellar liquid crystalline phase to be narrow in the
glycerol system.

Replacing decanol with the shorter chain length hexanol gives no
drastical changes of the aqueous system. All the phases are still present
albeit with some modifications in their regions. The diagram for the
glycerol system, on the other hand, is changed to a high degree. Now the
lamellar liquid crystalline phase is absent in the glycerol system and the
isotropic liquid solubility region extends uninterrupted from the hexanol
corner to the glycerol corner, Figure 7. The conditions in the hexanol

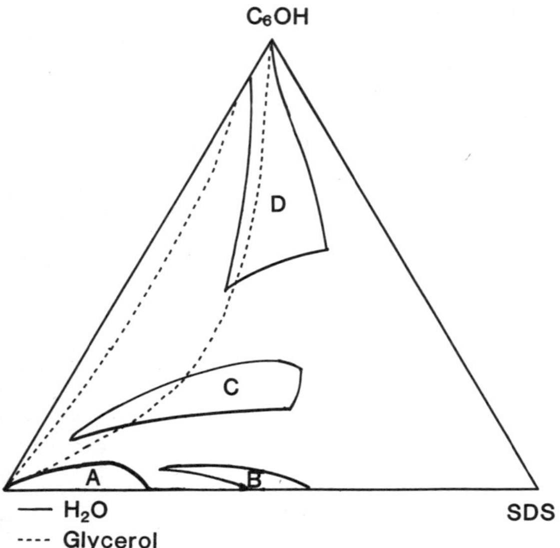

Fig. 7.   A comparison of phase regions for water (——) and glycerol (---)
          systems with sodium dodecyl sulfate (SDS) and hexanol (C_6OH).   A =
          aqueous micellar solution; B = liquid crystal of hexagonally
          close-packed cylinders; C = lamellar liquid crystal; D = inverse
          micellar solution.

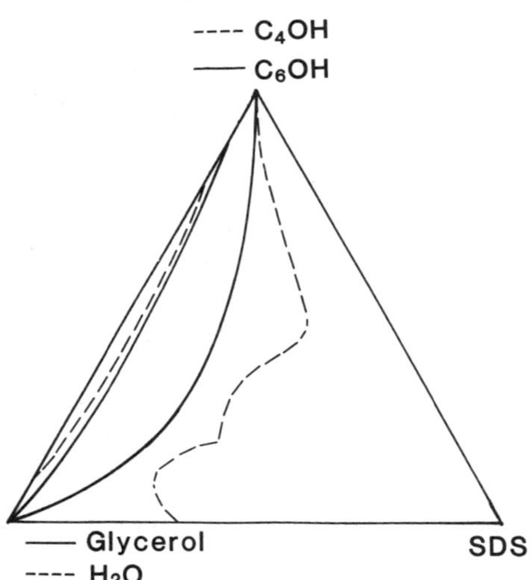

Fig. 8.   A comparison of the phase diagrams of sodium dodecyl sulfate (SDS)
          combined with glycerol plus hexanol (C_6OH) (——) or water plus
          butanol (C_4OH) (---).

glycerol system are similar to those found in the corresponding aqueous
system with butanol, Figure 8, except for the solubility of the surfactant
in the aqueous system.

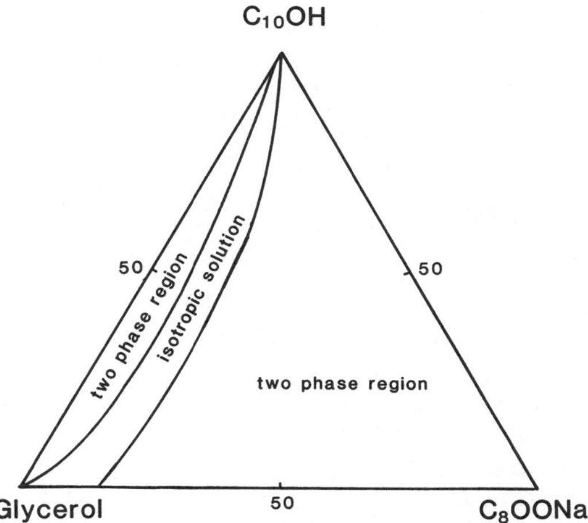

Fig. 9.   The system glycerol, sodium octanoate (C$_8$OONa) and decanol (C$_{10}$OH) gave no liquid crystal in spite of enhanced surfactant solubility in the glycerol in comparison with the sodium dodecyl sulfate system, Figure 7.

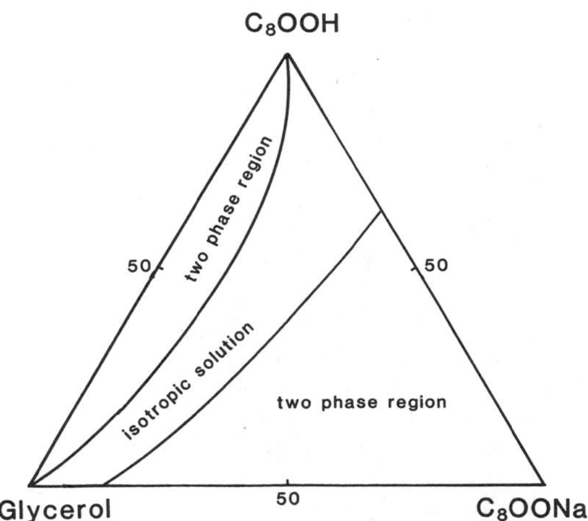

Fig. 10.   The combined solubility of sodium octanoate (C$_8$OONa) in both the glycerol and in the octanoic acid (C$_8$OOH) was not sufficient to give a lamellar liquid crystal.

     The glycerol lamellar liquid crystal shows less order than the aqueous system: the disorder induced by hexanol is sufficient to destabilize the lamellar liquid crystal while the aqueous system requires the shorter chain of butanol.

     One alternate explanation must be discarded. The lack of solubility of the surfactant in the glycerol is not the reason for the change in the phase diagrams as shown by the following examples. Sodium octanoate displays a pronounced solubility in glycerol, Figure 9, but there is no lamellar liquid crystal present. Instead, the solubility region reaches uninterrupted from the glycerol corner to the aqueous corner. An extension of the surfactant solubility in the hydrophobic corner showed as small an

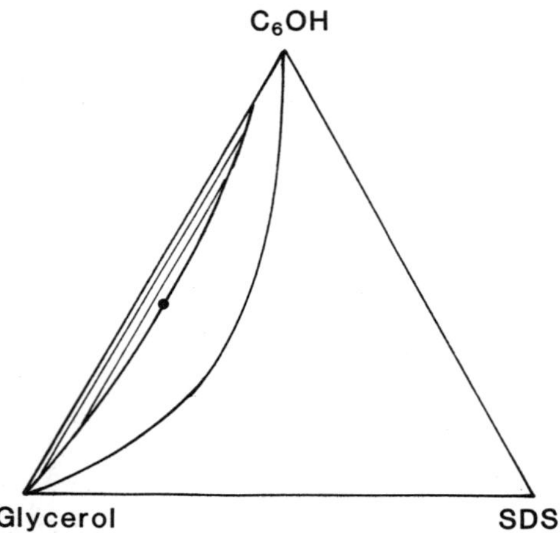

Fig. 11. The critical point in the solubility region in the glycerol, hexanol (C$_6$OH) and sodium dodecyl sulfate (SDS) was found at approximately equal amount of glycerol and hexanol, (●).

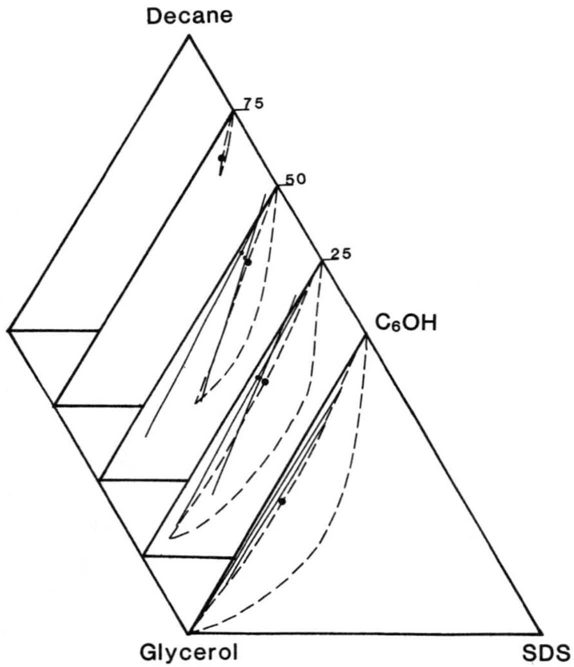

Fig. 12. Corresponding critical points were found in systems of glycerol, sodium dodecyl sulfate (SDS) and hexanol (C$_6$OH)/decane mixtures to 75% decane.

effect. Sodium octanoate is soluble in octanoic acid to a maximum soap/acid molecular ratio of 0.5[25]. Figure 10. but the combination of enhanced solubility in glycerol and in the hydrophobic amphiphile are not sufficient to give a lamellar liquid crystal.

The solubility region in the hexanol system necessitates the existence of a plait point an it is marked together with a few tie-lines in Figure 11. This solubility region is similar to the basis for the aqueous microemulsions, Figure 2, and addition of hydrocarbon to it should give solubility regions similar to those of W/O microemulsions in the aqueous system. Figure 12 shows these solubility regions for higher and higher amounts of hydrocarbon. The maximum solubility of glycerol is sufficiently high for the solution to be called glycerol-in-oil microemulsions and structure of these solutions is a pronounced interest. The main question is: Are they molecular solutions or do they contain discret glycerol aggregates in hydrocarbon? A preliminary answer is found in the tie-lines which showed convergence to one point in each of the solubility regions (solid circles, Figure 12); even in the one with 75% hydrocarbon.

Hence, these results indicate the glycerol-in-oil (G/O) microemulsions to be molecular solutions with critical points and that no long-life colloidal aggregates are present. These results have been confirmed by recent NMR investigations[26].

Conclusion

A comparison was made between water and glycerol systems combined with hydrocarbons, sodium dodecyl sulfate and a medium chain length alcohol. The results gave a significant difference between the two systems. The water-in-oil systems give microemulsions with well defined colloidal aggregates, the W/O microemulsion droplets. The glycerol system, on the other hand, gave no indication of long lived aggregates: the tie-lines coverged to a plait point for the glycerol-in-oil systems indicating them to be solutions with critical behavior.

Acknowledgement

This research has been supported by a grant from Dow Corporation.

REFERENCES

1. S. E. Friberg and R. L. Venable, Encydlopedia of Emulsion Technology $\underline{1}$ (4), 287 (1983).
2. I. D. Robb (Ed.) "Microemulsions", Plenum Press, New York, (1982).
3. D. O. Shah and R. S. Schechter (Eds.), "Improved Oil Recovery by Surfactant and Polymer Flooding", Academic Press, New York (1967).
4. D. O. Shah (Ed.), "Surface Pheonomena in Enhanced Oil Recovery", Plenum Press, New York (1981).
5. P. H. Doe, M. El-Amary, W. H. Wade and R. S. Schechter, J. Am. Oil Chem. Soc., $\underline{55}$, 505 (1978).
6. P. Ekwall in "Advances in Liquid Crystals", (G. H. Brown, Ed.), Academic Press, New York, (1975) P 1.
7. S. E. Friberg and I. Buraszewska, Prog. Colloid Polym. Sci. $\underline{63}$, 1 (1978).
8. M. Podzimek and S. E. Friberg, J. Dispersion Sci. & Technol., $\underline{1}$, 341 (1980).
9. J. Biais, P. Bothorel, B. Clin and P. Lalanne, J. Colloid Interface Sci. $\underline{80}$, 136 (1981).
10. I. Rico and A. Lattes, Nouveau J. De Chemie, $\underline{8}$, 424 (1984).
11. S. E. Friberg and M. Podzimek, Colloid & Polymer Sci., $\underline{262}$, 252 (1984).
12. P. D. I. Fletcher, M. F. Galal and B. H. Robinson, J. Chem. Soc. Faraday Trans. I, $\underline{80}$, 3307 (1984).
13. D. F. Evans, S. H. Chen, G. W. Schriver and E. M. Arnett, J. Am. Chem. Soc., $\underline{103}$, 481 (1981).

14. D. F. Evans, A. Yrmauchi and R. E. Casassa, J. Colloid Interface Sci., 88, 89 (1982).

15. D. F. Evans and B. W. Ninham, J. Phys. Chem., 87, 5025 (1983).

16. N. Moucharafieh and S. E. Friberg, Mol. Cryst. Liq. Cryst. 49, 231 (1979).

17. M. A. El-Nokaly, L. D. Ford, S. E. Friberg and D. W. Larsen, J. Colloid Interface Sci., 84, 228 (1981).

18. D. W. Larsen, S. E. Friberg and H. Christenson, J. Amer. Chem. Soc., 102, 6565 (1980).

19. L. Ganzuo, M. A. El-Nokaly rnd S. E. Friberg, Mol. Cryst. Liq. Cryst., 72, 183 (1982).

20. S. E. Friberg and Y.-Ch. Liang, To be published.

21. E. Sjoblom and S. E. Friberg, J. Colloid Interface Sci., 67, 16 (1978).

22. D. O. Shah and R. M. Hamlin, Science 171, 484 (1971).

23. M. Clausse, R. J. Sheppard, C. Boned and C. G. Essex in "Colloid and Interface Science", Vol. II, p. 233 Academic Press, New York (1976).

24. S. E. Friberg, T. D. Flaim and P. L. M. Plummer, ACS Symp. Proceedings, Washington, D.C., Series #272, pages 33-39 (1985).

25. S. Friberg, L. Mandell and P. Ekwall, Kolloid Z.u.Z., Polymere 233, 955 (1969).

26. B. Lindman, To be published.

THEORIES AND MODELS

SELF-ASSEMBLY: A BIASED RANDOM WALK THROUGH THE LITERATURE

B. Lindman and B. Ninham*

Physical Chemistry, 1, Chemical Center, University of Lund
P.O. Box 124, S-221 00 Lund, Sweden
*Department of Applied Mathematics, The Australian National
University, P.O. Box 4, Canberra 2601, Australia

INTRODUCTION

In his peroration the prophet of the Book of Ecclesiastes pronounced
that "of the making of books there is no end, and in too much study there
is much weariness of the spirit". That stern injunction is valid still. It
is an oft-repeated favourite of these authors. Then why bother to write
yet another review on micelles, vesicles, microemulsions and emulsions and
what more can usefully be said? The proper answer is that the sages of the
Ettore Majorana Centre for Scientific Culture decreed no paper and no
travel money. But on reflection there was some wisdom in that demand.
That is precisely because there are enough articles on our subject to fill
a large library, or two; and the technical literature is so very obstruse
and scattered that to the uninitiated it really is almost impossible to
extract any sense whatever. And in the end, since some better unity and
perspective on the business of association colloids did indeed emerge, our
obligation as privileged participants was to try to express that developing
consensus.

Others of our colleagues better practised in the real world will have
more to say on applications of microemulsion and vesicle technology, or
should: applications which range from enhanced oil recovery to the phar-
maceutical industry and modern cell biology. What we have tried to do is
rather to trace one path through the literature. Admittedly and necessarily that path reflects a particular world view. In so doing we hope to
throw into focus the key problems as we perceive them at this time. We have
borrowed freely from earlier papers in attempting to put together a coherent story.

WHY BOTHER?

From one point of view the exercise is unnecessary. The systems we
find under close scrutiny often appear arcane and academic. After all any
good product formulation chemist <u>knows</u> how to find a system which fits his
requirements best, and cost requirements impose limitations on what can
usefully be tried. If the problem is to solubilize the maximum amount of
oil and water with minimal surfactant, or to make a solution of particular
physical properties, or a system of ultralow surface tension, the pragmatic

approach is self evidently acceptable. One simply varies surfactant, (head
group or hydrocarbon chain), cosurfactant, oil, solute, counterion, salt,
temperature, pressure, whatever, until the right combination emerges. Or
more probably does not emerge, as the long experience of the oil industry
in enhanced oil recovery processes most readily attests. This, the seren-
dipitous approach, can be optimised only if we understand how changes in
these variables affect solution properties and microstructure.

There is another, more important reason for interest in our subject.
The present time is a period of transition during which physical chemistry
is beginning to move from a mastery of matter at a molecular level to the
design and understanding of hierarchies of aggregates of molecules. Biochemistry and materials science have been concerned into such systems from
the very beginning, but physical chemistry is just beginning to reach the
point where a rigorous background for understanding intermolecular forces
can begin to develop towards the same kind of understanding it has had for
interatomic forces and the covalent bonds that make up molecules. Some
very good progress has been made in quantifying the mechanisms that drive
the formation and set the structure of multimolecular aggregates. Parallel
advances have been made in understanding the nature of molecular forces
between surfaces below distances ~ 100 Å. The interplay between these two
areas provides new insights into the way that the physical chemistry of
self-assembly affects many biological processes. And that probably is
important.

APPROACHES TO SELF-ASSEMBLY

One approach to understanding is that of physicists, i.e., one ignores
detailed chemistry and attempts a global description. The powerful mathematical apparatus provided by the theory of phase transitions allows a
test for universality of critical exponents as a first guess. When that
approach fails, as it seems to for both surfactant-water and three phase
microemulsions, we can exploit new results which derive from the developing
area of mathematics and physics of random media; percolation theory and
fractal random walks provide a framework for description instead of the
more usual self-avoiding random walks which underlie the Ising model.
"Anomalous" critical behaviour which emerges from that description has
analogues in amorphous metal studies and transport in semi-conductors. The
problem with this approach is that it does not say anything about microstructure necessary for design, in the same way that knowledge of critical
exponents says nothing about molecular mechanisms (nucleation) in the usual
theory of phase transitions. For work along this line see other papers in
this book.

A different line of attack is as follows: One calculates the free
energy for micelles, droplets of oil in water, or vice versa, lamellar
phase, vesicles, liposomes, hexagonal, cubic and the bicontinuous phases.
At least when one component is present in excess, it is possible in principle to use the Maxwell construction to find tie lines, i.e. the equilibrium between two or more phases. This approach also suffers great
difficulties. The problems are several. Microemulsions are often bicontinuous, with a curvature set by a competition between surfactant tails
which oppose head group repulsion, and by volume fractions of components.
One has no idea of how to write down the entropy of such a system. In a
statistical mechanical description based on models, instead of thermodynamics, the partition function must average over size and <u>shape</u> of each
aggregate of a given class. There are some emerging signs of progress in
this area, particularly with inputs which will come to the field from work
on bicontinuous structures in inorganic chemistry and phospholipid systems.
The resurrection of some beautiful work from the last century on surfaces

of zero curvature suggests that the approach is not impossible in the long
term. Particularly is this so for ionic surfactant systems, where as we
shall see, the better understanding of intra- and interaggregate forces
operating which has emerged from both theory and experiment is really quite
hopeful.

Dilute Micellar Systems – Elementary Theories

Consider first surfactants in water. The foundations of theory were
first laid out by Tanford, who began with the law of mass action: If μ_N^o,
μ_1^o are the chemical potentials of a monomer in an aggregate of size N and
a monomer respectively, and X_N, X_1 the corresponding concentrations, then

$$\mu_N^0 + \frac{kT}{N} \ln \frac{X_N}{N} = \mu_1^0 + kT \ln X_1 . \tag{1}$$

The cmc is defined as

$$X_1 = \sum_{N>1}^{\infty} X_N$$

and the distribution of aggregates will be polydisperse unless μ_N^o is
sharply distributed. The problems with this approach are fairly severe.
For one, the <u>definition</u> of what we mean by a micelle is extremely difficult. It turns out to be possible to characterize self-assembly using
the law of mass action if (a) interactions between aggregates (and
monomers) are negligible, and (b) the hydrophobic interaction between
monomers which drives self assembly is short ranged and strongly
attractive. It is. However, there are further problems. The choice of
units of concentration is an awkward matter, and the problem of entropy for
a monomer confined to an aggregate is unresolved. There is also the
question of how the interface between the micelle "surface" and water is to
be defined. This is a question of small-system thermodynamics which has
not been properly resolved. In the end it can be shown, at least for
micelles, that dilute solution theory + statistical mechanics give a firm
basis for writing $N\mu_N^o$ = Bulk Helmholtz free energy + $P_o V$(aggregate) +
surface terms + terms due to curvature + molecular packing constants. The
pressure at which the chemical potential is to be evaluated is P_o (1
atmosphere), not the internal pressure, which may differ by terms $2\gamma/R$ due
to the Laplace pressure, where γ is the "surface tension" and R the radius.

If these caveats are accepted, then our problem reduces to the cal-
culation of $\mu_N^o - \mu_1^o = g_{HP} + g_S$, where g_{HP} is the (bulk) hydrophobic free
energy of transfer of a monomer from water to an assumed oil-like interior
of the micelle, and g_S is the surface free energy per monomer. In a zeroth
order theory we assume that $g_S = \gamma(a + \frac{a_o^2}{a})$; i.e. there is an attractive
head group interaction due to, e.g. exposure of hydrocarbon to water,
balanced by a repulsive interaction which could be due to hydration shells
around the head group, or electrostatics. This form is not to be taken
seriously as a quantitative description but contains the essential physics.
Then the equilibrium micelle is given by minimizing μ_N^o. Such a theory can
be extended to include oil (microemulsion droplets), curvature, packing
constraints and local packing criteria. It fits ionics quite well for
measured cmcs and aggregation numbers as a function of salt, as it must,
because the theory involves at least two hidden parameters among them the
head group area a_o in a bilayer, the unknown term γ, and the surface tension. But a useful criterion does emerge. This characterization predicts
that if a dimensionless parameter v/al be defined, where v is the known
hydrocarbon chain volume, a the area of the head group, and l a chain

length about 80% of the fully extended chain length, then spherical micelles form for $\frac{v}{al} < \frac{1}{3}$, polydisperse rods or cylindrical structures for $\frac{1}{3} < \frac{v}{al} < \frac{1}{2}$, and vesicles or bilayers for $\frac{1}{2} < \frac{v}{al} < 1$. The latter condition is satisfied by almost any double-chained surfactant, e.g. phospholipids. When $\frac{v}{al} > 1$, reverse structures form.

To illustrate how these rules work, consider the double-chained ionic surfactant dodecyl dimethyl ammonium bromide, for which $\frac{v}{al} \approx 0.83$ from X-ray analysis. The surfactant forms a cloudy lamellar phase in water. If we add to such a dispersion the single-chained micelle-forming surfactant dodecyl trimethyl ammonium bromide (same head group area, same chain length), with $v/al \approx \frac{1}{3}$, we expect the solution to clear, i.e., go over to small vesicles or rods which do not scatter light at $\frac{v}{al}$ (effective) =

$$\frac{(X_{DDAB} \times 0.83) + (X_{DTAB} \times (1/3))}{X_{DDAB} + X_{DTAB}} \approx 0.5,$$

where X are mole fractions of each. Indeed it does. This indicates the principle of bacteriocide, or toxicity of surfactants to cells. Single-chained surfactants at their cmc disrupt the double-chained phospholipid membrane of cells by inducing curvature changes by exactly the same principle.

Ion Binding and Chain Packing

The problem of an apparent conflict between the notion of a fluid-like core and the evident need to carry out properly the statistical mechanics of chain packing has been the subject of a considerable debate. The two are not inconsistent, and the debate has been resolved through the work of Gruen and Marcelja. A related question is how one reconciles the simple theoretical ideas above with the phenomenological description of ionic micelles in terms of measured 'ion binding' parameters. In this picture the micelle, together with a fraction of 'bound' counterions, is viewed as an entity in equilibrium with non-interacting monomers and free counter-ions. The micellization law is apparently very different here, and described by an equation of the form

$$K \, C_N = C_1^N \, C_2^Q; \quad C = C_1 + NC_N = C_2 + QC_N - C_3.$$

Here C_N, C_1, C_2, C, C_3 are concentrations of micelles, monomer, counterion, surfactant, co-ion respectively, and Q is the number of counterions 'bound' to the micelle. In this picture it can be shown that

$$kT \ln X_{cmc} = g_{HP} - kT \, \beta \ln (X_{cmc} + X_3), \tag{2}$$

where g_{HP} is as before, and $\beta = Q/N$ is the fraction of bound charge, which is supposed to be constant with salt. Indeed it is. The zeroth order theory (1) and the phenomenological model (2) appear in conflict.

However, the matter is resolved if we write $g_S = g_{es} + \gamma a$, where now γ subsumes all forces except the electrostatic free energy g_{es}, i.e., those

unknown forces due to steric effects, hydration, chain packing, exposure of
hydrocarbon to water. The electrostatic free energy - at the cmc, i.e.,
the limit of a low concentration of micelles - can be explicitly calculated
from double-layer theory. Then the optimal micelle is determined by

$$\left(\frac{\partial g_s}{\partial a}\right) = 0 \text{ or } \gamma = (\partial g_{es}/\partial a),$$

and γ is assumed independent of salt concentration. In fact it is, and
agrees with the value obtained from detailed statistical mechanical
calculations. Further, it can be shown that β emerges naturally from such
a calculation, at least for (and only for) those systems most studied when
β is large ~ 0.8.

Interactions

This development is very important. Its import is that if we can use
the Poisson-Boltzmann equation as the basis of a description for dilute
ionic micelles which subsumes all other unknown forces, then we can hope to
elucidate the phase behaviour of surfactant-water, and later surfactant-
water-oil systems with some confidence. However, questions still remain.
The important fact for colloid science generally is the following. What is
the effective Debye length in a strongly interacting assembly of micelles
or liposomes? And what the effective charge? The cell model of
Wennerström and Jönsson shows that the problem of phase behaviour for
ionic surfactants with monovalent cations is essentially a solved problem -
provided the DLVO theory is applicable. Theoretical work on new fluctuation forces due to correlations in the inhomogeneous double layer shows
that the older intuition does provide an acceptable base for quantitative
theory - except for divalent counterions where the forces operating between
aggregates can be very different. Further, direct force measurements
confirm, at least for some surfactant systems, that the DLVO theory holds
down to contact; i.e. there is no need to invoke new"hydration" forces
which would render analytic or computational procedures impossible at a
quantitative level. At least, the theory based on the primitive model of
electrolytes appears to be on the right track. For bulk electrolyte solutions it is known that the primitive model whereby ions are treated as
hydrated with a fixed hydration shell works well for the alkali halides,
providing a consistent picture of activity coefficients. The primitive
model works for alkali halide ions but fails for Cs^+, NO_3^-, and a range of
other monovalents. For 2:1 electrolytes like $CaCl_2$, Debye-Huckel theory
extended to include hard hydration shells also fails. Therefore we can
expect a description of ionic surfactant systems based on the
Poisson-Boltzmann distribution to fail also for such counterions.

A good deal of nonsense has been propagated in the literature on
inferences concerning the role of interactions between micelles as adduced
from quasi-elastic light scattering with ionic micelles. The debate here,
and on neutron scattering experiments, appears to be settling down to a
fairly sensible consensus as physicists begin to understand the subtleties
of chemistry.

Probably the central question which confronts ionic systems is how far
one can uncouple the notion of a well-defined aggregate or pseudophase from
interactions between these aggregates. With increasing surfactant concentration, the idea of an individual aggregate which can be considered as
an entity which exists without its fellows, simply disappears.

Counterion Specificity

The problem, and that of water structure about ions in mediating head group repulsion, hence aggregate nature, is thrown into sharp relief if we consider some peculiar phenomena exhibited by double chained cationic surfactants (DDAB) which form cloudy lamellar phases in water. The system at low surfactant concentration, e.g. 1% weight, forms metastable vesicles on sonication. These collapse back to lamellar phase over a period of time at a rate depending on concentration. The suspensions are cloudy because they contain large liposomes which scatter light (size $\geq$ 5000 Å). Take 1 ml of 10^{-1} M DDAB, and titrate with 10 M NaOH or concentrated acetic acid. The suspension abruptly clears after addition of 1.4 x 10^{-3} moles NaOH or 2 x 10^{-3} moles HAC. This is extraordinary. The suspension clears to a freely flowing solution, i.e. it consists of small vesicles at a salt concentration of the order of 2 Molar! DLVO theory would have led us to expect collapse of the lamellar phase.

To explore this counterion specificity further, we can ion exchange the surfactant to hydroxide, acetate or other carboxylate forms. These systems form spontaneous vesicles, and are soluble at extremely high concentrations while the halide forms give lamellar phase. The vesicles have been studied by cold stage and scanning electron microscopy, video enhanced differential interference microscopy, electrophoresis, fluorescence probe and other techniques. The vesicles are fairly monodisperse and thermo-dynamically stable. Their size depends on counterions, varying from $\sim$800 Å diameter at 10^{-5} M to 300 Å at 10^{-2} M. At 10^{-1} M only micelles exist in solution with aggregation numbers of the order of 50. For the corresponding single chained analogues cmcs are twice as large and aggregation numbers half those of the bromides. Further, these micelles do not increase in size with added salt until extremely high concentration.

It turns out that this behaviour can probably be accommodated for in terms of the Poisson-Boltzmann description if hydrated counterion size is taken into account and admitted as a working hypothesis. Strongly hydrated counterions would then sit relatively far out from the head groups and impose large curvature on the outer interface, as is required for vesicle formation. (Inside electrical neutrality demands that counterion con-centration is very high, so that head group forces are screened.) The stability and existence of such systems also requires the existence of large repulsive forces between such aggregates. Indeed, direct force measurements on bilayers absorbed onto mica confirm this. Acetate is 100% dissociated, while bromide behaves as if 90% bound, so that the forces between aggregates with different counterions are very different. A proper theory of ionic vesicles is still lacking.

The essential point which can be made is that at an elementary level the shape and size of aggregates is set by a balance of head group repul-sion and tail interactions. The net curvature can be changed by salt, counterion, chain length, mixed surfactants or mixed solents, or cosur-factant – long chain alcohols swell chain volume without substantially affecting head group area. (Short chain alcohols behave differently and are more complex). Awareness of these effects is useful in microemulsion design where we have to balance head group forces against those due to oil in the interface.

Sometimes these simple principles appear not to work e.g. for our model mixed surfactant system DDAB plus DTAB which reverts from lamellar phase to vesicles to rods and micelles as v/al (effective) decreases on successive addition of DTAB, the system can sometimes reorganise to ves-icles or rods in the presence of salt at a lower concentration of the single chained surfactant than without salt. This is because substantial uptake of DTAB into the bilayer phase occurs at the cmc, which is lower in the presence of salt, than without.

Surface Tension and Microemulsion Droplet Models

The same kind of zeroth order theory can be applied to model oil-
in-water and water-in-oil droplet models. We shall see why these models
for microemulsions are not always tenable. The Poisson-Boltzmann approach
gives a good description of the surface tension of a swollen micelle solution - oil interface. For SDS/heptane the predicted values are γ (interface) = 5.4, 4.6, 3.4, 2.3, 1 dyne /cm at salt concentrations of 0, 0.01,
0.03, 0.1, 0.3 M respectively, whereas a measured value is 5 at 0.01 M
NaCl. The tension is logarithmic in salt concentration. This means that
for ionic microemulsions - where the surface tension depends on curvature
in the same way - we expect a slow dependence of the surface tension with
added salt, as is observed. For sodium dodecyl sulphate/butanol/water/
toluene microemulsions the measured "droplet" size from QELS is $R \sim 100$ Å at
0.6 M NaCl. Theory gives γ 1 dyne/cm whereas the experimental value is
$\gamma \sim 0.1$ dyne/cm. The disagreement is worse at higher salt concentrations.
While interaggregate interactions will certainly affect the picture, the
inference is fairly clear. The 'fit' of QELS data to spherical droplets is
inadmissable. The microemulsions (and all ultralow tension systems) are
not droplets but bicontinuous phases of low curvature. This problem, of an
apparently excellent 'fit' to an assumed structure from neutron of QELS
scattering plagues the literature. The fit is non-unique.

IONIC SURFACTANT MICELLES AND LIQUID CRYSTALS: MORE REFINED STUDIES

To summarize, for ionic surfactants with simple head-groups and monovalent inorganic counterions, the actual formation of micelles is rather
insensitive to head-group and counterion. The same applies to a wide range
of processes on the molecular level, molecular packing and conformation,
hydration, molecular dynamics etc. On the other hand, the choice of head-
group and counterion may dramatically affect aggregate properties as dis-
played in micelle size and shape, intermicellar interactions and phase
behaviour. While the framework outlined above provides a useful benchmark
for qualitative thinking, the real subtleties are revealed only by more
refined experiments. In this section we wish to briefly review various
aspects of ionic surfactant micelles and liquid crystals as elucidated by
NMR spectroscopy.

Phase Diagrams

Knowledge of the phase diagram of a surfactant system is the most
basic information, and at least a rough knowledge of phase equilibria is a
prerequisite for meaningful physico-chemical investigations. There are
many examples in the literature where the inadvertent neglect of phase
boundaries has lead to fundamental misinterpretation.

The extensive phase diagram determinations by Ekwall, Fontell and
Mandell laid the foundation. These authors investited mainly three-
component systems of water, an ionic surfactant and a hydrophobic or amphiphilic compound. While only systems with monovalent counterions were
investigated by Ekwall et al., the divalent counterion systems have a great
fundamental and applied interest. Khan and co-workers have in recent years
investigated the phase behaviour of a number of calcium and magnesium
surfactants using the ^{2}H NMR method. This technique is based on the
appearance of quadrupole splittings for anisotropic phases but not for
isotropic ones. It allows identification of phases in a heterogeneous
sample without macroscopic separation. The most significant observation in
the work with Ca and Mg surfactants is that of a much decreased stability
range of the lamellar phase on substituting a monovalent counterion by a

divalent one; the maximum number of water molecules per surfactant ion is
reduced by a factor of 10-20. Khan et al. recently demonstrated for a
mixed sodium + calcium system the coexistence of two lamellar phases, thus
providing evidence for attractive electrostatic interactions between similarly charged surfaces.

Surfactant Self-association

The Fourier transform NMR technique for studying self-diffusion provides a simple way to a rather complete characterization of micelle-non-
micelle distributions of molecules and ions. In fact, it gives directly
the free and micellar concentrations provided the bulk and micellar self-
diffusion coefficients are known, and these can be obtained experimentally.
The example of decylammonium dichloroacetate illustrates well the behaviour
generally found for surfactants with monovalent counterions. (But not for
the 'peculiar' counterions already discussed.) We note the expected
decrease in free surfactant concentration above the cmc. The phase separation model of micelle formation is inadequate. This decrease, which is
less marked with a divalent counterion (calcium octylsulphate) and absent
for zwitterionic systems, is explained from the theory of micelle formation treating the electrostatic effects using the Poisson-Boltzmann
equation. Thus the inhomogeneous counterion distribution corresponds to an
unfavourable negative entropy contribution to micelle formation, but as the
total surfactant concentration increases this entropy term is reduced due
to partial counterion dissociation.

Micelle Size

Micelle size is an essential requirement for the description of a
micellar system, but its determination presents a number of problems.
Firstly, micellar solutions are polydisperse, and secondly, different
experimental techniques define the micelle size in different ways: hydrodynamic radius, radius of dry micelle, micelle aggregation number. In the
determination of the micelle radius from transport or scattering studies
the influence of intermicellar interactions must be taken into account.
Common ways of studying micelle size are dynamic and static light scattering, neutron scattering and self-diffusion. The micelle reorientational
correlation time determined from NMR relaxation studies gives a hydrodynamic radius which is less influenced by intermicellar interactions. To
determine the micelle aggregation number directly is not without problems,
and one of few possibilities is provided by the fluorescence quenching
method which gives a count of the number of monomers in a micelle.

For most ionic surfactants, micelles are close to spherical at low
concentrations. According to kinetic studies these micelles are fairly
monodisperse and there is a very deep minimum in the size distribution
curve. The radius of the spherical micelles is close to the length of the
extended surfactant molecule.

A striking observation for a number of ionic surfactant systems is the
very marked increase in viscosity observed on (sometimes relatively minor)
changes in surfactant concentration, on addition of electrolyte, on lowering the temperature and on adding a solubilizate. These viscosity
increases are due to micelle growth. A simple and direct way of observing
micelle growth is based on NMR relaxation in particular on observing the ^{1}H
NMR spectra, where line-broadening is direct evidence of micelle growth.

Micelle growth is strongly dependent on the counterion. For the alkyl
sulfate systems, micelle growth increases from Li^+ to Cs^+. $C_{16}N^+(CH_3)_3Cl^-$
does not grow appreciably at room temperature while the corresponding
bromide grows to large micelles. At higher temperature the micelles of the

bromide remain spherical up to high concentrations. With certain organic counterions this surfactant ion gives dramatic micelle growth already at quite low concentrations. Often these solutions are viscoelastic.

Micelle growth is thus dependent on subtle chemical changes in the system. We do understand the underlying factors at a global level, but not in detail.

Micelle Shape

According to our present understanding of the processes which drive surfactant self-association, any substantial micelle growth beyond the limit of a sphere with a radius equalling the length of the extended surfactant molecule corresponds to a micelle shape change. While it is easy to establish a growth into nonspherical aggregates, it is not trivial to determine unambiguously the actual shape. One is generally left with fitting variable concentration data with predictions of models based on certain assumptions regarding the packing conditions. However, phenomena such as micelle flexibility and polydispersity complicate such analyses.

It is generally accepted that in simple two-component solutions ionic micelles grow to rod micelles rather than to disc micelles. One important case exists where the aggregate shape has been unequivocally established using NMR. At high concentrations $C_{16}N(CH_3)_3Br$ solutions are very viscous. Using ^1H and ^2H NMR it is found that the NMR signals in the most concentrated isotropic solutions are virtually identical to the ones obtained in the hexagonal liquid crystalline phase. Since the signal is sensitive to micelle shape, this demonstrates that the aggregates in the isotropic system are rod shaped, with a long persistence length, since this shape has been established for the hexagonal phase.

There is, however, one procedure which in many cases is capable of giving a clear-cut distinction between different shapes over wide concentration ranges. This is based on the observation that different shapes of obstruction volumes give very different effects on the transport of small molecules. This approach has been used to establish the shape of nonionic micelles under different conditions (cf. below) but has not been applied to ionic systems. It should be a good approach also for many existing problems on ionic micelle shape such as micelle shape at high solubilisate content (towards the lamellar phase).

Chain Packing and Dynamics

For hexagonal and lamellar liquid crystalline phases, the ^2H quadrupole splitting technique has been applied already for quite some time to obtain characterizations, via order parameters, of the surfactant chain packing in the aggregated state. ^{13}C chemical shift studies of micellar solutions have also been applied for some time to show that there is a slightly increased proportion of trans conformers (downfield shift) of the alkyl chains on micellization.

In parallel with the theoretical studies by Gruen and others which attempt to characterize the internal structure of micelles, NMR methods have been devised which allow such a characterization also for isotropic solutions.

Two characteristic observations in ^2H and ^{13}C relaxation studies of micellar solutions are quite long spin-lattice relaxation times, T_1 (of the order of magnitude as those of simple liquids) and an inequality of T_1 and T_2, the spin-spin relaxation time. According to standard equations the first observation would give motional correlation times (τ_c) of the surfactant molecules in the range 10^{-10}–10^{-11} sec while the second one would give τ_c values in the range of 10^{-8}–10^{-9} sec. These observations are, therefore, interpreted to mean that there are at least two motions on widely different time scales that contribute to the relaxation. According to Wennerström's two-step model of relaxation there is a fast local motion (which is anisotropic and has a correlation time (τ_c^t) that averages part of the interaction as well as a slow overall motion (τ_c^s) that averages the remainder of the interaction. The interaction not averaged by the fast motion is characterized by an order parameter (S) defined as for liquid crystals with respect to the normal of the interface. The three quantities which characterize chain packing and dynamics of micelles can be obtained from three independent measures of the relaxation (relaxation at different magnetic fields, of different nuclei, T_1 and T_2 etc.).

For spherical micelles of ionic surfactants some general observations are the following:

- fast motions (slightly anisotropic) are of the order of 2.10^{-11} sec. and close to values of the corresponding alkanes or substituted alkanes

- the hydrophilic-hydrophobic interface in micelles imposes an order which along the chain is close to that of hexagonal and lamellar liquid crystals. The order parameter values are in agreement with Gruen's calculations.

- the slow motion can be understood from micelle rotation and surfactant lateral diffusion along the micelle surface with no need to introduce other motions. The τ_c^s values give a correct hydrodynamic radius of the micelles.

Counterion Binding

In a first approximation we can describe the inhomogeneous counterion distribution in terms of a two-site bound-free model and define a degree of counterion binding (β) as the ratio between bound counterions and aggregated surfactant molecules. The appropriateness of such an approximation will depend on the experimental technique. It is reasonable as shown from Poisson-Boltzmann calculations, but only when counterion binding is large. This is the case most studied so far.

The value of β is directly obtainable from self-diffusion data. Studies of a number of systems with monovalent counterions suggest that the observation of an approximately concentration independent β value is a quite general one. This so-called ion condensation behaviour has also been observed inter alia in counterion NMR chemical shift and quadrupole relaxation studies of micellar solutions and in counterion quadrupole splitting studies of liquid crystals and is theoretically predicted inter alia in Poisson-Boltzmann calculations.

While degrees of counterion binding are similar for different systems there are also significant differences. It is at these differences and at spectroscopic observations of counterion location that one has to look to understand the important counterion specificity in many properties of micellar solutions. This remains an incompletely understood field, but a number of significant observations can be made:

- Alkali ions remain hydrated when bound to micelles.

- Alkali ions penetrate between carboxylate head-groups but are located outside sulfonate and sulfate headgroups with an approximately symmetric location with respect to the oxygens there. This is attributed to

differences in interaction between counterion hydration water and
headgroups.

- β is slightly higher for Br^- than for Cl^- counterions, and Br^- ions are
  'bound' closer to the micelle interface than Cl^- ions, i.e., adsorption
  excesses are higher. The more interesting counterions H^+, Ac^-, OH^-, car-
  boxylates, NO_3^- have not been much studied.

<u>Hydration</u>

Water self-diffusion is much more affected by surfactant monomers than
by micelles. Interpreted in terms of the two-state model, such investigations give 10-15 water molecules per surfactant molecule diffusing with
the micelles as a kinetic entity. Accounting for head-group and bound
counterion hydration, this allows for very little contact between water
molecules and the alkyl chains in micelles. From water ^{17}O relaxation
studies it is deduced that less than two methylene groups in the alkyl
chain are exposed to water in micelles, that bound water molecules reorient
quite rapidly (less than a factor of 10 slower than in bulk water) with a
low degree of anisotropy, and that only one or tow layers of water molecules are perturbed appreciably with no indications of long-range
hydration structures.

Water ^2H quadrupole splitting studies have been performed quite exten-
sively, and a number of conclusions on hydration in liquid crystals can be
drawn from such work:

- there is often (but now always) an "ideal swelling" behaviour where water
  molecules in the vicinity of the surfactant aggregates have their order
  parameters almost constant on increasing the water content and where
  water addition only increases the amount of free water;

- there is no appreciable long-range ordering effect on the water
  molecules. Thus only about one layer of water molecules (hydration
  number <10) is significantly perturbed;

- even the water molecules in direct contact with the surfactant aggregates
  are very mobile and have a quite low anisotropy in their motion;
  generally S_b is of the order of 5.10^{-12}, thus much below the value of the
  polar head-group.

NONIONIC MICELLES

While the combination of experimental and theoretical approaches has
given us a generally accepted understanding of basic aspects of ionic
surfactant self-association, fundamental points remain unclear and controversial in the field of nonionic surfactants. In particular this is true
of the most investigated and technologically important class of nonionic
surfactants, i.e. that of oligo (ethyleneoxide) alkylethers.

These surfactants, here referred to as C_xE_y, x being the number of
carbons in an n-alkyl chain and y the number of ethyleneoxide (EO) groups,
show a quite complex phase behaviour in two-component systems with water.
An important feature is the appearance of a lower consolute temperature
often referred to as the cloud-point. Much effort has been devoted to the
rationalization of this phenomenon and to investigate critical effects in
the vicinity of the cloud point. Here we wish to discuss basic problems
related to the micelles, in particular micellar size and shape and intermicellar interactions, on the basis of experimental studies of the $C_{12}E_y$
series.

<u>Micelle Size</u>

The cmc's of these surfactants are so low that the micelle self-diffusion coefficient is directly obtained from the surfactant self-diffusion coefficient. A considerable advantage in these studies as compared to scattering studies is that self-diffusion coefficients are not influenced by critical fluctuations and thus we are able to study micelle size also in the vicinity of critical points.

A striking observation is that even a relatively small change in the surfactant chemical structure can have dramatic effects on the self-diffusion behaviour. For $C_{12}E_8$, D_{mic} increases with temperature and is very slightly dependent on concentration. The hydrodynamic radius is found to be roughly constant but increases slightly with temperature and concentration. This value corresponds to the length of the extended alkyl chain plus around 2/3 of the length of the extended EO chain.

For $C_{12}E_5$ the behaviour is very different, with D decreasing strongly both with temperature and concentration. There is very marked micelle growth both with temperature and concentration over the entire ranges investigated. There is no indication of special growth in the vicinity of the cloud point, and shifting the cloud point by small additions of ionic surfactant (sodium dodecylsulfate) does not have any measurable effect on micelle size. The results for $C_{12}E_5$ (and to a smaller extent for $C_{12}E_6$) are in contrast to those of ionic surfactant micelles, where micelle size has been observed to decrease quite strongly with increasing temperature.

The observed increase in micelle hydrodynamic radius could be due to a real micelle growth where the hydrocarbon cores grow, or to an aggregation of intact micelles. Since the observed decreases in micelle self-diffusion coefficient, at least at not too high temperatures, are accompanied by an increased rate of alkyl chain ^{1}H NMR relaxation, the first model appears to be the correct one. A direct support of this is given by recent fluorescence quenching results. Even so, this question deserves further attention in more specific relaxation experiments. Such experiments would also provide information on chain packing and dynamics under different conditions. A striking observation in the ^{1}H NMR spectra is the more rapid relaxation of the alkyl than of the EO chains. This provides evidence for lower order parameters of the EO chains than of the hydrocarbon cores. Furthermore, there appears to be a decreased order in the alkyl chains as temperature is increased.

<u>Micelle Hydration</u>

From studies of the water self-diffusion coefficient as a function of temperature it is found that the hydration number decreases progressively with increasing temperature. Furthermore, the hydration number is higher for $C_{12}D_5$ than for C_2E_5. A striking observation is that for a given temperature, hydration is the same per EO group for spherical micelles of $C_{12}E_8$, for nonspherical micelles of $C_{12}E_5$ and for poly(ethylene glycol.)

<u>Micelle Shape</u>

In view of the large hydrodynamic radii (>>length of extended surfactant molecule) observed for $C_{12}E_5$ at higher temperatures and concentrations, these micelles must be strongly anisometric; however, the question is, what nonspherical shape is formed? We will discuss here only rod- and disc-shaped micelles. There can be little doubt that micelles grow to rods with concentration at lower temperatures in the presence of a hexagonal phase, but it is not obvious which micelle shape occurs at lower concentrations and higher temperatures.

Another interesting subject in these systems is that of micelle shape in the "anomalous" or L_3 phase, which is an additional isotropic solution phase formed at temperatures much above the cloud point for certain systems. Proton relaxation and light scattering studies demonstrate the existence of very large micelles in the anomalous phase.

To discuss these problems we can make use of the very important difference in obstruction effect in the diffusion of small molecules between large prolate and large oblate objects. From water self-diffusion data, obstruction effects were calculated taking the hydration to be that of the same number of EO groups in solutions of poly (ethylene glycol). The observed obstruction effects clearly demonstrate the existence of disc rather than rod micelles in the L_3 phase, while rod shape is suggested for the normal micellar (L_1 solutions). The observation of different shapes in the two phases is not unexpected. It would instead have been remarkable to observe the same shape in two phases separated by a number of two-phase regions and three-phase lines.

Intermicellar Interactions

Considering surfactant self-diffusion results at higher concentrations, we note a further strong difference between the $C_{12}E_5$ and $C_{12}E_8$ systems. For $C_{12}E_5$, D starts to increase at higher concentrations while for $C_{12}E_8$, D continues to decrease. The latter behaviour is the expected one for repulsive intermicellar interactions and is the one observed for ionic surfactant micelles.

The facilitated surfactant molecule diffusion over macroscopic distances, observed for $C_{12}E_5$ at 25°C, is attributed to attractive intermicellar interactions allowing intermicellar exchange in, for example, fusion-fission processes. The high concentration self-diffusion behaviour is very strongly dependent on temperature, and it can be deduced that intermicellar interactions change from repulsive to attractive with increasing temperature. For a longer EO chain, a higher temperature is needed to give attractive interactions.

Qualitatively, the same behaviour is observed in recent direct force measurements, where the interactions between $C_{12}E_5$ layers has been found to change from repulsive to attractive in the range 15-30°C.

In the anomalous phase, the interactions between the disc micelles appear to be strongly attractive, as surfactant self-diffusion is very rapid down to low concentrations (a few % by weight). The surfactant self-diffusion is close to 2/3 of the neat surfactant self-diffusion over the entire concentration range investigated, which is an unexpected observation which remains to be explained.

Addition of sodium dodecylsulfate to $C_{12}E_5$ micelles strongly decreases the intermicellar exchange, as expected, as the introduction of charges in the micelles provides a repulsive term in the interaction. For the L_1 phase in C_xE_y-hydrocarbon-water systems the interaction may be repulsive throughout a broad concentration range, may change from repulsive to attractive with concentration, or be attractive already at low concentrations. A striking observation is the more rapid diffusion of hydrocarbons compared with surfactants, as observed in the $C_{10}E_4$-$C_{16}H_{34}$-water system. The intermicellar interactions as obtained from this self-diffusion behaviour appear to be related to the distance in temperature from the HLB temperature, but more systematic studies are needed for a meaningful discussion of these problems.

<u>Mechanisms</u>

The various observations described above seem all to be easily under-
standable from the model of Israelachvili, Mitchell and Ninham, where
aggregate shape is taken to depend on the effective surfactant molecular
shape. For these EO surfactants <u>a</u> increases with increasing hydration and
with increasing number of EO groups. As an example, since hydration
decreases with increasing temperature, we expect micelle shape to change
sphere→rod→disc. For a sufficiently long EO chain steric effects are large
enough to keep <u>a</u> high so that the micelles remain spherical also at high
temperatures.

The phase behaviour of EO surfactants has been investigated in detail
by Mitchell and Tiddy and others and could be well rationalized from the
same model. The observed decreasing hydration with increasing temperature
would favour a transition from hexagonal to lamellar liquid crystal-line
phase with increasing temperature.

STRUCTURE AND DYNAMICS OF MICROEMULSIONS

Microemulsions form at widely different positions in the phase diagram for
different surfactants or for different combinations of surfactants, and it
is clear that microstructure must be different for different systems.
However, it has not been straightforward to unambiguously establish these
structures. Further problems in the field of microemulsions relate to
demonstrating whether surfactant films exist or not; if they do, we wish to
characterize these inter alia in relation to other phases in surfactant
systems.

<u>Self-diffusion and Solution Structure</u>

Molecular self-diffusion over macroscopic distances is very sensitive to
confinement into closed domains. Therefore, multicomponent self-diffusion
studies, as obtained most conveniently in Fourier transform spin-echo NMR
work, can easily distinguish between different structural models in many
cases. The following limiting cases may form a basis of our discussion:

a)   <u>Water-in-oil droplet structure</u>. With confinement in closed domains
water diffusion will be slow. If surfactant molecules occur only at the
interfaces their diffusion will also be the same as that of the droplets
themselves. Oil diffusion will be high as it forms continuous domains. It
will only be retarded with respect to the neat oil by an obstruction effect
due to the droplets and by penetration between surfactant alkyl chains.
Therefore, we predict $D_{oil} \gg D_{water} \simeq D$ surfactant $\simeq D_{droplet}$.

b)   <u>Oil -in-water droplet structure</u>. Now water diffusion occurs in a continuous medium and is retarded compared to heat water only by surfactant
solvation and obstruction. Thus $D_{water} \gg D_{oil} \simeq D_{surfactant} \simeq D$ droplet.

c)   <u>Bicontinuous structure</u>. A large number of structures are possible in
which both water and oil form domains which are continuous over macroscopic
distances and where surfactant molecules are located at the interfaces
between these domains. In particular we may think of layered or channel-
type structures, but we will not in this discussion consider any distinction between different possible bicontinuous structures. D_{water} and D_{oil}
will both be high, lowered from the values of the neat liquids only by
obstruction and solvation/penetration effects. Surfactant diffusion will
be hindered by the location at interfacial films but will be unrestricted.
We expect $D_{surfactant}$ to be intermediate between the value for nonassoci-

ated surfactant molecules (of the order of $10^{-9} m^2 s^{-1}$) and its value for droplet type structures (of the order of $10^{-11} m^2 s^{-1}$).

d) <u>Molecule disperse solutions</u>. When the different components diffuse as single molecules all D values will be high ($5.10^{-9} - 5.10^{-10} m^2 s^{-1}$).

The concentration dependences can also be informative, since both medium and droplet molecules are expected to diffuse more slowly with increasing volume fraction (ϕ) of the dispersed phase. For example for spherical droplets:

$$D_{medium}/D^0_{medium} = 1/(1 + \phi/2); \quad D_{droplet}/D^0_{droplet} \alpha$$

$$1 - 2\phi.$$

<u>Examples of Self-diffusion Behaviour</u>

By now a large number of systems have been investigated and a few typical examples may be given;

a) <u>Solubilized micellar solutions</u>. Here $D_{water} \gg D_{surfactant} D_{oil}$. Depending on the system D_{water}/D_{oil} ratios lie in the range of 40-400 for high concentrations.

b) <u>Double-chain surfactant systems</u>. For the large isotropic phase of the double-chain anionic surfactant Aerolsol OT, $D_{oil} \gg D_{surfactant} D_{water}$. D_{oil}/D_{water} is around 20 in a wide concentration region up to above 50% water. For the double-chain cationic surfactant, didodecyldimethyl-ammonium bromide $D_{water} D_{oil}$ or $D_{oil} \gg D_{water}$ depending on composition and hydrocarbon. At higher water contents of the isotropic phase D_{oil}/D_{water} can exceed 100.

c) <u>Four-component systems of surfactant, cosurfactant, hydrocarbon and water</u>. With butanol or pentanol as cosurfactant D_{oil} and D_{water} differ only by a factor of 2-3 over wide concentration ranges both being high, while $D_{surfactant}$ is about $10^{-10} m^2 s^{-1}$. D_{water} decreases at the highest hydrocarbon contents and D_{oil} at the highest water contents. A very strik-ing feature is the strong dependence of self-diffusion behaviour on the cosurfactant. Changing cosurfactant from butanol to decanol at constant composition changes D_{water} by 2 orders of magnitude while D_{oil} is quite constant. On the other hand a change of oil has only marginal influence.

d) <u>Five-component systems of surfactant, cosurfactant, hydrocarbon, electrolyte and water</u>. Increasing the salinity may change the phase behaviour Winsor I→Winsor III→Winsor II and this is parallelled by dramatic changes in diffusion behaviour from a region where $D_{water} \gg D_{oil}$, via a region where $D_{water} \simeq D_{oil}$, both being high, to a region where $D_{oil} \gg D_{water}$. The change in the ratio D_{water}/D_{oil} is by a factor of around 10^4.

$D_{surfactant}$ is around $10^{-11}m^2s^{-1}$ or somewhat above in the low and high salinity regions, while it rises to $10^{-10}m^2s^{-1}$ in the intermediate region.

These examples suffice to demonstrate that different microemulsion systems can be very different in their self-diffusion behaviour. It may be mentioned that Shinoda's systems near hydrophilic-hydrophobic balance (like nonionic surfactant-hydrocarbon-water, Aerosol OT-NaCl-hydrocarbon-water and calcium surfactant-cosurfactant-hydrocarbon-water) may show very dramatic changes in D_{water}/D_{oil} over narrow ranges of temperature or of cosurfactant-to-surfactant ratio.

Microemulsion Structure

From these examples of self-diffusion behaviour we can deduce that microemulsions can have very different microstructures depending on surfactant chemical structure, cosurfactant chemical structure, temperature, salinity and composition. For the aqueous microemulsions investigated so far it appears that the surfactant molecules are almost entirely in the aggregated state, i.e., they form interfacial films between the hydrophilic and hydrophobic domains. The structure of the microemulsions can be either of the water-in-oil droplet type, of the oil-in-water droplet type, or of the bicontinuous type. Cosurfactant microemulsions may change structure from the bicontinuous type to the W/O type with increasing salinity and with increasing length of the cosurfactant. Double-chained surfactants predominantly form the W/O type structure. The bicontinuous structure formed in the didodecyldimethylammonium bromide systems under certain conditions is probably rather different from the one formed in systems with butanol or pentanol as cosurfactant. (For example, the viscosities are quite different in the two cases). An important aspect in future work, in addition to the study of many more microemulsion systems, will be to characterize the bicontinuous structures in more detail, most importantly the geometry of the domains.

Molecular Dynamics of Microemulsions

The low viscosity bicontinuous microemulsions necessarily have very flexible and dynamic surfactant interfacial films, but little is known about time-scales of motion either at long or at short molecular ranges. NMR relaxation times for both surfactant, cosurfactant, water, hydrocarbon and counterions are long and, therefore, there are important rapid motions. A closer analysis requires a multi-field/multi-nucleus investigation as described above for micelles. Such studies are yet quite limited for microemulsions.

For the system sodium octylbenzenesulfonate-butanol-toluene-water, $^{13}CT_1$'s and nuclear Overhauser enhancement factors were studied at different magnetic fields, whereby contributions to relaxation from motions on widely different time-scales could be identified. An analysis of such data using the two-step model of relaxation gives order parameters and fast motion correlation times. For a wide range of compositions the order parameters have a magnitude and variation along the chain which is very similar to order parameters measured for lamellar liquid crystals. This provides direct evidence for the presence of interfacial surfactant films with an organization similar to that of bilayers in lamellar liquid crystals.

The fast motion correlation times are rather close to those of the corresponding liquid hydrocarbon, thus giving evidence for a highly mobile state within the interfacial films.

The slow motion correlation time is found to be quite short, a few nanoseconds. This means that the organisation of the surfactant films is very short-lived. The motion occurring on the nanosecond timescale has not yet been identified; there are several possibilities such as film flexibility, surfactant diffusion along highly curved interfaces or a short surfactant residence time in the interfacial films.

These studies of the dynamics of microemulsions are still in an early stage, and we need to investigate more systems in a systematic way. Furthermore, measurements have to span a wider frequency range, and to include ^{2}H T_2 measurements, to provide a more detailed picture. For example, it cannot be ruled out that the nanosecond time-scale motion is not completely isotropic but leaves a small interaction to be averaged by motions on a longer time-scale.

Microemulsions with Insoluble Double-Chained Surfactants

In contrast with the broad spectrum of behaviour outlined above, there are some revealing insights which emerge if one works with double chained ionics that are insoluble in both oil and water. The surfactant must then reside at the oil-water interface and the microstructure is much easier to elucidate. The system described below provides a useful benchmark for other studies, as it can be successively perturbed by addition of cosurfactant, surfactant, salt, etc. Recall that according to zeroth order theories, provided that oil uptake in the surfactant tails is ignored, the problem of solubilising a large amount of oil or water is reduced to the statement that we require $v/al \approx 1$, and weak interactions, i.e. double chained surfactants. (Alternatively, for a single chained micelle forming surfactants $v/al \approx (1/3)$, we require a cosurfactant which increases the effective chain volume to make the effective $v/al \approx 1$). Alcohols hardly affect head group area, and pentanol and butanol especially reduce interaggregate interactions. For ionics, head group repulsion is set by electrostatics, depends on counterion, and depends weakly, logarithmically, on salt concentration. We have already seen that for ionic surfactant-water systems the nature of the aggregate is set by the double-layer free energy which opposes chain repulsion and other forces operating at the interface. With oil as a third component oil uptake in the tails forces the chains apart and imposes a curvature which opposes head group repulsion. As a rough rule deduced from studies of alkane uptake in black films, it can be expected that when alkane chain length is less than the surfactant tail length the oil penetrates strongly, imparting a high negative curvature ($v/al \gg 1$); whereas when the alkane has a chain length greater than the surfactant tail length, little oil is taken up. We expect normal (oil-in-water) curvature in this case. The ideal system is then one with weak bilayer interactions, with double chains, insoluble in water and oil, fluid chains at room temperature, and a chain length which allows a range of oils to be taken up. The microstructure can then be made to vary from strong reverse (water-in-oil) curvature to normal (oil-in water).

Our model surfactant DDAB is ideal for this purpose. It has the additional advantage that as we have seen, head group forces imposed by anions vary enormously with the anions. (Counterion specificity is much weaker for cations). We expect and observe substantial changes in the phase diagrams for Cl^-, Br^-, I^-, Ac^-. From the point of view of this approach to design, the Aerosol OT system is a silly choice, because the surfactant is significantly soluble in both oil and water, and because the surfactant chain length is such that most oils will be taken up equally in the chains. There has been a good deal of work on the DDAB, water, alkane or alkene system. It can be briefly summarised as follows:

(1) There is a high and systematic degree of oil specificity in the phase
diagrams. A single phase microemulsion forms with a highly penetrating oil
like cyclohexane or 1-hexene with as little as 1% water at most S/O ratios.

(2) The viscosity is extremely high for low alkanes or alkenes at lowest
water content, suggesting a high degree of structured tabular regions. The
viscosity decreases with increasing water content, i.e. the tubes grow in
diameter as expected on dilution of counterion.

(3) The single phase region is highly conducting at low water content.
With increasing water the conductivity decreases until at a certain point,
still within the single phase boundary, it exhibits a percolation
threshold, and changes by 8 orders of magnitude corresponding to disconnected water droplets. This behaviour is opposite to that shown by the
more usual systems studied. The conductivity goes to zero along a line in
the ternary phase diagram of constant surfactant/water ratio (curvature)
different for each oil.

(4) The microemulsions are bicontinuous in oil and water except beyond the
precolation threshold.

(5) For alkanes up to and including dodecane, reverse droplets extend to
the oil corner. Tetradecane has 'normal' curvature and other physical
properties.

(6) The systems are stable up to 100°C.

     Much of the behaviour observed can be rationalized in terms of packing
arguments first given by Lissant to explain the structure of low external
phase emulsions. At low surfactant and roughly equal water/oil ratios
there is a region of spontaneous emulsification. The emulsion is difficult
or impossible to break. This suggest that the older distinction between
emulsions and microemulsions in tenuous. Those spontaneous emulsions
probably have the same structure as those investigated first in the work of
Friberg, i.e., multilamellae of swollen bilayers surrounding large water or
oil droplets.

     A good deal of work on this system has been concerned with perturbing
the structure through systematically changing curvature; e.g., recall that
without oil, addition of the surfactant DTAB will change the curvature of
the lamellar phase formed from DDAB in water. Then the phase diagram,
conductivity and other properties of a microemulsion formed from the mixed
system DDAB + DTAB (v/a_l (effective) decreases), water, alkane will mimic
that of DDAB, water and less penetrating alkane. This is so. Again systems formed with the cosurfactant dodecanol as an additive (v/a_l effective
increases) will behave in the opposite manner. These correlations can be
made quantitative. Mixed oils again behave as might be expected from
elementary arguments.

     The problem of tuning up the structure by opposing head group repul-
sion against tail/oil interactions is then an easy matter to control.
Changes induced by counterion, mixed surfactants, cosurfactants, mixed
oils, mixed solvents (which affect interactions and head group repulsion,
e.g. urea-water or glycerol-water mixtures) can all be explored. At least
for ionic microemulsions formed with insoluble double chained surfactants
it seems as though a part of the puzzle is on the way of being sorted out.
It is satisfying that the behaviour observed correlates nicely with that of
the corresponding surfactant water systems. In this absurdly simple
approach curvature, which can be dictated at will, is the key to microstructure. A system of ultra low interfacial tension can only be formed if
the net curvature is 'normal', e.g. with tetradecane, which, on addition of

enough salt to swing the curvature back close to zero will achieve that
condition. Progress in our understanding of bicontinuous surfaces using
differential geometry is likely to have a big impact.

Forces

     It is also a matter of satisfaction that the properties observed for
micelles, vesicles and microemulsions correlate nicely with an emerging
series of direct force measurements of high accuracy, using the apparatus
of Israelchvili. Numerous reports on the status of these measurements
exist. We need for our purposes only to note that surfactants can be
deposited by Langmuir-Blodgett techniques or by direct adsorption of vesicles onto molecularly smooth mica as monolayers of bilayers of measureable
thickness. The forces between two such planar surfaces – oil, water, salt,
or whatever – are measured directly with a distance resolution of 1Å. The
surfactant layers are usually anchored firmly by electrostatic interactions
between head group and the (charged) mica substrate. Head group area is
usually fixed in this situation, and the interface between bilayer and
water, or surfactant tails and oil, is rigid. In the real situation of
concern here, especially with soluble single chained surfactants with a
very flexible interface, head group area and chain extension will change in
response to interaggregate separation i.e. the microstructure will therefore rearrange depending on distances, volume fractions of components, as
for example the sphere to rod transition for ionic micelles. Despite that
limitation, the force measurements provide a central tool for understanding. The following brief discussion provides a cursory survey of the
relevant highlights of what has emerged as a major useful developing program:

(1) Short range hydration forces responsible for phospholipid lamellar
phases are still not fully understood, and depend on the size of head
groups which are large in most situations. These interactions are much
less mysterious than was formerly thought.

(2) The forces between PEO surfactants in water exhibit a dramatic change
from repulsive to attractive at precisely the temperature at which those
surfactants exhibit a cloud point. This work complements very nicely
corresponding NMR work.

(3) Between sugar headed surfactants, galactolipids, in which the head
groups are small and lie flat, hydration forces are small or non-existent.
The forces appear to satisfy Lifshitz theory of van der Waals forces very
well.

(4) With fully hydrophobic monolayers (contact angle > 90°) immersed in
water, attractive forces are 10 to 100 times greater than those predicted
by continuum theory. Water structure imposed by a large oil surface is
unexpectedly long range. The implication, since the hydrophobic interaction between small molecules is normally a contact force, is that hydrophobic association can depend strongly on the surfactant, its chain flexibility and hydrocarbon-water area.

(5) Binding of divalent counterions has a real, chemical origin, in distinction to monovalents, which generally tend not to give up their associated water molecules on adsorption to nearly the same extent.

(6) With Br^-, Cl^-, Na^+, K^+ as counterion, force measurements are well
described by DLVO theory, if a phenomenological ion binding parameter in
agreement with that involved to characterize micellar systems is invoked.
This is a mystery. The mystery is about to be resolved, but its explanation will not be clear to any except those initiated into the arcane art

of modern liquid state physics. In fact the hypernetted chain equation
rather than Poisson-Boltzmann is a far more accurate theory as tested by
Monte Carlo simulations on model systems. The HNC equation for interactions of fully charged surfaces with electrolytes intervening agrees with
the Poisson Boltzmann result assuming binding. But the qualitative claims
we make still hold! Watch the literature for new developments. According
to theories based on the Poisson-Boltzmann description, no ion binding
should be necessary.

(7) At least for those counterions which induce "peculiar" surfactant
behaviour, e.g. Ac$^-$, OH$^-$, the forces are described very well by DLVO theory
with no ion binding. The repulsive forces are very much larger here.

(8) With pentanol in water, cationic surfactants experience very different
forces. The alcohol does not penetrate into the surfactant hydrocarbon
tail region, and adds an additional weak attractive hydrophobic force to
the double layer interaction. That latter finding is of much interest, and
deserves fuller explanation.

We are still at an early stage, but the expected correlation - that
weak forces must operate in order to form microemulsions, and certainly to
form three phase systems - does emerge. Emulsions remain an open question,
but the rediscovery by various authors (and much work is unreported being
constrained by patent applications) of spontaneous emulsification and of
Fribergs' work in this area will probably see a disappearance of the older
distinction between emulsions and microemulsions.

CONCLUSION

We have said little on how studies on surfactants and vesicles and
microemulsions fit into modern biology. They do very directly. After all
phospholipid bilayers are surfactant aggregates in part. The beginnings of
a new role for physical chemistry have been documented elsewhere, and there
is emerging a new role for surface chemistry in immunology and drug design.
Cationic surfactants, widely used in household products and as bacterio-
sides and disinfectants, both double and single chained, but not all of
them, appear to operate through precisely the same principles as outlined
above. They are also, at concentrations far below the cmc, potent immuno-
suppressants, better indeed than the new wonder drug cyclosporin A which is
used for controlling organ transplant rejection, is something to wonder at,
especially since we understand why, but this is in fact no cause for con-
sumer alarm.

At this stage it is also clear that the prophet of the Book of
Ecclesiastes was correct in his analysis. We conclude by again quoting
James Morris, who, writing on the British Empire at the time of Queen
Victoria's Jubilee celebration, said that "If to the Queen herself all the
myriad peoples of the Empire really did seem one, to the outsider their
unity seemed less than apparent. Part of the Jubilee jamboree was to give
the Empire a new sense of cohesion, but it was like wishing reason upon the
ocean, so enormous was that span of association, and so unimaginable its
contrasts and contradictions". Microemulsions we think are a lot like the
British Empire: elusive, full of contradiction, and yet with an underlying
thread of untiy. A sense of that unity is achievable if one were to read
all the papers quoted, and those in this book. We wish the reader well in
that monumental enterprise.

SELECTED REFERENCES

<u>Introduction</u>: The Bible, King James Version, Last Chapter of Ecclesiastes.
<u>Approaches to Self-Assembly</u>: Other papers cited in this book, especially
    associated with e.g. work of Talmon and Prager, de Gennes, Taupin,
    Widom, Scriven, Davis, Also see
Mathematics and Physics of Disordered Media, U. Minnesota Institute of
    Mathematics and its Applications Publications 1; B. Hughes and B.W.
    Ninham (eds.). Springer-Verlag publishers (1983).
J.N. Israelachvili, D.J. Mitchell, and B.W. Ninham, Theory of Self-Assembly
    of Hydrocarbon Amphiphiles into Micelles and Bilayers, <u>J.C.S., Faraday
    Transactions II</u> 72:1525 (1976).
J.N. Israelachvili, D.J. Mitchell, and B.W. Ninham, Theory of Self-Assembly
    of Lipid Bilayers and Vesicles, <u>Biochim. Biophys. Acta</u> 470:185 (1977).
S.L. Carnie, D.Y.C. Chan, D.J. Mitchell, and B.W. Ninham, The Structure of
    Electrolytes at Charged Surfaces: The Primitive Model, <u>J. Chem. Phys.</u>
    74: 1472 (1981).
B. Lindman and H. Wennerström, Micelles, Amphiphile Aggregation in Aqueous
    Solution, <u>Topics in Current Chemistry</u> 87:1 (1980)
Bengt Jönsson and H. Wennerström, <u>J. Colloid Interface Sci.</u> 80:482 (1980).
Bengt Jönsson and H. Wennerström, Phase Equilibria in a Three-Component
    Water- Soap- Alcohol System. A Thermodynamic Model, <u>J.Phys.Chem.</u>
    91:338 (1987).
For applications of minimal curvature from differential geometry see recent
    work of S. Hyde. S. Anderson, K. Larsson in Acta Cryst. and elsewhere.

<u>Ion Binding and Chain Packing</u>

<u>Chain Packing:</u>
A. Wulf, cited in Mitchell and Ninham has reconciled the apparent
    contradiction between chain packing and a fluid-like core.
D.W. Gruen, <u>J. Colloid Interface Sci.</u> 24:281 (1981).
D.W. Gruen and E.H.B. de Lacey, <u>in</u>: "Surfactants in Solution", K.L. Mittal
    and B. Lindman, eds., Plenum Press, New York (1984), vol. 1, p. 279.
D.W.R. Gruen, <u>J. Phys. Chem.</u> 89:146, 153 (1985).
D.W.R. Gruen, <u>Progr. Colloid Polymer Sci.</u> 70:6 (1985).

<u>Ion Binding:</u>
G. Gunnarsson, Bengt Jönsson and H. Wennerström, Surfactant Association
    into Micelles. An Electrostatic Approach, <u>J. Phys. Chem.</u> 84:3114
    (1980)
H. Wennerström, Bo Jönsson and P. Linse, The Cell Model for Polyelectrolyte
    Systems. Exact Statistical Mechanical Calculations, Monte Carlo Simu-
    lation, and the Poisson-Boltzmann Approach, <u>J. Phys. Chem.</u> 76:4665
    (1982).
D.F. Evans, D.J. Mitchell, and B.W. Ninham, Ion Binding and Dressed
    Micelles, <u>J. Phys. Chem.</u> 88:6344 (1984).
D.F. Evans and B.W. Ninham, Ion Binding and the Hydrophobic Effect, <u>J.
    Phys. Chem.</u> 87:5025 (1983).

<u>Interactions:</u>
See Forces below.
See Jonsson and Wennerström, in Approaches to Self-Assembly.
On bulk electrolytes activity coefficients see e.g.
B.W. Ninham, B.A. Pailthorpe, and D.W. Mitchell, Ion Solvent Interactions
    and the Activity Coefficients of Real Electrolyte Solutions, <u>J.Chem.
    Soc. Faraday II</u> 80:115 (1984).
D.J. Mitchell and B.W. Ninham, Range of the screened Coulomb Interaction in
    Electrolytes and Double Layer Problems, <u>Chemical Physics Letters</u>
    53:397 (1978)

There is a vast literature on the "civilised" model for bulk electrolytes
i.e., hard core ions in model hard sphere dipoles in Journals like J. Phys.
Chem., J. Chem. Phys., Mol. Phys. There is also a vast literature on real
electrolytes germane, mainly associated with and inspired by H. Friedmann,
with which the general informed public will be familiar.

Quasi-Elastic Light Scattering for Ionic Micelles:

See:

B.W. Ninham, D.J. Mitchell, S. Mukherjee, and D.F. Evans, Surfactant Dif-
    fusion: New Results and Interpretations, J. Colloid Interface Sci.
    93:184 (1983).
D.J. Mitchell, B.W. Ninham, and D.F. Evans, And again the Micelle Diffusion
    Coefficient, J. Colloid Interface Sci. 101:292 (1984).

The critique and reply of M. Corti and V. Degiorgio in the same Journal.
This is scientific dispute at its very best. In the end, both sides, as
always, are partly correct, and the resolution of the debate has apparently
been achieved by Drifford (yet to be published) J. de Physique.

Counterion Specificity:

Y. Talmon, D.F. Evans and B.W.: Ninham, Spontaneous Vesicles formed from
    Hydroxide Surfactants: Evidence from Electron Microscopy, Science
    221:1047 (1983).
B. Kachar, D.F. Evans, and B.W. Ninham, Rapid Characterization of Colloidal
    Systems by Video Enhanced Contrast Light Microscopy, J. Colloid
    Interface Sci. 99:593 (1984).
S. Hashimoto, J.K. Thomas, D.F. Evans, and B.W. Ninham, Unusual Behaviour
    of Hydroxide Surfactants, J. Colloid Interface Sci. 95:594 (1983).
J.E. Brady, D.F. Evans, B. Kachar, and B.W. Ninham, Spontaneous Vesicles,
    J. Am. Chem. Soc. 106:4279 (1984).
J. Brady, D. F. Evans, G. Warr, F. Gnessir, and B.W. Ninham, Counterion
    Specificity as a Determinant of Surfactant Aggregation, J. Phys. Chem.
    90:1853 (1986).
B. Kachar, D.F. Evans, and B.W. Ninham, Video Enhanced Contrast Differ-
    ential Interference Contrast Microscopy, J. Colloid Interface Sci.
    100:287 (1984).

Surface Tension and Microemulsion Droplet Models

See Mitchell and Ninham - Approaches to Self-Assembly
Evans, Mitchell, and Ninham, Ion Binding and Dressed Micelles (above)
D.J. Mitchell and B.W. Ninham, Electrostatic Curvature Contributions to
    interfacial Tension of Micellar and Microemulsion Phases, J. Phys.
    Chem. 87:2996 (1983), and work of Langevin and Cazabat there cited.

Ionic Surfactant Micelles and Liquid Crystals

For a general review on how to obtain structural and dynamic information
from NMR, see:

B. Lindman, H. Wennerström, and O. Söderman, NMR Studies of Surfactant
    Systems, in: "Surfactant Solutions. New Methods of Investigation", R.
    Zana, ed., Marcel Dekker, New York, 1987, p. 295.

This account also gives literature references to all the studies
referred to below. For NMR work on surfactant systems see also Specialist
Periodical Reports "Nuclear Magnetic Resonance", e.g. O. Söderman, B.

Lindman, and P. Stilbs, in vol. 12 (1983) and O. Söderman in vol. 14
(1985).

Phase Diagrams

P. Ekwall, Adv. Liquid Cryst. 1:1 (1975).
P. Ekwall, L. Mandell, and K. Fontell, Mol. Cryst. Liq. Cryst. 8:157
     (1969).
K. Fontell, Mol. Cryst. Liq Cryst. 63:59 (1981)
G.J.T. Tiddy, Surfactant-Water Liquid Crystal Phases, Phys. Rep. 57:1
     (1980).

Divalent Couterion Systems

A. Khan, K. Fontell, G. Lindblom, and B. Lindman, J. Phys. Chem. 86:4266
     (1982).
A. Khan, K. Fontell, and B. Lindman, Colloids & Surfaces 11:401 (1984).
A. Khan, K. Fontell, and B. Lindman, J. Colloid Interface Sci. 101:193
     (1984).
A. Khan, K. Fontell, and B. Lindman, Progr. Colloid & Polymer Sci. 70:30
     (1985).
A. Khan, B. Jönsson and H. Wennerström, J. Phys. Chem. 89:5180 (1985).

NMR Self-Diffusion Studies of Micellization, Solubilization and Microemulsions

     The progress in this field can be attributed mainly to the development
of the Fourier transform version of the pulsed gradient spin echo NMR technique by P. Stilbs.

P. Stilbs and M.E. Moseley, Chem. Scr. 15:176 (1980).
P. Stilbs, J. Colloid Interface Sci. 87:385 (1982).
P. Stilbs, Progress NMR Spectroscopy, 19:1 (1987).

Surfactant Self-Association

H. Wennerström and B. Lindman, Micelles. Physical Chemistry of Surfactant
     Association. Physics Reports 52:1 (1979).
P. Stilbs and B. Lindman, J.Phys. Chem. 85:2587 (1981).
B. Lindman, M.C. Puyal, N. Kamenka, R. Rymdén, and P. Stilbs, J. Phys.
     Chem. 88:5048 (1984).
B. Lindström, A. Khan, O. Söderman, N. Kamenka, and B. Lindman, J. Phys.
     Chem. 89:5313 (1985).
N. Kamenka, G. Haouche, B. Faucompré, B. Brun, and B. Lindman, J. Colloid
     Interface Sci. 108:451 (1985).
For theoretical rationalization see above cited paper by Gunnarsson,
     Jönsson and Wennerström.

Micelle shape

Fluorescence quenching:

N.J. Turro and A. Yekta, J. Amer. Chem. Soc. 100:5951 (1978).
M. Almgren and J.E. Löfroth, J. Colloid Interface Sci. 81:486 (1980).
P. Lianos and R. Zana, J. Phys. Chem. 84:3339 (1980).

Kinetic studies:

E.A.G. Aniansson, S.N. Wall, M. Almgren, H. Hoffmann, I. Kielmann, W.
     Ulbricht, R. Zana, J. Lang, and C. Tondre, J. Phys. Chem. 80:905
     (1976).

NMR work:

J. Ulmius and H. Wennerström, J. Magn. Resonance 28:309 (1977)

Dynamic light scattering:

P.J. Missel. N.A. Mazer, M.C. Carey, and G.B. Benedek, in: "Solution Behaviour of Surfactants", K.L. Mittal and E.J. Fendler, eds., Plenum, New York (1982), vol. 1. p. 373.

Viscoelastic systems:

J. Ulmius, H. Wennerström, L.B.A. Johansson, G. Lindblom and S. Gravsholt, J. Phys. Chem. 83:2232 (1979).

Micelle shape NMR studies:

J. Ulmius, H. Wennerström, J. Magn. Resonance 28:309 (1977).
U. Henriksson, R. Klason, L. Odberg, and J.C. Eriksson, Chem. Phys. Lett. 52:554 (1977).

The obstruction effect in self-diffusion is very useful:

B. Jönsson, H. Wennerström, P.G. Nilsson, and P. Linse, Colloid Polymer Sci. 264:77 (1986).

Chain packing and dynamics

Experiments on liquid crystals:

B. Mely, J. Charvolin, and P. Keller, Chem. Phys. Lipids 15:161 (1975).
J.N. Davis, Biochim. Biophys. Acta 737:117 (1983).
J. Seelig, Quart. Rev. Biophys. 10:353 (1977).

Experiments on micelles:

H. Wennerström, B. Lindman, O. Söderman,, T. Drakenberg, and J. Rosenholm, J. Amer. Chem. Soc. 101:6869 (1979).
H. Walderhaug, O. Söderman, and P. Stilbs, J. Phys. Chem. 88:1655 (1984).
O. Söderman, and P. Stilbs, J. Phys. Chem. 88:1655 (1984).
O. Söderman, H. Walderhaug, U. Henriksson, and P. Stilbs, J. Phys. Chem. 89:3693 (1985).
H. Néry, O. Söderman, D. Canet, H. Walderhaug, and B. Lindman, J. Phys. Chem. 90:580 (1986).

Two-step model of relaxation:

H. Wennerström, G. Lindblom, and B. Lindman, Chem. Scr. 6:97 (1974).
B. Halle and H. Wennerström, J. Chem. Phys. 75:1928 (1981).

Theory: see above cited work by D. Gruen.

Counterion binding

In addition to the review paper on NMR see:

H. Gustavsson and B. Lindman, J. Amer. Chem. Soc. 100:4647 (1978).
H. Wennerström, B. Lindman, S. Engström, O. Söderman, G. Lindblom, and G.J.T. Tiddy, in: "Magnetic Resonance in Colloid and Interface Science", J.P. Fraissard and H.A. Resing, eds., Reidel, New York (1980), p. 609.

B. Lindman, in: "NMR of Newly Accessible Nuclei", P. Laszlo, ed., Academic
   Press, New York (1983), vol. 1, p. 193.
G. Lindblom, B. Lindman, and G.J.T. Tiddy, J. Amer. Chem. Soc. 100:2299
   (1978).
H. Fabre, N. Kamenka, A. Khan, G. Lindblom, B. Lindman, and G.J. Tiddy, J.
   Phys. Chem. 84:3428 81980).

Hydration

B. Lindman, H. Wennerström, H. Gustavsson, N. Kamenka, and B. Brun, Pure
   Applied Chem. 52:1307 (1980).
B. Halle and G. Carlström, J. Phys. Chem. 85:2142 (1981).
N.O. Persson and B. Lindman, J. Phys. Chem. 79:1410 (1975).
H. Wennerström, N.O. Persson, and B. Lindman, Am. Chem. Soc. Symp. Ser.
   9:253 (1975).

Nonionic micelles

For phase diagram work see:

D.J. Mitchell, G.J.T. Tiddy, L. Waring, T. Bostock, and M.P. McDonald, J.
   Chem. Soc. Faraday 1, 79:975 (1983), and references therein.

NMR relaxation and self-diffusion work:

P.G. Nilsson, H. Wennerström, and B. Lindman, J. Phys. Chem. 87:1377
   (1983).
P.G. Nilsson and B. Lindman, J. Phys. Chem. 87:4756 (1983).
P.G. Nilsson and B. Lindman, J. Phys. Chem. 88:4764 (1984).
P.G. Nilsson and B. Lindman, J.Phys. Chem. 88:5391 (1984).
P.G. Nilsson and B. Lindman, J. Phys. Chem. 86:27 (1982).
P.G. Nilsson, H. Wennerström, and B. Lindman, Chem. Scr.25:67 (1985).

Fluorescence quenching studies:

A. Malliaris, J. LeMoigne, J. Sturm, and R. Zana, J. Phys. Chem. 89:2709
   (1985).
J.E. Löfroth and M. Almgren, in: "Surfactants in Solution", K.L. Mittal and
   B. Lindman, eds., Plenum, New York (1984), vol. 1, p. 627.
G.G. Warr, Thesis, Melbourne (1985).

Force measurements: P. Claesson et al. (see below)
See also other papers on nonionics in this book (Kahlweit, Degiorgio)

Structure and Dynamics of Microemulsions

An extensive review of self-diffusion is under publication:

B. Lindman and P. Stilbs, Molecular Diffusion in Microemulsions, in:
   "Microemulsions", S. Friberg and P. Bothorel, eds., CRC Boca Ralon Fl,
   1987, p. 119-

Selected studies are:

P. Stilbs and B. Lindman, J. Colloid Interface Sci. 99:290 (1984).
F.D. Blum. S. Pickup, B.W. Ninham, S.J. Chen, and D.F. Evans, J. Phys.
   Chem. 89:711 (1985).
K. Fontell, A. Ceglie, B. Lindman, and B.W. Ninham, Acta. Chem. Scand.
   A49:241 (1986).
B. Lindman, N. Kamenka, T.M. Kathopoulis, B. Brun, and P.G. Nilsson, J.
   Phys. Chem. 84:2485 (1980).

B. Lindman, P. Stilbs, and M.E. Moseley, J. Colloid Interface Sci. 83:569
     (1981).
P. Stilbs and B. Lindman, Progress Colloid & Polymer Sci. 69:391 (1984).
B. Lindman, T. Ahlnäs, O. Söderman, H. Walderhaug, K. Rapacki, and P.
     Stilbs, Faraday Disc. Chem. Soc. 76:317 (1983).
P. Stilbs, K. Rapacki, and B. Lindman, J. Colloid Interface Sci. 95:583
     (1983).
P. Guéring and B. Lindman, Langmuir 1:464 (1985).

For NMR relaxation work see:

B. Lindman, T. Ahlnäs, O. Söderman, H. Walderhaug, J. Rapacki, and P.
     Stilbs, Faraday Disc. Chem. Soc. 76:317 (1983).
O. Söderman and H. Walderhaug, Langmuir, 2:57 (1986).
K. Shinoda's surfactants with balanced hydrophile-lipophile properties
     which give microemulsions at low surfactant contents are interesting
     not the least from a structural point of view:
K. Shinoda, H. Kunieda, T. Arai, and H. Saijo, Principles of Attaining Very
     Large Solubilization (Microemulsion): Inclusive Understanding of the
     Solubilization of Oil and Water in Aqueous and Hydrocarbon Media, J.
     Phys. Chem. 88:5126 (1984).
K. Shinoda, The Significance and Characteristics of Organized Solutions, J.
     Phys. Chem. 89:2429 (1985).
K. Shinoda, Solution Behaviour of Surfactants: The Importance of Surfactant
     Phase and the Continuous Change in HLB of Surfactant, Progress in
     Colloid & Polymer Sci. 68:1 (1983).

For self-diffusion studies of these systems see publications of K. Shinoda
and the Lund group.

Microemulsions with Double-chained Surfactants

L.R. Angel, D.F. Evans, and B.W. Ninham, Three Component Ionic Micro-
     emulsions, J. Phys. Chem. 87:538 (1983).
S.J. Chen, D.F. Evans, and B.W. Ninham, Properties and Structure of Three
     Component Ionic Microemulsions, J. Phys. Chem. 87:538 (1983).
B.W. Ninham, D.F. Evans, and S.J. Chen, Role of Oils and Other Factors in
     Microemulsion Design, J. Phys. Chem. 88:5855 (1984).
F.D. Blum, S. Pickup, B.W. Ninham, S.J. Chen, and D.F. Evans, Structure and
     Dynamics in Three Component Microemulsion, J.Phys. Chem. 89:711
     (1985).
S.J. Chen, D.F. Evans, B.W. Ninham, D.J. Mitchell, F.D. Blum, and S.
     Pickup, Curvature as a Determinant Microstructure in Microemulsions,
     J. Phys. Chem. 90:842 (1986).
D.F. Evans, D.J. Mithcell, and B.W. Ninham, Oil, Water and Surfactant.
     Properties and Conjectured Structure of Simple Microemulsions, J.
     Phys. Chem. (Feature article) 90:2817 (1986).
K. Fontell, A. Ceglie, B. Lindman, and B.W. Ninham, Acta Chem. Scand.
     A49:247 (1986).

Forces (Experimental) A large number of papers in J. Colloid Interface
Sci., J. Phys. Chem., Biophys, J. are associated with the work of J.N.
Israelachvili, R.M. Pashley, R.G. Horn, H. Christenson, J. Marra V.A.
Parsegion, P. Rand.

Much of the literature can be tracked down in "Hydration Forces". Chemica
Scripta vol. 25 (Orenäs Conference).

(1) See also J. Marra and J.N. Israelachvili, Biochemistry 24:4608 (1985).
(2) P. Claesson, R. Kjellander, P. Stenius and H. Christenson, J.C.S.
     Faraday 1,82:2935 (1986).

(3)   J. Marra, J. Phys. Chem. 90:2145 (1986).
(4)   R.M. Pashley, P. McGuiggan, B.W. Ninham, D.F. Evans, Attractive Forces
      between Unchanged Hydrophobic Surfaces: Direct Measurements in
      Aqueous Solution Science 229:1088 (1985).
(5)   see (1).
(6)   (7) R.M. Pashley, P.M. McGuiggan, B.W. Ninham, D.F. Evans, and J.
      Brady, Direct Measurements of Surface Forces between Bilayers of
      Double-chained Quaternary Ammonium Acetate and Bromide Surfactant, J.
      Phys. Chem. (1986). 90:1637. See Ion Binding.

The apparent excellent agreement with DLVO may or may not be an
artefact. The hypernetted chain or higher approximations beyond the
Poisson-Boltzmann equation probably is necessary.

See Bo Jönsson et al. and R. Kjellander and S. Marcelja:

B. Jönsson, P. Linse, and T. Åkesson, Breakdown of the Poisson-Boltzmann
      Approximation in Polyelectrolyte Systems: A Monte Carlo Simulation
      STudy, in: "Surfactants in Solution", K.L. Mittal and B. Lindman,
      eds., Plenum, New York (1984), p. 2023.
L. Guldbrand, B. Jönsson, H. Wennerström, and P. Linse, Electrical Double
      Layer Forces. A Monte Carlo Study, J. Chem. Phys. 80:2221 (1984).
B. Svensson and B. Jönsson, The Interaction between Charged Aggregates in
      Electrolyge Solution. A Monte Carlo Simulation Study, Chem. Phys.
      Letters 108:580 (1984).
R. Kjellander and S. Marcelja, Inhomogeneous Coulomb Fluids with Image
      Interactions between Planar Surfaces. I, J. Chem. Phys. 82:2122
      (1985).
R. Kjellander and S. Marcelja, Correlation and Image Charge Effects in
      Electric Double Layers, Chem. Phys. Letters, 112:49 (1984).

Conclusions

R.B. Ashman and B.W. Ninham, Immunosuppression Induced by Cationic Sur-
      factants, Molec. Immunology 22:609 (1985).
R.B. Ashman, R.V. Blanden, B.W. Ninham, and D.F. Evans, A Role for Surface
      Chemistry in Immunology, Immunology Today, (to appear).
D.F. Evans and B.W. Ninham, Molecular Forces in the Self-Organization of
      Amphiphiles, J. Phys. Chem. 90:2817 (1986).

NEW IDEAS FOR MICROEMULSION STRUCTURE:

THE TALMON-PRAGER AND DE GENNES MODELS

C. Taupin

Physique de la Matière Condensée
Collège de France
11 Place Marcelin-Berthelot
75231 Paris Cedex 05, France

1. INTRODUCTION

In the preceding lectures, we have heard about a variety of experimental investigations which are tightly bound with the "droplet model". This model, which appeared in the early studies of microemulsions, has been very successful and illuminating in many monophasic microemulsions, but it seems more questionable for the three phase microemulsion systems (Figure 1).

A different approach is proposed in two recent models[1,2,3] where the statistics of geometrical repartition as a function of the three components (water, oil and amphiphilic film) are studied without focusing on the existence of well-defined structural elements such as droplets.

In these models, which are still under experimental investigation, the volume is randomly divided into three different parts; the interfacial film is supposed to contain all the surfactant molecules and occupy a negligible volume. Then the statistical mechanics of phase equilibria in these pseudo-ternary systems are developed taking into account the following elements:

a) entropy of the geometrical repartition;
b) interfacial energy (i.e., interfacial tension in the film), which is unusually small;
c) contribution of the curvature of the film to the energy;
d) Van der Waals attraction between similar layers.

The original idea for these models resides in a paper by Scriven[4]. This author pointed out that as the oil/water ratio is increased, three types of structure will evolve in succession. Starting from the oil in water emulsion to be expected at low oil content, there should be a smooth transition to a bicontinuous structure, in which neither phase can be said to surround the other, followed by a second transition to a disconnected water in oil emulsion. In this early paper, special attention is given to the variation of the interfacial area which is an important factor in the energy function. The two models (Talmon-Prager and de Gennes) differ mainly by the way the random geometry is generated. One of them (TP) is the Voronoi tesselation for which many properties have been calculated;

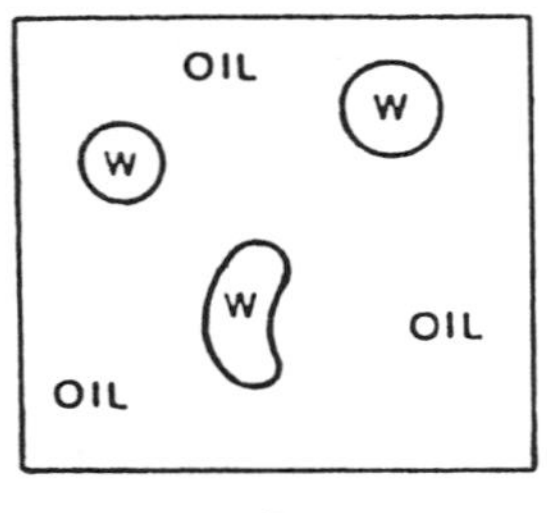

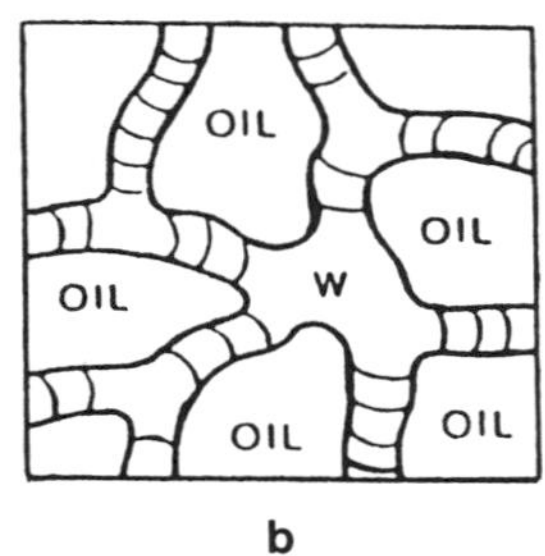

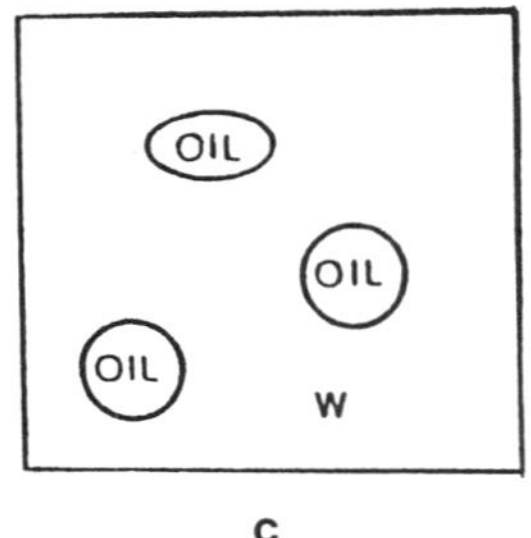

Fig. 1. Structure of microemulsions as a function of the water-to-oil
ratio (w/o): (a and c) swollen micelles (respectively w/o <<1 and
w/o >>1; (b) "bicontinuous" structures, first proposed by Scriven.
The example shown here can be described as a network of water
tubes in an oil matrix (in this figure the individual surfactant
molecules are not shown; the surfactant film is represented as a
continuous sheet).

the other one is a division of space into cubes of edge ξ_K, ξ_K being
determined by the flexibility of the film. Talmon-Prager focuses mainly
on phase equilibria, whereas de Gennes also considers the competition
between microemulsions and lamellar phases.

2.   TALMON-PRAGER MODEL

A simple procedure is proposed to generate random two-fluid
geometries, with the main attention given to entropy and curvature. The
entropy effects were mentioned first by Ruckenstein and Chi[5]. They are
usually difficult to calculate. In their model, in order to determine at
least within a constant of proportionality the number of configurations
available to the oil and water regions in a microemulsion of specified
composition, Talmon and Prager consider the class of geometries resulting
from a two step process, consisting of a Voronoi tesselation followed by
random segregation into oil and water domains. N points are first
distributed completely at random over the volume V to be occupied by the
microemulsion; with each of these Poisson points is associated the
polyhedral region which lies closer to it than to any of the other (N - 1)
points. In the second step, N_O of these polyhedra are selected at random
to be occupied by oil, and the remainder filled with water; the result is
a random two-fluid geometry in which the oil occupies the volumic fraction
v_O and the water v_w of the total volume V.

The number of distinct positions available to a Poisson point is
proportional to the volume V, the proportionality constant 1/w being one
of the model parameters. It is the size of w that determines the scale of
the microemulsion structure, and it is through w that information about
the molecular nature of the fluids enters into the models. Taking into
account the number of configurations for the placement of all N points
and for each configuration the number of ways of assigning the oil and
water domains, the configurational entropy is calculated:

$$S = - Vkc(\ln(\frac{wc}{e}) + v_o \ln v_o + v_w \ln_w).$$

This equation must be supplemented by a relation between $c \equiv N/V$, the
concentration of Poisson points, and the true composition variables, in
particular the surfactant concentration σ.

114

The mean area S of internal oil/water boundary per unit volume is given by

$$S = \eta_1 c^{1/3} v_o v_w \quad \text{with } \eta_1 = 5.82$$

and the surfactant concentration σ will be taken proportional to S, $\sigma = \alpha S$ where the surfactant capacity α is a second parameter of the system. The phase equilibria of such a system, which is entirely controlled by the entropy, shows that above a certain critical value of surfactant there is only one phase; below that concentration, there is a mixture of two microemulsions in equilibrium with one another. No three phase equilibria exist with only the entropic term.

Talmon and Prager added the effect of curvature in the following way: a curved interface implies a lowering of the surfactant capacity by a factor $(1 - \beta\lambda)$, λ being the edge length per unit area. The curvature is evaluated by the length of the edges of the Voronoi polyhedra which lie on the oil/water interface: $5.83\ c^{2/3} \times 3v_o v_w$. The new expression for σ is:

$$\sigma = 5.82 \quad \sigma c^{1/3} v_o v_w (1 - 3\beta c^{1/3}).$$

The corresponding expression for the free energy gives rise to interesting new features: a triple tangent plane appears which lies entirely below the free energy surface, proving the possibility of three phase equilibria in this model. This is a very new and remarkable result; many ingredients of the model are very simple and reflect the fundamental characteristics of the three phase systems:

a)   very low interfacial tension;
b)   effect of entropy;
c)   curvature terms.

It is interesting to note that the curvature energy term in the Talmon-Prager model is symmetrical as regards the sense of curvature, in contrast to many other theoretical studies[6]. In spite of the fact that it is clear that such an hypothesis is not true in many systems, it could be realistic in the vicinity of the "optimal composition", where the symmetry of the properties of the surfactant molecule relative to oil and water has been frequently emphasized[7].

Nevertheless, some physical phenomena are omitted in this model: (i) because in such geometry the interfacial film is wrinkled, a necessary condition is the flexibility of the interfacial film, which is known to be very small in the stiff lamellar phases; (ii) Talmon-Prager considered only zero interfacial tension state: phase equilibria could occur close to (but not exactly at) a situation of zero interfacial tension; and (iii) the Van der Waals interaction, which have been shown to be important in microemulsion systems are neglected.

The de Gennes-Taupin[3] model studies the role of these ingredients with a different way of dividing space, linking the statistics of this problem to that of the well-known Ising model.

3.   THE ROLE OF INTERFACIAL FLEXIBILITY: DE GENNES MODEL

1)   The Existence of a Well Defined Interfacial Area per Molecule can be Correlated with the Experimental Ultralow Interfacial Tensions in the Following Manner (Schulman Description)

Let us consider a single interface of arbitrary shape separating the

oil from the water: the total area of interface is a. It contains a
number n_s of surfactant molecules, each of them covering an area $\Sigma = a/n_s$.
In the simple case where: (i) the surfactant is insoluble in bulk oil or
water; (ii) interactions between different portions of the interface are
negligible; and (iii) curvature energies are omitted, the free energy may
be written:

$$f = f_{bulk} + \gamma_{ow}a + n_s G(\Sigma), \tag{1}$$

where the second term corresponds to the bare (i.e., without surfactant)
interface (with an interfacial tension γ_{ow}) and $G(\Sigma)$ is a surfactant free
energy, depending on the area Σ and containing in particular the effect
of surfactant/surfactant repulsions. The Langmuir surface pressure of
the film is

$$\Pi(\Sigma) = - \frac{\partial G}{\partial \Sigma} \tag{2}$$

and the actual interfacial tension is

$$\gamma = \gamma_{ow} - \Pi. \tag{3}$$

If (1) is minimized, at fixed n_s, with respect to Σ, the condition (4) is
obtained:

$$0 = \frac{df}{d\Sigma} = n_s(\gamma_{ow} - \Pi) = n_s\gamma. \tag{4}$$

Thus the system will adopt a well defined area per surfactant which
is called Σ^*: it is defined by the implicit equation

$$\Pi(\Sigma^*) = \gamma_{ow}. \tag{5}$$

The state $\Sigma = \Sigma^*$ will be called the <u>saturated state</u>. It can be
reached only if other possible states of the surfactant (such as pure
surfactant micelles in water, or in oil) are of higher free energy.

In the saturated state, with a system of zero surface tension, as
shown by Eq. (4), the area a is entirely defined by the number of
surfactants available:

$$a^* = n_s\Sigma^*. \tag{6}$$

It has to be noted that if $\gamma = 0$, several factors become non-
negligible:

a)   entropy of mixing oil and water (see recent published calculations –
     Refs. [1,5]);
b)   interaction forces (electrostatic and Van der Waals);
c)   curvature energy and flexibility.

2.   Rigidity and Spontaneous Curvature

The Schulman description ignores all energies associated with the
curvature of the interface. Indeed, for many problems involving
fluid/fluid interfaces, curvature energies represent only a very minor
correction, and the interfacial energy γ dominates the behavior. However,
with interfaces where $\gamma \to 0$, curvature effects become relevant.

The two basic ingredients have been defined most clearly in a paper
by W. Helfrich[8]. For a curvature $1/R$ one expects an energy contribution
per unit area of the form

$$F = \gamma - \frac{K}{R_o R} + \frac{K}{2R^2} , \qquad (7)$$

where $1/R_o$ is the spontaneous curvature, and can be of either sign (R_o is counted as positive when the trend is towards direct micelles). The parameter K has the dimensions of an energy, and may be called the rigidity of the interface. Equation (7) holds only if R and R_o are much larger than the interfacial film thickness L. How can one estimate K and R_o?

(i) For ionic surfactants the steric considerations of Ninham and Mitchell[6] may give an estimate of the spontaneous curvature (Figure 2).

(ii) The addition of a cosurfactant may act strongly on $1/R_o$ and also on K. For instance, in the (different but related) soap water systems, Charvolin and Mely showed that a certain mixture of C_{18} and C_{10} soaps could give a cubic phase which is not present with the pure C_{14} soap[9]. The cubic phase is believed to be an array of rod portions (with positive curvature) related by branching regions (with negative curvature).

(iii) For films of non-ionic surfactants, Robbins has considered curvature effects in some detail[10].

As experiments indicate a small contribution of spontaneous curvature in microemulsions, we will focus on the role of the flexibility (K^{-1}).

3. Fluctuations of a Flexible Interface

The general idea is that with $\gamma = 0$ there may be giant fluctuations of the film shape. This approach follows the lines of a previous study of red blood cell scintillation[11].

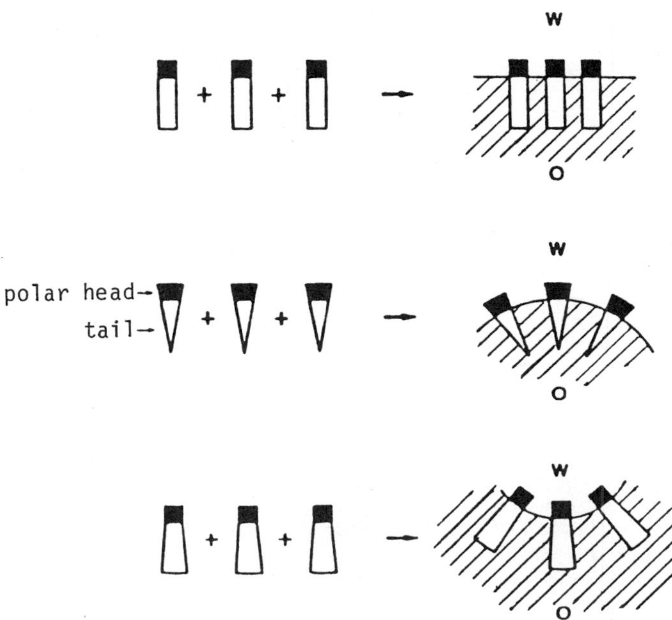

Fig. 2. A naive steric model correlating the shape of the amphiphile to the spontaneous curvature of the interface.

Let us assume now that the interface has a negligible spontaneous curvature $((1/R_o) \to 0)$ and is close to a certain reference plane (xy). The distances between the plane and the interface will be called $\zeta(xy)$. The curvature is then:

$$\frac{1}{R} = \frac{\partial^2 \zeta}{\partial x^2} + \frac{\partial^2 \zeta}{\partial y^2} \equiv \Delta_\perp \zeta. \tag{8}$$

The free energy (7) becomes

$$f_K = \int \frac{1}{2} k (\Delta_\perp \zeta)^2 = \sum_q \frac{1}{2} K q^4 |\zeta_q|^2, \tag{9}$$

where we have used the two-dimensional Fourier transforms

$$\zeta_q = \int dx dy \, \zeta(xy) \exp[i(q_x x + q_y y)]. \tag{10}$$

We shall be mainly interested in the local orientation of the surface, defined by a unit vector $\underset{\sim}{n}$ normal to it:

$$n_x = - \frac{\partial \zeta}{\partial x}, \quad n_y = - \frac{\partial \zeta}{\partial y}, \quad n_z \simeq 1. \tag{11}$$

For small fluctuations $\partial \underset{\sim}{n} = (n_x, n_y)$ we can write

$$|\partial \underset{\sim}{n}_q|^2 = q^2 |z_q|^2. \tag{12}$$

Applying the equipartition theorem to all modes in Eq. (10) we obtain the thermal average of these fluctuations:

$$<|\partial n_q|^2> = \frac{kT}{Kq^2}, \tag{13}$$

where T is the temperature and k is Boltzmann's constant. We can now look at the angular correlations between two points (0 and $\underset{\sim}{r}$) on the surface:

$$\theta^2(r) = <|\underset{\sim}{\delta}n(o) - \delta\underset{\sim}{n}(r)|^2> = \sum_q 2[1 - \cos(q \cdot \underset{\sim}{r})]<|\delta n_q|^2>$$

$$= \frac{kT}{\pi K} \int_o^{1/a} [1 - J_o(qr)] \frac{dq}{q}, \tag{14}$$

where $1/a$ is a high q cut-off - a microscopic length related to the detergent size. $J_o(x)$ is a Bessel function; the factor $1 - J_o(x)$ is essentially equal to 1 for $x \gg 1$, and to 0 for $x \ll 1$. Omitting uninteresting constants, the result is thus

$$<\theta^2> = \frac{kT}{\pi K} \ln(\frac{r}{a}). \tag{15}$$

For small θ we may also present it in the form

$$<\cos\theta> \simeq <1 - \frac{\theta^2}{2}> \simeq \exp(- \frac{<\theta^2>}{2}) = (\frac{a}{r})^{(kT/2\pi K)}. \tag{16}$$

The law (16) shows an exponent $kT/2\pi K$ which is continuously variable with T - a frequent feature of two-dimensional fluctuations[12]. We have derived it here only for θ small or not too large r. The question of its validity for larger r values is not known at present. For our purposes, however, we know enough with Eq. (16). In particular, we can define a persistence length ξ_K by the following procedure: at distances r smaller than ξ_K the angle θ is small on the average, while at distance $r > \xi_K$ it is large. Choosing for instance $<\cos\theta> = 1/e$ as the crossover value, we obtain from Eq. (16)

$$\xi_K = a \, \exp\left(\frac{2\pi K}{kT} \right). \tag{17}$$

Thus the persistence length ξ_K is extremely sensitive to the value of the rigidity constant K. If, following Helfrich[13] we assume that a simple monolayer (without any cosurfactant) has a rigidity comparable to that of a thermotropic liquid crystal, we arrive at values $K \sim 10^{-13}$ erg, corresponding to $2\pi K/kT \sim 12$. In such a case ξ_K is exponentially large, $\xi_K \sim 10^3 a$, and the interface is <u>stiff</u>. On the other hand, if, by addition of a suitable cosurfactant, we can decrease K by a factor of 5 ($K \sim 2.10^{-14}$ erg) then $\xi_K \sim 10a \sim 100$ Å and the interface, observed at scales r larger than 100 Å, is <u>strongly wrinkled</u>. Clearly this distinction must play an important role for the selection of disordered (rather than ordered) structures.

4. SOME APPLICATIONS OF THE CORRELATION LENGTH ξ_K

1) <u>The Multisurface Problem</u>

We now consider a situation where the interface is present in all the sample volume Ω. Numerically, this can be characterized by an <u>amount of surface per unit volume</u> which we call 1/d (since it has the dimensions of an inverse length). If the number of surfactant molecules per cm^3 is n_s, and if they are all located at an interface, we have

$$\frac{1}{d} = n_s \Sigma. \tag{18}$$

Physically d represents a certain average distance between consecutive sheets of the interface: for instance, if we had a lamellar structure, 2d would be the repeat period (each period containing <u>two</u> interfaces).

If our sheets were completely ideal, i.e., if there were no interaction between them, each could show a persistence length ξ_K: it would be rigid at short scales ($r < \xi_K$) and wrinkled at large scales ($r > \xi_K$). However, all sheets interact. We have long range Van der Waals attractions and also repulsive forces: electrostatic forces in the water phase, and steric forces appearing whenever the aliphatic tails from two neighboring sheets begin to overlap.

All these forces tend to restore order in the sheet system. In the following paragraphs, we try to give a qualitative analysis of three very complex effects: steric, long-range interactions and entropic.

2) <u>Steric Effects</u>

Let us start with non-interacting sheets; it is probably correct to visualize any sheet as a system of adjacent platelets, each with a certain typical size ξ_K and areas ξ_K^2. Let us think of them as independent units. Each platelet has a number ξ_K^2/Σ^* of detergent molecules, and thus the number of platelets per unit volume is

$$c_p = n_s \frac{\Sigma^*}{\xi_K^2} = \frac{1}{d\xi_K^2}. \tag{19}$$

If different platelets cannot intersect each other, they may tend to stack and form a nematic phase of flat objects; this type of liquid has indeed been observed recently with suitable organic molecules, and is currently called a <u>discotic</u> phase[14]. Clearly, the discotic phase can exist only if the <u>distance d</u> between platelets is not too large. We can make this

statement slightly more quantitative by a transposition of the Onsager
argument concerning the nematic alignment of rod-like molecules[15].   In
the present case we may say that each platelet is associated with an
interaction volume of order ξ_K^3 (Figure 4). Whenever two platelets
have a finite overlap of their interaction volumes, they are
strongly correlated in their orientation. Thus the criterion for
nematic order is of the form

$$c_p \xi_K^3 \gg 1.\tag{20}$$

Returning now to Eq. (19), we see that there are two limits (see
Figure 3):

a)   if $\xi_K > d$ the sheets tend to be parallel to each other;
b)   if $\xi_K < d$ the sheets are wrinkled and can build up an isotropic,
     disordered liquid phase.

Thus for a given surfactant concentration n_s, there is in this case a
critical value of the rigidity constant $K = K_c$ corresponding roughly to

$$a \exp\left(\frac{2\pi K_c}{kT}\right) = d = \frac{1}{n_s \Sigma^*},\tag{21}$$

where we have used Eqs. (17) and (18).

3)   Long Range Interaction Effects

Our approach here will start from an ordered lamellar phase, adding
fluctuations to the average order, and then looking for their amplitude.
This line of thought has already been followed by Chun Huh[16], but there
are important differences. Chun Huh's interfacial energy γ is different
from zero (in fact, for the fluctuation calculation, he uses the bare
oil/water surface tension). In the present paper we start from a
saturated interface with $\gamma = 0$ and thus the fluctuations are
automatically much more important.

In particular, let us look at the motions of one interface in the
presence of neighboring layers which are fixed at their equilibrium
position. We use an augmented version of Eq. (9) where we include a
potential energy term $1/2\, U'' z^2$ (per unit area). Here U'' is the curvature
of the potential due to the neighboring layers, and is positive (stable
equilibrium). The conclusion of the calculations which were developed
elsewhere[17] is that if

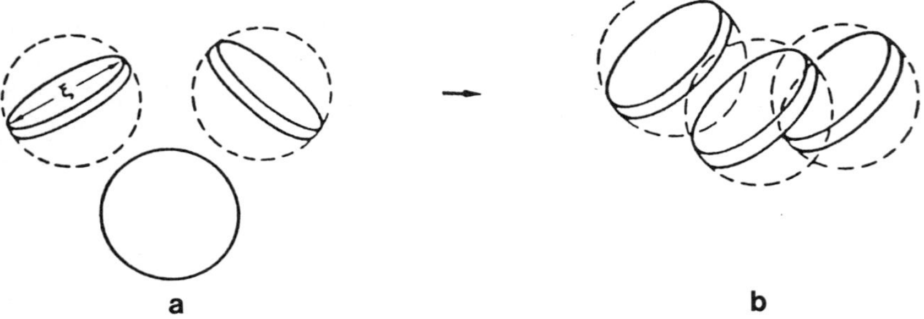

a   b

Fig. 3.   The stacking of disks: (a) at low concentration the disks are
disordered; (b) as soon as their "envelope spheres" (dashed)
overlap significantly they tend to stack. A "discotic phase"
appears.

$$K < \frac{(kT)^2}{64 \ U'' \ d^4} \ ,$$

a disordered state is preferred.

4) Entropy Effects for Flexible Interfaces

We discuss these effects in a model which is related to but somewhat different from the original proposal by Talmon and Prager[1]. We observe first that the interface must have a certain <u>persistence length</u> ξ_K: (i) it is essentially flat at scales smaller than ξ_K; and (ii) consecutive "pieces of interface", with an area ξ_K^2, have independent orientations. A rough but convenient model is then obtained by dividing all space into consecutive cubes, each of linear size ξ_K. Each cube is either filled with oil or with water. The overall proportion of cubes filled with oil (water) is called ϕ_O (ϕ_w). Two adjacent cubes will have no interface, and no energy, if they are of the same type. But if they are different, we must count a free energy contribution $\gamma \xi_K^2$, γ being the interfacial tension. For the moment, we do not assume $\gamma = 0$. Rather, we say that a given chemical potential μ_s of the surfactant imposes a certain area per surfactant Σ_s, through the thermodynamic condition (obtained from differentiating (1) with respect to n_s and using (2) and $\Sigma_s = a/n_s$):

$$\mu_s = G(\Sigma_s) + \Pi(\Sigma_s)\Sigma_s . \tag{22}$$

The result is then a certain $\gamma(\Sigma_s)$ which depends ultimately on μ_s.

We have now reduced the statistics of the interface to a "lattice gas model". Clearly, the description is very crude; the oil (or water) regions in a microemulsion do not look like an assembly of cubes. However, the lattice gas model keeps some essential features of a random surface. Also, the resulting statistical behavior is well-known. When the coupling between adjacent cubes ($\gamma \xi_K^2$) is weaker than kT, or more precisely when

$$\gamma < \gamma_c = \alpha kT/\xi_K^2 \tag{23}$$

(where $\alpha = 0.44$ for a simple cubic lattice), we expect a single phase with the oil and water mixed down to the scale ξ_K. However, when the inequality (23) is reversed, we may have <u>phase separation</u> (Figure 4).

A number of significant properties emerge from the model, and are probably of more general validity:

a) The values of γ involved are weak. As shown by (23), the range of interest is $\gamma \sim kT/\xi_K^2$. The persistence length ξ_K is expected to be rather large, and thus γ should be small; we are not very far from the Schulman criterion.

b)   Phase separation occurs not because of specific interactions between the droplets, but purely from a balance between interfacial entropy and interfacial energy.

c) In practice one often observes more complex phase diagrams. In particular, certain microemulsions can coexist simultaneously with an oil phase and a water phase. The above lattice gas model generates two-phase equilibria only. However, the model is highly degenerate. Small perturbations on the structure of the free energy (induced by curvature

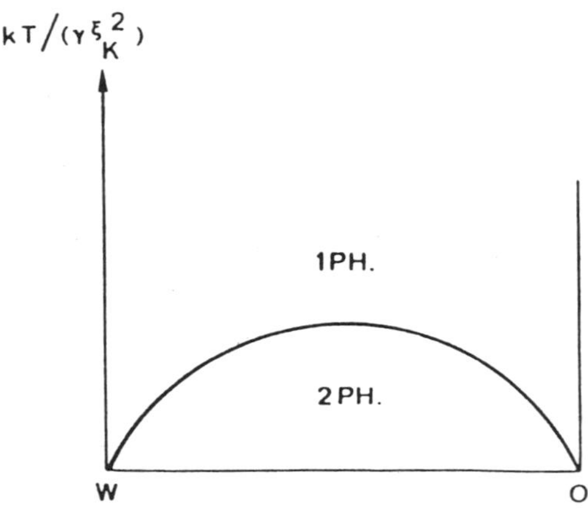

Fig. 4. Phase diagram in the modified Talmon-Prager model. At high
surfactant contents (low γ) the microemulsion is stable at all
water fractions. At lower surfactant contents, a phase separation
occurs. The present model does not include the spontaneous
curvature of the interface associated with Brancroft's rule. Then
(and only then) the plot is symmetrical.

effects or other corrections) might lead to three-phase equilibria. A
first attempt in this direction is described in Ref.[1]. Some difficulties
are pointed out in Ref.[18].

5. CONCLUSIONS

Microemulsions were discovered thirty years ago and described as
"extraordinary emulsions". Their structure do not obey to a unique model.
Two recent models focused on a random structure were reviewed.
Experimental studies[19] are still in progress to investigate these models
and their dynamical implications.

REFERENCES

1. Y. Talmon and S. Prager, J. Chem. Phys., 69:2984 (1978).
2. Y. Talmon and S. Prager, J. Chem. Phys., 76:1534 (1982).
3. P. G. de Gennes and C. Taupin, J. Phys. Chem., 86:2294 (1982).
4. L. E. Scriven, Nature, 263:123 (1976).
5. E. Ruckenstein and J. Chi, J. Chem. Soc., Faraday Trans. II, 71:1690
    (1975).
6. D. J. Mitchell and B. W. Ninham, J. Chem. Soc., Faraday Trans. II,
    77:601 (1981).
7. D. O. Shah and R. S. Schechter, "Improved Oil Recovery by Surfactants
    and Polymers Flooding", Academic Press Inc. (1977).
8. W. Helfrich, Z. Naturforsch., 28C:693 (1973).
9. J. Charvolin and B. Mely, Mol. Cryst. Liquid Cryst. Lett., 41:209
    (1978).
10. M. L. Robbins, "Micellization, Solubilization and Microemulsions", K.
    Mittal, ed., 2:273, Plenum, NY (1977).
11. F. Brochard and J. F. Lennon, J. de Phys., 36:1035 (1975).
12. J. G. Dash and J. Ruvalds, "Phase Transition in Surface Films", NATO
    Advances Study Series (Physics), eds., Plenum Press, NY (1980).
13. W. Helfrich, Z. Naturforsch., 33a:305 (1978).

14. J. Billard, in: " Chem. Phys. Ser. Vol. II, pp 383, 395, Springer, Berlin (1980).
15. L. Onsager, Ann. N.Y. Acad. Sci., 51:627 (1949).
16. Chun Huh, J. Colloid Interface Sci., 71:408 (1979).
17. P. G. de Gennes and C. Taupin, J. Phys. Chem., 86:2294 (1982).
18. J. Jouffroy, P. Levinson and P. G. de Gennes, J. Physique, 43:1241 (1982).
19. J. M. di Meglio, M. Dvolaitsky, R. Ober and C. Taupin, J. Physique (Lettres), 44:L-229 (1983).

WINSOR MICROEMULSIONS: EVIDENCE FOR BICONTINUOUS

STRUCTURE BY X-RAYS AND NEUTRON SCATTERING

C. Taupin[1], R. Ober[1], J.-P. Cotton[2]
and L. Auvray[1,2]

[1]Laboratoire de Physique de la Matière Condensée
College de France, 11 Place Marcelin-Berthelot
75231 Paris Cedex 05, France
[2]Laboratoire Léon Brillouin, CEA-CEN Saclay
91191 Gif-sur-Yvette Cedex, France

1. RANDOM BICONTINUOUS MICROEMULSIONS

A) Introduction

Since the early model of ordered bicontinuous structures generated by
minimal surfaces proposed by Scriven[1], two models of random bicontinuous
structures[2-4] have been published in which the local curvature of the
surfactant film strongly fluctuates.

These random bicontinuous structures and the statistical
configurations of the oil, water and surfactant are described by simple
mean-field models[2,4,5] in which the volume of the microemulsion is
divided into elementary cells of mean size ξ, randomly filled by oil and
water, the surfactant being distributed at the oil water interface.

The models differ mainly by the way the random geometry is generated.
In the original Talmon-Prager model[2], the cells are generated by a
random Voronoi tesselation and their size distribution is large. In
contrast to Talmon and Prager, de Gennes-Jouffroy-Levinson[4] and also
Widom[5] assume that the film stiffness forbids the curvature fluctuations
at spatial scales smaller than ξ_K, the persistence length of the
surfactant layer. This length is the basic size of the cells, which are
taken to be identical and cubic.

In both models, by contrast to the well-known equation giving the
radius R of water in oil spheres:

$$R = 3\phi_w/C_S\Sigma, \tag{1}$$

the mean size ξ of the oil and water elementary volumes, which is the
length scale ($\sim$ 100 Å) of the random structure, is related to the
microemulsion composition by the geometrical constraint:

$$\xi = 6\phi_o\phi_w/C_S\Sigma, \tag{2}$$

in which ϕ_o and ϕ_w are the oil and water volume fractions ($\phi_o + \phi_w = 1$), C_S

the surfactant concentration (number of molecules per unit volume), and Σ the area per surfactant molecule in the film ($\Sigma \sim 60$ Å^2).

Equation (2) interpolates continuously between the case of water in oil droplets ($\phi_w \ll 1$) and the case of oil in water droplets ($\phi_o \ll 1$). In the models of random microemulsions, the average mean curvature of the film <C> (by convention positive for W/O droplets) increases continuously with ϕ_o at constant interfacial area (C_S constant); <C> is negative for $\phi_o < 0.5$ and positive for $\phi_o > 0.5$, and it vanishes by symmetry when the microemulsions contain as much oil than water (inversion point, $\phi_o = \phi_w$).

A particularly interesting case of a microemulsion system is "the Winsor III Microemulsion" which appears when an oil, surfactant and brine system separates in three phases. The "middle phase" microemulsion is associated with extremely low interfacial tensions ($\sim 10^{-3}$ dynes/cm) and has many industrial applications in enhanced oil recovery and phase transfer. Its structural organization is not well understood. Several features are particularly interesting:

a) such microemulsions are associated with extremely low interfacial tensions;
b) they appear in a range of salinity of the brine where the spontaneous curvature of the amphiphilic film (which is strongly dependent on electrostatic repulsions) could be very low;
c) if they also correspond to highly flexible films they could be good candidates for testing the random bicontinuous models which were proposed recently[2,4].

To test this prediction, different techniques have been used: conductivity measurements[6], self-diffusion coefficient measurements [7,8], electron microscopy[9] and scattering techniques[10-16]. However, as the measurement of transport coefficients only yields very indirect information on the microemulsion structure and as it has not yet been possible to obtain artefact-free electronmicroscopy pictures of the middle-phases[17], the main information on these systems comes presently from small angle X-rays and neutron scattering experiments.

By using these techniques and the method of contrast variation with deuterated molecules[18] and studying a very representative and well-known system[19], we have obtained four main experimental results[12,13,15]:

i) A well-defined surfactant film, evidenced by an asymptotic behavior of the scattered intensities, exists even in "critical" Winsor microemulsions.

ii) The characteristic microemulsion size ξ, defined as a mean radius of curvature and drawn from the spectra in the intermediate range of scattering vector, q, follows experimentally from the prediction (2).

iii) When $\phi_o = \phi_w$, the macroscopic concentration fluctuations of water and surfactant (measured from the scattering at zero angle for different contrasts) are not correlated.

iv) The intensity scattered only by the oil and the water exhibits a pronounced peak at a given scattering vector q* proportional to the surfactant concentration C_S.

Let us recall that in any scattering experiment the important parameter is the scattering vector q, related to the scattering angle 2θ and wavelength λ by

$$q = \frac{4\pi}{\lambda} \sin\theta. \tag{3}$$

By varying q, it is possible to test the structure at different length scales. In our experiments q varies between 10^{-2} and $0.25 \; \mathring{A}^{-1}$; we explore the microemulsion structure between roughly 15 and 300 $\mathring{A}$. Three different q ranges can be distinguished, in correlation with the characteristic sizes ξ (around 100 $\mathring{A}$) of the microemulsions:

1) A large q-range (q > 0.1 $\mathring{A}$, small spatial scale < 30 $\mathring{A}$), which corresponds to the effects due to the surfactant film. This range is called the asymptotic range since in this domain the scattered intensity follows general laws which are independent of the large scale structure.

2) A "medium q-range" which corresponds to the size of the water and oil domains.

3) A very small q domain which describes the large scale structure of the microemulsion and depends on correlations between domains.

Another important possibility is to use neutron scattering to vary the contrast, i.e., to enhance selectively the scattering power of the various domains[18] (oil, water or film) of the system.

Assuming that a microemulsion reduces itself to three incompressible geometrical parts; the oil, the water and the film (volume fraction ϕ_o, ϕ_w, and ϕ_f, scattering length density n_o, n_w and n_f) the intensity scattered by a unit volume of the sample is written[13]:

$$i(q) = (n_w - n_o)^2 \chi_{ww}(q) + (n_f - n_o)^2 \chi_{ff}(q) + \\ + 2(n_w - n_o)(n_f - n_o)\chi_{wf}(q). \tag{4}$$

The partial structure factors are defined by the relation:

$$\chi_{ij}(q) = \int d^3 r < \delta\phi_i(0)\delta\phi_j(r) > e^{i\vec{q}\cdot\vec{r}} \tag{5}$$

in which $\delta\phi_i(r)$ is the local fluctuation of the ith component volume fraction (i = w,f). The brackets mean average value.

The direct structure factors $\chi_{ww}(q)$ and $\chi_{ff}(q)$ can be measured in a single experiment by using respectively an oil-water contrast ($n_o = n_f \neq \neq n_w$) and a film contrast ($n_o = n_w \neq n_f$). On the contrary, a complete variation of the contrasts is necessary to determine the cross structure factor $\chi_{wf}(q)$.

B)   Results

1) Large q measurements. The observed signal is due to scattering density variations at small scale sizes. Asymptotic laws were established long ago, and this domain has been widely investigated experimentally to check the presence of interfaces in finely divided matter (the measurement of specific area). The case of microemulsions is particularly interesting since the presence of a well defined interface, which was clearly demonstrated in droplet microemulsions, is not evident in Winsor microemulsions which are frequently in the vicinity of critical points [5,19].

These experiments revealed the existence of a well defined surfactant film even in Winsor microemulsions. More precisely, the general laws of large angle scattering intensity $i(q)$ have been predicted by Porod[20] under the general form

$$i(q) = \frac{S}{V} f(q),$$ (6)

S/V being the interfacial area per unit volume of the sample $((S/V) = C_S\Sigma$ with C_S the surfactant concentration (number of molecules per unit volume) and Σ the area per surfactant molecule in the film $(\Sigma \sim 60 \text{ Å}^2))$ and $f(q)$ is a function of the type of contrast. One obtains the following: (a) in the case of two densities (the usual case of one finely divided component, or D_2O and C_7H_8 contrast in neutron scattering):

$$i(q) = 2\pi(n_w - n_o)^2 C_S\Sigma q^{-4};$$ (7)

b) contrast in a pure film:

$$i(q) = 2\pi(n_f - n_o)^2 C_S\Sigma d^2 q^{-2};$$ (8)

c) in the more complex contrast observed with X-rays:

$$i(q) \sim C_S\Sigma(Aq^{-4} + Bq^{-2}).$$ (9)

These various behaviors have been experimentally observed[12,13], in particular the proportionality of the asymptotic intensity to C_S which proves the existence of the interfacial film. The area per molecule Σ is found to be around $70 \text{ Å}^2 \pm 10 \text{ Å}^2$ and the thickness of the film $10 \text{ Å} \pm 5 \text{ Å}$.

2) <u>Semi-local scale</u> $(q\xi \sim 1)$. Expressions[14-16] of Porod's law are valid as long as the interface looks flat at the scale defined by the scattering angle. Our idea is that the location of the q_1 crossover zone between the asymptotic behavior and the small angle behavior gives the size scale (πq_1^{-1}) at which the film becomes curved, i.e., an estimation of the persistence length of the film ξ_K introduced in the model.

Empirically, we have chosen as a definition of the crossover scattering vector q_1, the abscissa of the minimum of the $q^4 I$ versus q plot. The detailed discussion of the practical and theoretical limitations of this choice has been given in [12].

A very interesting point about this phenomenological procedure is that, for all the samples, the intensities in the medium q-range can be written as

$$I(q) = I_o f(q/q_1),$$ (10)

thus proving that q_1 is a good choice to determine a characteristic length of the structure.

Recall that the main prediction of the models of bicontinuous microemulsions is (2)

$$\xi = \frac{6\phi_o\phi_w}{C_S\Sigma}.$$

In agreement with our assumption that q_1^{-1} reflects ξ, it has been experimentally verified that q_1 is proportional to C_S. Moreover $C_S q_1^{-1}$ is proportional to the product $\phi_o\phi_w \sim \phi_o(1 - \phi_o)$: Figure 1 exhibits a parabolic shape around $\phi_o \sim 0.5$. At low water or oil volumic fractions $(\phi_o$ or $\phi_w < 0.3)$ the behavior is similar to that of spherical droplets. A semi-quantitative calibration of q_1^{-1} is possible when observing by this procedure an otherwise known system.

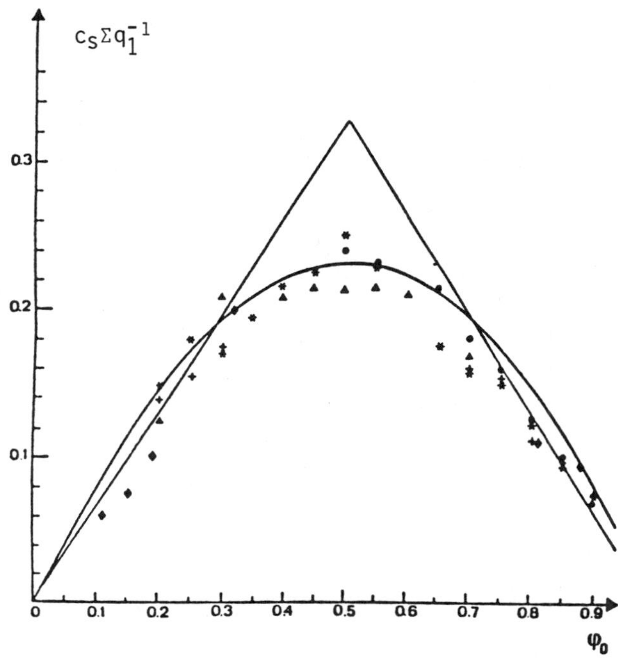

Fig. 1.   Product $C_S \Sigma q_1^{-1}$ plotted versus the toluene volume fraction ϕ_0.
The straight line corresponds to $C_S \Sigma q_1^{-1}$ = 0.65 (ϕ_w or ϕ_0).
The parabola corresponds to $C_S \Sigma q_1^{-1}$ = 0.9 ϕ_0 ϕ_w.

The observed behavior indicates that as ϕ_0 increases, one passes
progressively from a structure made of distinct oil in water spheres to a
bicontinuous structure described by the Talmon-Prager or de Gennes model;
the curvature of the film is inverted at ϕ_0 = 0.5 and as ϕ_0 increases
again, the reverse process takes place.

3) <u>Very small q-range.</u> In this domain, long-distance correlations
or interactions are observed.

We previously noticed that the structure of the Winsor microemulsions
are very similar to within a scale factor which has been discussed in other
papers, at scales which are of the same order of magnitude or smaller than
the characteristic size. The differences between the samples appear at
larger scales ($q < 10^{-2}$ Å^{-1}).

In the inversion zone ($0.4 < \phi_0 < 0.7$), the samples exhibit a diffuse
band centered around q*, the corresponding Bragg spacing being $2\pi q^{*-1}$
(Figure 2). q* does not move as one goes far away from the inversion
point but varies proportionally to the amount of surfactant C_S. The
comparison of various systems reveals a strong correlation between the
position of the peak and q_1.

In this inversion zone, the intensity scattered by a two phase random
oil and water partition in Voronoi polyhedra has been calculated by Kaler
and Prager[22]. As this partition is random, they predict that at small q
the intensities approximately follow a Guinier law where the equivalent
radius of gyration R_G is proportional to the characteristic length of the
microemulsion; within this model R_G = 0.55 ξ_K. This prediction does not
agree with the experimental results, which exhibit clearly a peak.

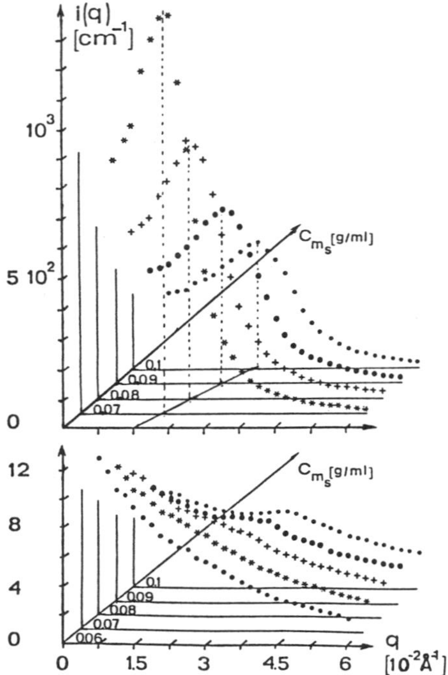

Fig. 2. Intensity of the two-phases spectra (top) and of the film spectra
(bottom) as a function of q and C_{m_s} (same q-scale on the two
plots, q^* (peak scattering vector) versus C_{m_s} in the q–C_{m_s} plane).

Quantitatively the peak is not very different from what it would be if the
microemulsions were dispersions of droplets. In fact the peak is clearly
visible in the case of a water-oil contrast but it disappears in the case
of a pure film contrast (Figure 2), in contradiction with a droplet model.
The reason for these effects is pictured schematically on Figure 3.

The differences between the two structures appear either at the semi-
local scale (studied in Ref. [12]) or in the film geometry. The study of
the film fluctuations and correlations through the two functions $\chi_{wf}(q)$
and $\chi_{ff}(q)$ is particularly interesting. Here, we will focus mainly on the
zero angle limit of $\chi_{wf}(q)$, which is related to the film curvature.

4) <u>Scattering at zero angle and zero curvature of the film.</u> If $\delta\phi_w$
and $\delta\phi_f$ designate the fluctuations of water and film volume fraction
respectively in a macroscopic volume V, $\chi_{wf}(q)$ and $\chi_{ff}(q)$ are given in the
thermodynamic limit, q going to zero, by the two relations:

$$\chi_{wf}(0) = V \ <\delta\phi_w \delta\phi_f> \tag{11}$$

$$\chi_{ff}(0) = V \ <\delta\phi_f^2> \tag{12}$$

Because of the Schulman's condition, the area per surfactant in the
interfacial film does not fluctuate, to a first approximation. A
surfactant (or film) fluctuation in a macroscopic volume V is proportional
to the fluctuation of the interfacial area in the volume V. As first
noticed by Widom[5] and developed in Refs. [13] and [15], $\chi_{wf}(0)$ and
$\chi_{ff}(0)$ (calculable from the thermodynamical models of microemulsions
[2,4,5]) then have a simple geometrical interpretation.

For a microemulsion made of water in oil droplets, a water excess in

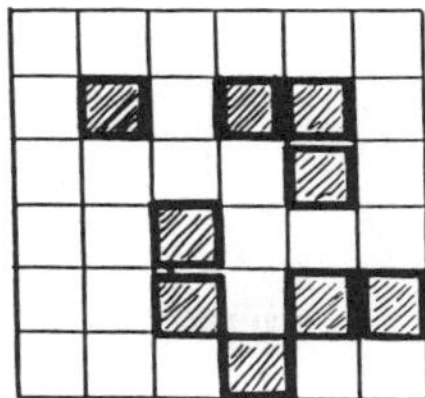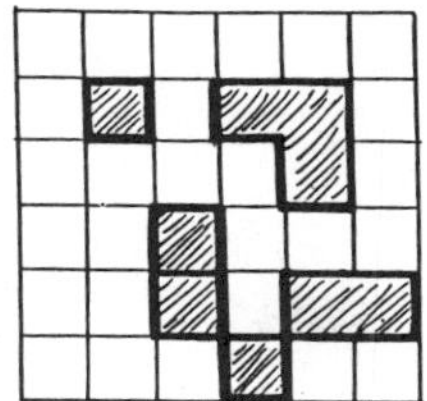

Fig. 3. Parallel between a lattice-gas model of water in oil droplets and
a model of random microemulsion with a well-defined length scale.
If the surface and curvature energies of the interfacial film are
neglected before the entropy of dispersion of the oil and the
water, the statistical configurations and correlations of the oil
and the water are the same in the two models. The differences
between the two models only appear at the local scale ($r < \xi$) or
on the geometry of the interfacial film (thick line).

a volume V means an excess of droplets, and hence a surfactant excess:
$\chi_{wf}(0)$ is positive. It is the reverse for oil in water microemulsions.
In the past, this distinction (formulated in another way) has been used to
evidence the inversion of microemulsions[23]. Up to a numerical factor
and to the first order in d (d is the thickness of the film and $<C>$ the
average mean curvature of the film) one obtains

$$\chi_{wf}(0)/\chi_{ww}(0) \; \propto \; <C>d \tag{13}$$

by geometrical considerations.

The Widom's remark[5] enables this relation between $\chi_{wf}(0)$ and the
film curvature to be generalized to the random microemulsions. In this
case one finds that the average curvature of the film and $\chi_{wf}(0)$ are
both proportional to $(1 - 2\phi_w)$. As was predicted by Widom[5], when $\phi_O =$
$= \phi_w = 0.5$, the average curvature of the microemulsion vanishes and the
fluctuations of water and film are not correlated.

Figure 4 represents the experimental variation of the ratio $\chi_{wf}(0)/$
$\chi_{ww}(0)$ as a function of ϕ_O (at constant C_S). This ratio increases
progressively when ϕ_O increases, and changes its sign for $\phi_O = 0.5 \pm 0.03$.
This shows directly that the mean film curvature is progressively inverted
as the oil and water proportions in the samples are progressively
inverted, in agreement with the predictions of the models of random
bicontinuous microemulsions (the straight continuous line on Figure 4 is
calculated from the model of de Gennes-Jouffroy-Levison).

In order to appreciate the sensitivity of the method of contrast
variation, it is interesting to compare the experimental results with a
model of water in oil droplets. In this model:

$$\chi_{wf}(0)\chi_{ww}(0) = 3d/R_w + O(d/R_w)^2 \tag{14}$$

(d is the film thickness and R_w is given by Eq. (1). With d = 10 $\overset{\circ}{A}$ [25]
and Σ = 60 $\overset{\circ}{A}{}^2$, the value of $3d/R_w$ is 0.25 for the sample $\phi_O = 0.5$. This
value is much higher than the almost zero experimental value. A
description of the $\phi_O = 0.5$ sample in terms of a droplet model is clearly
not possible. Let us consider now the sample which contains the largest
proportion of oil, $\phi_O = 0.7$. If $\phi_O = 0.7$, $3d/R_w = 0.42$. This value is
larger than the observed value of $\chi_{wf}(0)/\chi_{ww}(0) = 0.33$. The curvature of
the interfacial film is smaller in this sample than it would be if this
microemulsion was made of distinct droplets. This means that the
structure of this microemulsion probably remains highly connected.

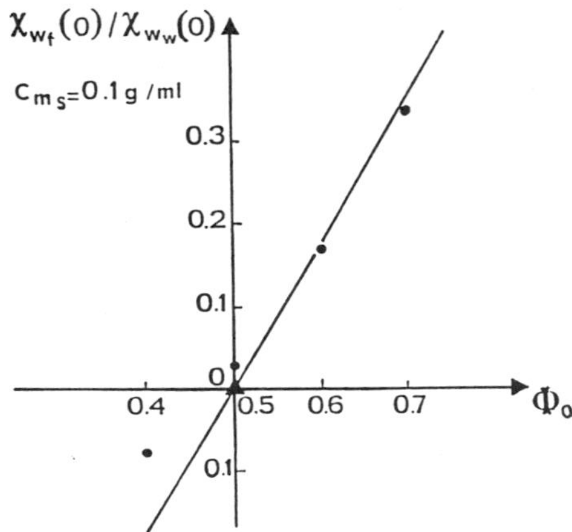

Fig. 4. Variation of the ratio $\chi_{wf}(q)/\chi_{ww}(q)$ extrapolated at zero angle as a function of the oil volume fraction at constant surfactant concentration.

2. SPONTANEOUS CURVATURE EFFECT ON THE MICROEMULSION INTERFACIAL PROPERTIES

The ultra-low interfacial tensions $\gamma_{m/o}$ and $\gamma_{m/w}$ between a middle-phase microemulsion (m), the upper oil phase (o), and the lower brine phase (w) have been much studied experimentally[19,23] and theoretically [2-23]. However, their relationship to the middle-phase structure is still unclear. In terms of models of random bicontinuous microemulsions, we present here a description of the structure of the m/o and m/w interfaces, fully developed in Ref. [25], which interprets the variations of the middle-phase interfacial tensions as a function of the brine salinity.

To be specific, let us consider the middle-phase-oil-phase interface. Because the structure of the microemulsion is random and bicontinuous, it seems at first sight that the microscopic interface between the oil and the water (occupied by the surfactant film) in the oil-microemulsion interfacial region cannot be flat at the scale of a few hundred Angstroms (a few ξ). There should be continuous paths connecting the oil channels in the bulk of the microemulsion and the oil in the upper phase. In this case, we say that the o/m interface is open and we notice (Figure 5a) that the surfactant film in the o/m interfacial zone is on the average bent towards water.

If the film spontaneous curvature C_O is positive (favoring water in oil microemulsion, highly salinity case), this structure is energetically favorable, but this is not the case if C_O is negative (favoring the film bending towards oil, i.e., oil in water microemulsion, low salinity case).

If $C_O < 0$, an alternative structure of the o/m interface may be that of a closed interface (Figure 5c): the closed o/m interface is the superposition of a microemulsion-water open interface, where the surfactant layer is on the average bent towards oil, a water wetting layer (whose thickness is of order ξ) and a flat Langmuir surfactant film separating the water wetting layer and the oil upper phase.

Limiting the discussion to these two extreme structures of the microemulsion interfaces, the main results of the model, which are supported by recent ellipsometry experiments[25], are the following:

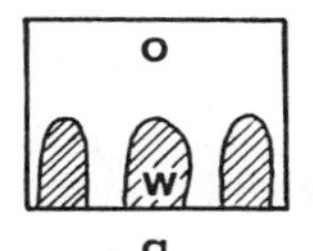 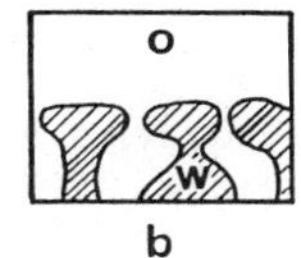 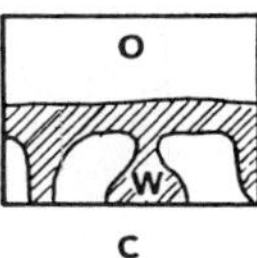

a   b   c

Fig. 5. Microscopic structures of the interface between a Winsor middle-phase and the oil upper phase in excess at the scale of a few hundred Angströms. The water salinity (spontaneous curvature) decreases from (a) to (c). In (a), the spontaneous curvature of the surfactant film, C_O, is positive or very small, the film is bent towards water. In (c), C_O is smaller than $-C_O^*$ ($\sim -\xi^{-1}$), the microemulsion surfactant film is preferrentially bent towards oil, the interface is closed by a flat Langmuir film and a water wetting layer.

1) There exists a characteristic value of the spontaneous curvature, C_O^*, which is of the order of ξ^{-1}, ξ being the microemulsion characteristic length.

2) Depending on the relative value of C_O with respect to C_O^*, there are three different situations: (i) if $C_O < -C_O^*$ (low salinity), the middle phase minimizes its surface free energy with an open w/m interface and a closed m/o interface; (ii) if $-C_O^* < C_O < C_O^*$, the two interfaces are open; and (iii) if $C_O > C_O^*$ (high salinity), the w/m interface is closed whereas the m/o interface is open.

Because of the free energy excess of the wetting layer and the Langmuir surfactant film, $\gamma_{m/o}$ is larger than $\gamma_{m/w}$ in case (i) and smaller in case (iii). The intermediate case (ii), which corresponds to a small spontaneous curvature, is particularly interesting; the difference $\Delta\gamma = \gamma_{m/o} - \gamma_{m/w}$ only depends on the curvature energy and is directly related to C_O:

$$\Delta\gamma \propto -K C_O C_S \Sigma. \tag{15}$$

Taking $K \sim 10^{-14}$ erg{3,26], $1/C_O \sim 100$ Å and $1/C_S\Sigma \sim 100$ Å, one estimates $\Delta\gamma \sim 10^{-2}$ dyne/cm, which is the right order of magnitude. Note that if $C_O = 0$, one expects that the middle-phases contain as much oil as water[5] and that $\Delta\gamma = 0$. This is indeed what is observed in many Winsor systems at the optimal salinity[23].

3. CONCLUSIONS

We believe from the results of the above simplified model and from the scattering experiments that the main features of the phase behavior, structure and interfacial properties of the Winsor middle-phases have now been distinguished and are well described by the recent theories presenting them as random bicontinuous dispersions resulting from a competition between the oil-water dispersion entropy and the surfactant film surface and curvature energies.

Acknowledgements

This work has received partial financial support from PIRSEM (CNRS) under A.12P.no. 2004.

REFERENCES

1. L. E. Scriven, in: "Micellization, Solubilization and Microemulsions", K. L. Mittal, ed., 2:77, Plenum Press (1977); Nature, 263:123 (1976).
2. Y. Talmon and S. Prager, J. Chem. Phys., 69:2984 (1978).
3. P. G. de Gennes and C. Taupin, J. Phys. Chem., 86:2294 (1982).
4. P. G. de Gennes, J. Jouffroy and P. Levinson, J. Physique, 43:1241 (1982).
5. B. Widom, J. Chem. Phys., 81:1030 (1984).
6. K. E. Bennett, J. C. Hatfield, H. T. Davis, C. M. Macosko and L. E. Scriven, in: "Microemulsions", I. D. Robb, ed., 65, Plenum Press (1982).
7. M. T. Clarkson, D. Beaglehole and P. T. Callaghan, Phys. Rev. Lett., 54:1722 (1985).
8. P. Guering and B. Lindman, Langmuir, 1:464 (1985).
9. J. Biais, M. Mercier, P. Bothorel, B. Clin, B. Lalanne and B. Lemanceau, J. Microsc., 121:169 (1981).
10. E. W. Kaler, K. E. Bennett, H. T. Davis and L. E. Scriven, J. Chem. Phys., 79:5673 (1983).
11. E. W. Kaler, H. T. Davis and L. E. Scriven, J. Chem. Phys., 79:5685 (1983).
12. L. Auvray, J. P. Cotton, R. Ober and C. Taupin, J. Physique, 45:913 (1984).
13. L. Auvray, J. P. Cotton, R. Ober and C. Taupin, J. Phys. Chem., 88:4586 (1984).
14. A. de Geyer and J. Tabony, Chem. Phys. Lett., to appear.
15. L. Auvray, Ph.D. Thesis, Université de Paris-Sud, France (1985), available on request.
16. J. N. Chang, R. T. Hamilton, J. F. Billman and E. W. Kaler, poster at this conference.
17. Y. Talmon, Proceedings of the 5th International Symposium on Surfactants in Solutions, Bordeaux, K. L. Mittal and P. Bothorel, eds., July 9th–13th (1984), to appear.
18. H. B. Stuhrmann, J. Appl.. Crystallogr., 7:173 (1974).
19. A. M. Cazabat, D. Langevin, J. Meunier and A. Pouchelon, Adv. Coll. Int. Sci., 16:175 (1982).
20. R. Kirste and G. Porod, Koll. Z. u. Z. für Polym., 184:1 (1962).
21. L. Auvray, Thesis of Doctorat d'Etat, Orsay (1985).
22. E. W. Kaler and S. Prager, J. Coll. Int. Sci., 86:359 (1982).
23. M. Laguës, R. Ober and C. Taupin, J. Physique Lett., Orsay, France, 39:L-487 (1978).
24. R. L. Reed and R. N. Healy, in: "Improved Oil Recovery by Surfactant and Polymer Flooding", D. O. Shah and R. S. Schechter, eds., Academic Press, New York (1977).
25. L. Auvray, J. Physique Lett., 46:L163 (1985).
26. D. Beaglehole, M. T. Clarkson and A. Upton, J. Coll. Int. Sci., 101:330 (1984).
27. J. M. di Meglio, M. Dvolaitzky and C. Taupin, J. Phys. Chem., 89:871 (1985).

LIQUID STATE THEORY AND ITS APPLICATIONS TO MICELLAR

SOLUTIONS AND TO OTHER COMPLEX LIQUIDS

L. Reatto

Dipartimento di Fisica
Università di Milano
Italy

I. INTRODUCTION

The liquid state theory for simple one component fluids has greatly
advanced in recent years and we have reached a good understanding of their
thermodynamic properties and of some basic aspects of the static correlations of the system, those which involve the pair distribution function.
It is natural to ask if the theories so developed can find application in
other fields. In fact recently these theories have often been applied to
complex mixtures like micellar solutions[1-5], water in oil microemulsions[6] and dispersions of colloidal particles[7].

The first step is the identification of the entity which corresponds
to the single particle in a simple fluid, for instance a single micelle
with its coordinated water in the case of a micellar solution. Then one
must specify the effective interaction between these entities as function
of the thermodynamic state. This interaction is the sum of a direct interaction between the entities and of an indirect one mediated by the solvent.
In this way the computation of the properties of the complex mixture is
reduced[8] to a statistical mechanical problem of a fictitious one component fluid and the theory of liquid state can be applied.

Here I do not discuss the important issue of the extent to which such
one component models are appropriate to describe specific complex mixtures.
My intention is to show that even if this first step is appropriate one has
to be very careful in applying the liquid state theory, because the character of the effective interaction can be very different with respect to the
one of a simple liquid. In these applications, for instance, it has been
always assumed that the effective interaction is pairwise additive

$$V(\vec{r}_1,\ldots,\vec{r}_N) = \frac{1}{2} \sum_{i \neq j} v(\vec{r}_i - \vec{r}_j),$$

(1)

where $\vec{r}_i$ represents the coordinate of the i-th entity. In the liquid state
theory a variety of approximations[9] have been introduced and it is well
established that their accuracy depends on the shape of $v(r)$. For instance
the Percus-Yevick equation for the radial distribution function $g(r)$ gives
quite accurate results for the hard sphere system, but this is not so for
softer repulsive potentials and if attractive forces are also present.

Therefore one has to consider is the shape of v(r) arising in the problem
at hand is compatible with the approximate theory of liquid state one plans
to use.

The previous observation is quite relevant. For instance, in model-
ling the effective interaction v(r) for the complex mixtures that I have
mentioned above it is believed that v(r) contains two pieces[1-7]. There
is an essentially impenetrable core at short distance which prevents the
overlap between different entities. At larger distance there is an attractive interaction. In the case of ionic mixtures a screened coulombic
interaction is also present[5,7]. Thus for model computations a convenient
parametrization of v(r) is

$$\phi(r) \equiv V(r)/k_B T = \begin{cases} \infty & r < \sigma \\ -\dfrac{K}{(r/\sigma)} \exp(-z(\dfrac{r}{\sigma} - 1)) & r > \sigma. \end{cases} \qquad (2)$$

Here T is the absolute temperature of the system and we are assuming that
the basic entities have an essentially spherical shape. This form contains
three parameters. σ represents the diameter of the basic entity. Notice
that this does not necessarily coincide[3] with the size, for instance, of
the micelle, K gives the amplitude of the attractive force (or of the
repulsive one in the ionic case) in units of k T, and z is the inverse
range of these forces in units of the diameter σ. It is typical to find
[1,3,4,6] that this range is very small compared with σ, so that $z \gg 1$.
On the other hand, the existence in many complex mixtures of phase
separation implies that the model has to be studied in the strongly
interacting regime, i.e. $K \gg 1$. This is not the situation prevalent in
simple fluids, and there has been no assessment of the different
approximate theories under such conditions. In fact, in simple fluids the
range of the attractive forces typically has the same range of σ or longer
and, for instance, the minimum of the Lennard-Jones potential is ~ 0.75
$k_B T_c$, where T_c is the temperature of the critical point.

In simple liquids it is known that three-body interactions v(3) are
present[9] in addition to the two-body ones. However, these three-body
forces have only a minor role and their effects are usually included just
by a small modification of the two-body v(r). There is no guarantee that
the same is true in the case of complex mixtures. The solvent induced
interaction might lead to important contributions to v(3) and, in any case,
the role of v(3) could be different in presence of strong and very short
range two-body forces.

The purpose of this paper is twofold. On one hand I present the
results of an investigation[10,11] of the accuracy of some liquid state
theories for a model system with interaction (2) when z and K are large
quantities. The theories considered are the mean spherical approximation
(MSA) and a generalized version of it (GMSA) and the high temperature
approximation (HTA). Secondly, I consider the effect of the presence of
three-body forces on the critical point and on the coexistence curve. This
is in view of explaining the low value of the critical concentration for
phase separation found in many non-ionic micellar solutions[12].

II. PHASE DIAGRAM AND CORRELATIONS OF FLUIDS WITH VERY STEEP ATTRACTIVE
     FORCES

In applications to complex mixtures[1,3-6] the most widely used theory
of liquid state appears to be MSA[13]. The reason is that this approximation is known to be reasonably accurate for dense simple liquids and that

the resulting integral equation has been solved in closed form for certain
forms of τ interaction, for instance[14] for the Yukawa form (2).

MSA is the equation resulting from the Orstein-Zernike relation

$$g(r) - 1 = c(r) + \rho \int d^3 r c(r')[g(|\vec{r} - \vec{r}'|) - 1] \tag{3}$$

between radial distribution function $g(r)$ and the direct correlation function $C(r)$ and the two statements:

$$c(r) = -v(r)/k_B T \qquad r > \sigma \tag{4}$$

$$g(r) = 0 \qquad r < \sigma. \tag{5}$$

Here one is assuming that the system has a hard core of diameter σ. the
accuracy of this equation in presence of very steep and strong attractive
forces was not known, but recently we have studied[10] this equation for
systems of hard spheres with an attractive Yukawa tail (2) with large
values of z and K. Usually the accuracy of an integral equation is studied
by comparison with the result of simulation performed for the same interaction. Unfortunately no simulation has been performed yet for such steep
interactions. In this situation we have investigated the accuracy of MSA
on the basis of its internal consistency. It is well known[9] that from
$g(r)$ one can compute the equation of state in different ways. The pressure
is obtained directly from the virial relation

$$p/\rho k_B T = 1 - \frac{2\pi}{3} \frac{\rho}{k_B T} \int_o^\infty dr r^3 g(r) \frac{dv(r)}{dr} \tag{6}$$

or it can be obtained by a density integration of the compressibility sum
rule

$$S(o) = \rho k_B T \, K_T = 1 + \rho \int d^3 r[g(r) - 1], \tag{7}$$

where K_T is the isothermal compressibility. If $g(r)$ were exact these two
routes would give the same result. An approximate $g(r)$ gives, in general,
two different equations of state and the amount of their deviation is a
measure of the accuracy of the approximate theory. The virial relation (6)
is true if $v(r)$ is density independent. In applications to complex mixtures this is not necessarily true but $v(r)$ might depend on concentration.
This, however, is inessential for the present purpose and we take $v(r)$ as
state _independent_ since we want to establish the accuracy of MSA per se.

This one component model with $v(r)$ attractive outside the core has a
phase transition with a critical point. In the application of this model
to a micellar solution, for instance, this phase transition corresponds to
the cloud point transition[12] between a low and a higher concentration
phase. In the first place we have computed[10] the spinodal line, i.e. the
locus of diverging compressibility, directly from (7) and by numerical
differentiation of (6). When the inverse range z of $v(r)$ is small, $z \gtrsim 1$,
the two spinodal lines turn out to be very similar indicating that MSA is a
good approximation. On the contrary large deviations are present when z is
large and the results for z=5 are shown in Fig. 1. The critical temperatures estimated by the two routes differ by almost a factor of three and
this factor becomes about eight when z = 15 (see Table 1). In addition,
the critical densities disagree. It is clear that MSA is not acceptable in

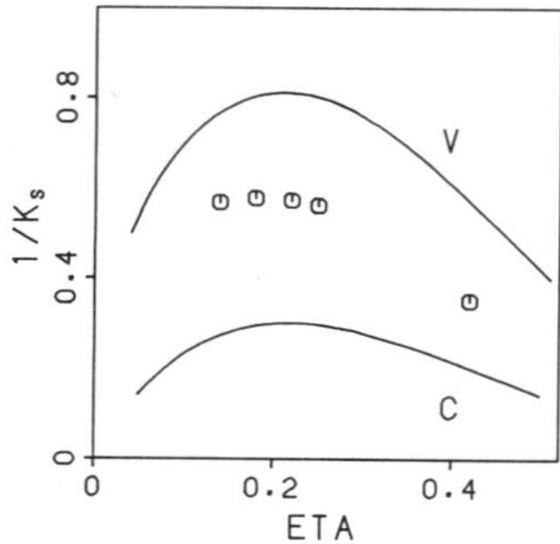

Fig. 1. Temperature of the spinodal line as function of the packing
fraction $\eta = \pi\rho\sigma^3/6$ in GMSA (dots) and in MSA from the virial (V)
and from the compressibility (C) relation. The range parameter is
z = 5.

this regime. This breakdown of MSA is not limited to the phase transition
region but also at higher density one finds a strong discrepancy between
the two equations of state[10].

It is know that MSA becomes unreliable also when the interaction is
repulsive and rather large immediately outside the core. This can happen,
for instance, in coulombic systems, and in this case a generalized mean
spherical approximation[15,14] (GMSA) has been introduced. This consists
in replacing the approximate statement (4) by

$$c(r) = - v(r)/K_B T + \lambda(r), \quad r > \sigma. \tag{8}$$

$\lambda(r)$ is a correction term which for convenience is taken of the form[14]

$$\lambda(r) = K_2 \exp[-z_2(\frac{r}{\sigma} - 1)]/(r/\sigma) \tag{9}$$

and the two parameters K_2 and z_2 are determined by imposing suitable conditions on g(r). We have extended[16,10] this GMSA to the case of an
attractive potential and the two parameters k_2 and z_2 are fixed by
requiring that the two equations of states (6) and (7) give the same result
and, in addition, that (6) is thermodynamically consistent with the
internal energy:

$$U = 2\pi\rho \int_\sigma^\infty g(r)v(r)r^2 dr. \tag{10}$$

This gives the condition

$$\frac{\partial(\beta p/\rho)}{\partial\beta})_\rho = \rho \frac{\partial U}{\partial\rho})_\beta, \quad \beta = (k_B T)^{-1}. \tag{11}$$

In the present case c(r) for $r>\sigma$ is the sum of two Yukawa functions
and the resulting integral equation has also been solved[14] in closed form
In terms of one of the roots of two quartic coupled algebraic equations.
The two free parameters K_2 and z_2 have been determined so that the compat-
ibility of (16) and (7) is verified and condition (11) is approximately
verified in the region of the phase transition. In this way we obtain a
unique critical point and the result is shown in Fig. 1 and in Table 1.
The critical temperature falls between the two values given by MSA but it
is closer to the virial value of MSA. The critical density given by GMSA
is much smaller than both values, the virial and the compressibility one,

Table 1. Inverse temperature (K_c) and packing fraction (η_c) of the Critical Point for the Yukawa Interaction given by HTA, MSA and GMSA.

	HTA		MSA, virial		MSA, compressibility		GMSA	
z	K_c	η_c	K_c	η_c	K_c	η_c	K_c	η_c
1	0.391	0.141	0.373	0.148	0.425	0.143	0.383	0.143
5	1.794	0.225	1.233	0.213	3.309	0.218	1.759	0.180
15	3.421	0.360	1.299	0.320	9.921	0.287	2.938	0.218

given by MSA. Since we do not have the exact results of simulation with which to compare, we cannot say what is the absolute accuracy of these GMSA results but experience with other systems leaves no doubt that it is better than MSA.

As mentioned above MSA is inadequate also at large density if z is large. This means that the MSA structure factor S(q)

$$S(q) = 1 + \rho \int d^3r[g(r) - 1]e^{i\vec{q}\cdot\vec{r}} \qquad (12)$$

has significant deviations (see Fig. 2) from the one given by GMSA which is reputedly more accurate. As a consequence, if one tries [1,4,6] to extract information on the effective interparticle interaction starting from experimental scattering results by using the MSA S(q), large errors can arise. In fact we found that the MSA S(q) can be made to match very closely the S(q) given by GMSA, at the price of a large modification of the parameters of the interaction (2). For instance when z=5 and for a packing fraction η= 0.42 this matching requires a decrease of the diameter by about 1%, an increase of the strength K of the attractive interaction by a factor larger than two and a decrease of its range z^{-1} by 25%.

Another popular approach to liquid state is the high temperature approximation[9], i.e., a thermodynamic perturbation theory with respect to the attractive part of the interaction limited to first order. HTA has been applied[7], for instance, to study the possibility of a "liquid-vapor" phase transition in charge stabilized, acqueous dispersions of large colloidal particles. Comparison[11] (see Table 1) of the results of HTA for the interaction (2) with those of GMSA indicates that HTA is better than MSA as far as the critical temperature is concerned but the critical density is very wrong. Therefore also HTA can have limited applications when the attractive force is very strong and of very short range.

III. THREE-BODY FORCES AND THE CLOUD POINT TRANSITION IN NON-IONIC MICELLAR SOLUTIONS

The phase separation associated with the cloud point transition in non-ionic micellar solutions has many peculiarities[12] in comparison with other phase separation transitions. For instance the critical concentration can be very small, even at the level of 1-2%, the coexistence curve is very asymmetric and the critical exponents for the divergence of the osmotic compressibility and of the correlation length are non universal. These features have not yet found a general explanation. Consider for instance the low value of c_c for which two explanations have been advanced. One[2] assumes that a micelle has a rod shape with a very large axis ratio so that it behaves essentially as a random coil. There is some controversy

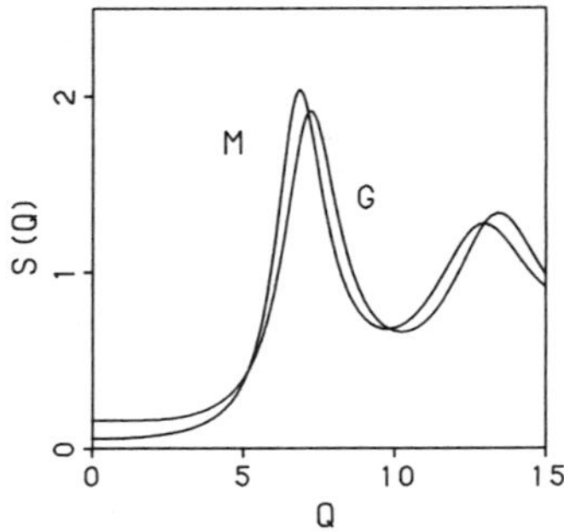

Fig. 2. Structure factor in MSA (M) and in GMSA (G) for z = 15, K = 2.90 and η = 0.42.

on the subject of the shape of a micelle but most experiments do not show evidence of such rod shape and usually a globular shape adequately explains the data[12]. The other explanation[3] assumes that a micelle has, in addition to some strongly bound water, a rather large region of loosely bound water, and this is assumed to give rise to a temperature dependent effective repulsive interaction between micelles. This explanation can be reasonable when c_c is of order of 5-10% but runs into difficulties if c_c is smaller because the required amount of the loosely bound water becomes extremely large and there is no experimental evidence of this. Here I suggest another explanation, that the low value of c_c is due to effective three body forces of repulsive nature.

Recently I have shown[17] that <u>repulsive</u> three body forces are very effective in reducing the value of the critical concentration when the <u>attractive</u> two-body forces have a very short range. I have considered a one component system of particles with interaction

$$V(\vec{r},..,\vec{r}_N) = \frac{1}{2} \sum_{i \neq j} v(|\vec{r}_i - \vec{r}_j|) + \frac{1}{6} \sum_{i\,j\,l} v^{(3)}(r_i,r_j,r_l). \tag{13}$$

The two-body interaction is assumed to be the square well potential

$$v(r) = \begin{cases} \infty & r < \sigma \\ -\epsilon & \sigma < r < \sigma(1 + b) \\ 0 & r > \sigma(1 + b). \end{cases} \tag{14}$$

For the three-body term the Axilrod - Teller form[18]

$$v^{(3)}(\vec{r}_1,\vec{r}_2,\vec{r}_3) = \nu(1 + 3\cos\theta_1\cos\theta_2\cos\theta_3)/R_1^3 R_2^3 R_3^3 \tag{15}$$

has been assumed. R_i and θ_i are the sides and angles of the triangle formed by the three particles. This interaction is the leading non-additive contribution for non-ionic molecules and it is predominantly repulsive v(3) favoring linear configurations of the three particles with respect to equilateral triangle configurations.

The equation of state and the critical point has been determined from the virial expansion truncated to third order:

$$pV/NK_BT = 1 + \frac{B}{V_m} + \frac{C}{V_m^2} . \tag{16}$$

140

V_m is the molar volume. $v(3)$ contributes to the virial coefficients starting from third order and this reads[18]

$$C = C^{(2)} + C^{(3)}, \tag{17}$$

$$C^{(3)} = \frac{N_m^2}{3K_BT} \int d^3r_2 d^3r_3\; v^{(3)}(\vec{r}_1,\vec{r}_2,\vec{r}_3)e_{13}e_{23}, \tag{18}$$

$$e_{ij} = \exp(-\,v(|\vec{r}_i - \vec{r}_j|)/K_BT). \tag{19}$$

N_m is Avogadro's number and $C(2)$ is the usual third virial coefficient[18] for the two-body interaction. B and $C(2)$ for the interaction (14) are known in closed form[19]. When $b \ll 1$ $C(3)$ can be approximated by the contribution due to configurations for which all three distances or at least two fall in the range σ-σ (1+b) of the attractive forces. The position of the critical point has been determined in a standard way from the conditions $\partial p/\partial \rho = \partial^2 p/\partial \rho^2 = 0$. In Fig. 3 the variation of the critical packing fraction

$$\eta_c = (\pi/6)\,\rho_c\sigma^3$$

and of the critical temperature T_c are shown as function of the adimensional strength

$$V_3 = v/\sigma^9 K_B T_c \tag{20}$$

of the three-body interaction. We see that this interaction is very effective in depressing the value of η_c. η_c is reduced by a factor of ten when V_3 changes from 0 to 0.5. At the same time T_c is reduced only by 45%. In order to appreciate the strength of V_3, I notice that in Argon one finds[18] $V_3 \approx 0.06$.

The previous computation is based on a severe truncation of the virial expansion. On the other hand it is known for many different forms of $v(r)$ that the virial expansion truncated to third order gives a reasonable estimate of the critical point[19]. I believe that also in this case this approximation gives results at least of semiquantitative value. It should be noticed that qualitatively these results do not depend on the specific form[15] of $v(3)$ as long as $v(3)$ is predominantly a repulsive interaction so that $C(3){>}0$.

The presence of $v(3)$ has two additional effects on the phase diagram. The diameter ρ_d of the coexistence curve close to the critical point is expected to have the form[20]

$$\rho_d \equiv \frac{\rho_+ + \rho_-}{2} = \rho_c(1 + bt + dt^{1-\alpha}), \qquad t = |(T-T_c)|/T_c. \tag{21}$$

$\rho \pm$ are the densities of the two coexisting phases and α is the exponent characterizing the divergence of the specific heat at constant volume along the critical isocore. The term linear in t in (21) represents the "rectilinear diameter" of the phase boundary and the additional contribution is a non-analytic one due to critical fluctuations. We do not have a theoretical estimate of the coefficients b and c. In simple dielectric fluids the asymmetry of the measured coexistence curve is very small. If one neglects the $t^{1-\alpha}$ terms in (21) the coefficient b is of order of unity. In such fluids the presence of the $t^{1-\alpha}$ term is still controversial and evidence for it has been found only in SF_6[21]. Also in this case the asymmetry is

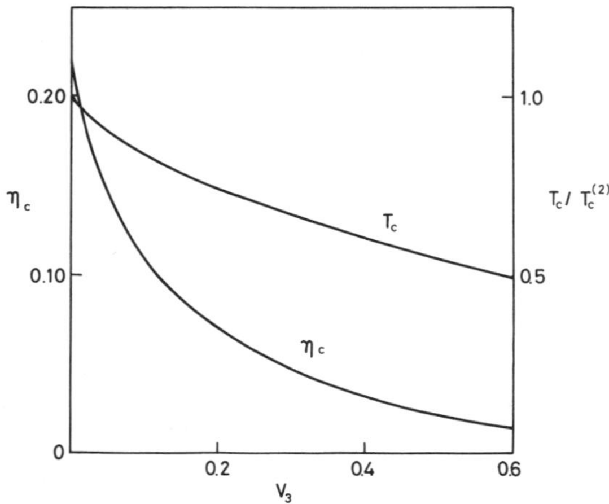

Fig. 3. Packing fraction and critical temperature of the critical point as function of the strength of the three body interaction V_3 with b = 0.1. $T_c(2)$ is the critical temperature when $V_3 = 0$ ($T_c(2) = 2.0\varepsilon$).

very small and from the fit of the data one finds[22] a strong cancellation between the t and the $t^{1-\alpha}$ term. I find that the presence of the three-body interaction gives an additional contribution proportional to V_3 both to the coefficients b and d. These contributions have the same sign and become large when V_3 is large enough to depress strongly the value of c_c.

In application of this theory to complex mixtures like a micellar solution (16) becomes[2] the expression for the excess chemical potential and the packing fraction η becomes the volume fraction. Therefore the previous results can explain the low value of the critical concentration in micellar solutions and of the strong asymmetry of the line of cloud points. it is clear that the physical mechanism at the basis of the Axilord-Teller interaction is present also in the micellar system but the analytic form (15) might be inappropriate for extended objects. In addition also the solvent mediated intermicellar interaction should be expected to give a non-additive contribution. Therefore the specific form (15) for v(3) might be inappropriate for micelles but, as I have already mentioned, the basic effect does not depend on the shape of v(3).

If this explanation of the low value of c_c is correct, the theory predicts the presence of a large coefficient for the singular contribution to the diameter of the coexistence curve. This is a difficult measurement but it is clear that it could give an important contribution to our understanding of these mixtures. At the same time a measurement of the specific heat in order to determine α is very important.

REFERENCES

1. J. B. Hayter and M. Zulauf, Colloid Polymer Sci., 260:1023 (1982).
2. R. Kjellander, J.Chem.Soc.Faraday Trans.II 78:2025 (1982).
3. L. Reatto and M. Tau, Chem.Phys.Lett. 108:292 (1984).
4. M. Corti, V. Degiorgio, J. B. Hayter amd M. Zulauf, Chem.Phys.Lett. 109:579 (1984).
5. J. B. Hayter and J. Penfold, J.Chem.Soc.Faraday Trans. I 77:1851 (1981).

6. J. S. Huang, et. al., Phys.Rev.Lett. 53:592 (1984).
7. J. M. Victor and J. P. Hansen, J. de Physique Lett. 45: L-307 (1984).
8. J. S. Rowlinson, Mol.Phys. 52:567 (1984).
9. See, for instance, J. P. Hansen and I. R. McDonald, "Theory of Simple Liquids," Academic Press, New York, (1976).
10. M. Tau and L. Reatto, J.Chem.Phys. 83:1921 (1985).
11. J. P. Hansen, L. Reatto, M. Tau and J. M. Victor, Mol.Phys. 56:385 (1985).
12. V. Degiorgio in: "Physics of amphiphiles, Micelles, Vesicles and Microemulsions," V. Degiorgio and M. Corti eds., North Holland, Amsterdam (1985).
13. J. L. Lebowitz and J. K. Percus, Phys.Rev., 144:251 (1966).
14. E. Waisman, Mol.Phys. 25:45 (1973).
15. J. S. Høye, J. L. Lebowitz and G. Stell, J.Chem.Phys. 61:3253 (1974).
16. J. S. Høye and G. Stell, Mol.Phys. 52:1071 (1984).
17. L. Reatto and M. Tau, Europhys. Lett. 3:527 (1987).
18. See, for instance, J. A. Barker and D. Henderson, Rev.Mod.Phys., 48:587 (1976).
19. J. O. Hirschfelder, C. F. Curtiss, and R. B. Bird, "Molecular Theory of Gases and Liquids," Wiley, New York, (1954).
20. J. J. Rehr and N. D. Mermin, Phys.Rev. A8 472 (1973).
21. J. Weiner, K. H. Langley and N. C. Ford, Jr., Phys.Rev.Lett., 32:879 (1974).
22. M. Ley-Koo and M. S. Green, Phys.Rev., A16 2483 (1977).

A SCALING THEORY OF SURFACE TENSION

NEAR A CRITICAL POINT

L. Mistura

Dipartimento di Energetica-Sezione Fisica
Università "La Sapienza" Roma
Via A. Scarpa, 14 Roma, Italy 00161

In recent years a number of generalizations and alternative derivations
of the van der Waals[1] theory of surface tension have appeared. (For
recent reviews see Refs. [2] to [5].) The present work is focused on the
development of the theory for near critical fluid interfaces.

As is well-known, the basic assumption of the van der Waals theory,
justified on the basis of a mean field calculation, is that the free density
of a non-uniform state can be obtained by adding a term proportional to
the square of the density gradient to the local equilibrium free energy
density, namely

$$f = f_o(T,\rho) + \frac{1}{2} m|\nabla\rho|^2 \tag{1}$$

where m is a positive constant, independent of the state variables T and ρ.

A priori Eq. (1) should not provide an accurate representation of the
free energy on the entire coexistence curve and in particular near the
triple point, since here the density is varying too rapidly through the
interface to truncate the gradient expansion after the first non-vanishing
term. Better agreement is expected near the critical point since here the
density gradient is spread over a distance large with respect to the range
of intermolecular forces.

As a matter of fact, much of what has been done on the theory of the
near critical interface has been within the framework of the van der Waals
theory. Nevertheless, it is known that the theory, in its original form,
cannot be valid even near the critical point, in particular if one assumes,
consistent with the mean field approximation, that the local equilibrium
free energy density f_o is obtained by integrating the van der Waals or a
van der Waals-like classical equation of state.

A previous attempt to extend the validity of the theory to the critical
region was reported by Fisk and Widom[6]. Although they explicitly
recognize[7] that the basic assumptions of the theory are open to criticisms, they develop their approach on those assumptions, but they replace
the classical mean field type equation of state by one that incorporates
the non-analytic behavior of the thermodynamic properties near the critical
point.

Although this approach has a great heuristic value it cannot be considered as fully satisfactory from the theoretical point of view. Indeed, while for the equation of state one recognizes the failure of the mean field approximation, for the gradient term one fully accepts its consequences without any a priori justification. It is known however, from the theory of the correlation function, that the van der Waals coefficient m, which is also a measure of the second moment of the Ornstein-Zernike direct correlation function, is not constant but diverges as the critical point is approached along the critical isochore, as $|T - T_c|^{-\eta\nu}$ [8].

The divergence of m is a clear symptom of the failure of the gradient expansion. This failure is expected to be quantitatively negligible in three dimensions (for $d = 3$, $\eta\nu \simeq 0.02$), but its theoretical relevance is beyond any doubt (for the two-dimensional Ising model $\eta\nu = 0.25$).

There are, moreover, some experimental indications, recently reported by Moldover[9], that the theory in its present form does not account for the observed values of the surface tension. Although the reported discrepancy does not necessarily indicate a breakdown of the theory at a fundamental level, but might well be a consequence of somewhat more conjectural assumptions introduced in the latest stage, it appears to be worthwhile to develop the theory anew, without introducing a priori a term proportional to the square of the density gradient. To show how such a theory can be formulated, fully in accord with modern scaling, is the main purpose of this paper.

We adopt the point of view of classical thermodynamics[10]: to describe the non-uniform states of a simple fluid we introduce, besides the usual variables T and ρ, one further scalar quantity λ, defined as the norm of the density gradient

$$\lambda = \sqrt{\overline{g^{iJ}\rho_i\rho_J}}, \tag{2}$$

where the g^{iJ} are the contravariant components of the metric tensor and the $\rho_i \equiv \dfrac{\partial\rho}{\partial x^i}$ are the components of the density gradient. The total free energy of the system F is then given by the invariant integral

$$F = \int\sqrt{g}\, f\, dx \tag{3}$$

where $f(T,\rho,\lambda)$ is the (scalar) free energy density and g is the (positive) determinant of the matrix g_{iJ}. Note that, through λ, f depends on the g^{iJ} and the ρ_i.

We have decided to write the equations in general curvilinear coordinates, although this is not usual in thermodynamics or statistical mechanics, because, as we have shown in a recent paper[11], the calculation of the stress tensor is then greatly facilitated. On the other hand, we need the stress tensor components to calculate the surface tension according to the so-called mechanical definition.

The equilibrium density distribution is obtained in the usual way by means of the variational principle

$$\delta F - \mu\delta N = 0, \quad \delta T = 0, \quad \delta g^{iJ} = 0. \tag{4}$$

From (3) and (4) we get the Euler-Lagrange equation in the form

$$\mu' - \frac{1}{\sqrt{g}}\frac{\partial}{\partial x^i}\left(\sqrt{g}\,\nu'\,\frac{\partial\lambda}{\partial\rho_i}\right) = \mu \tag{5}$$

where we have introduced the notation $\mu' \equiv \left(\dfrac{\partial f}{\partial \rho} \right)_{T,\lambda\sigma}$, $\nu' \equiv \left(\dfrac{\partial f}{\partial \lambda} \right)_{T,\rho}$. In (5) $\dfrac{\partial \lambda}{\partial \rho_i}$ is obtained from (2) as

$$\frac{\partial \lambda}{\partial \rho_i} = \frac{g^{iJ}\rho_J}{\lambda} \; ; \tag{6}$$

however, the specification of μ' and ν' as functions of T, ρ and λ depends on the knowledge of the free energy, and this is a problem which is outside the framework of thermodynamics. Any further progress is therefore based on special assumptions concerning the function $f(T,\rho,\lambda)$.

Near the critical point, the mean field approximation is equivalent to a smoothness assumption for the free energy. A Taylor expansion then gives

$$\nu' = \left(\frac{\partial f}{\partial \lambda} \right)_{T,\rho} = m\lambda + \dots \quad m = \left(\frac{\partial \nu'}{\partial \lambda} \right)_{T,\rho} = \left(\frac{\partial^2 f}{\partial \lambda^2} \right)_{T,\rho} = \text{const} \tag{7}$$

and

$$\mu' = \left(\frac{\partial f}{\partial \rho} \right)_{T,\lambda} = \mu'(\rho_c) + A\Delta T\Delta\rho + C(\Delta\rho)^3 \tag{8}$$

where $\Delta T = T - T_c$, $\Delta\rho = \rho - \rho_c$ and $\mu'(\rho_c) = \mu'_c + \dfrac{\partial \mu'}{\partial T} \Delta T + \dots$

From these equations one derives, in the familiar way, the well-known classical, but ultimately incorrect, predictions for the density profile and the surface tension[5]. They also imply however a remarkable property of the free energy which provides a clue for developing a new theory. It follows indeed, within the approximation indicated by Eqs. (7) and (8), that the classical theory obeys a scaling law, namely that the classical free energy $f(T,\rho,\lambda)$ is a positive quasi-homogeneous function of degree d with exponents α_T, α_ρ and α_λ, i.e.,

$$f(L^{\alpha_T}\Delta T, L^{\alpha_\rho}\Delta\rho, L^{\alpha_\lambda}\lambda) = L^d f(\Delta T, \Delta\rho, \lambda) \tag{9}$$

for any positive L.

The classical values of the exponents are readily obtained from Eqs. (7) and (8) to be

$$\alpha_\rho = \frac{d}{4} \, , \; \alpha_T = \frac{d}{2} \, , \; \alpha_\lambda = \frac{d}{2} \, . \tag{10}$$

In conformity to what we know on the critical behavior of uniform systems, we believe that this homogeneity property of the classical theory has a more general validity than the theory itself, if we allow for the different values of the exponents. Accordingly, we can formulate our basic assumption as follows. Instead of dividing the free energy density as in Eq. (1), we divide it into a regular part f_{reg}, which may well be of the van der Waals type, and a singular part f_{sing} which we assume to obey a scaling law as in Eq. (9), but with non-classical values of the exponents. We will now develop the consequences of this assumption.

Let us start determining the value of the exponents. For the first two it follows immediately from the requirement that for $\lambda = 0$ in Eq. (9) we must recover the free energy of a homogeneous phase; therefore

$$\alpha_\rho = \frac{d}{\delta + 1} \, , \; \alpha_T = \frac{d}{\beta(\delta + 1)} \, . \tag{11}$$

As far as α_λ is concerned we will determine its value from the fact that now the coefficient m is defined as a thermodynamic derivative by the equation

$$m = \left(\frac{\partial^2 f}{\partial \lambda^2} \right)_{T,\rho} \tag{12}$$

and therefore it must obey the scaling law

$$m(L^{\alpha_T}\Delta T, L^{\alpha_\rho}\Delta\rho, L^{\alpha_\lambda}\lambda) = L^{d-2\alpha_\lambda}m(\Delta T, \Delta\rho, \lambda). \tag{13}$$

To compare this scaling law with the prediction of the theory of critical correlations mentioned above, we observe that λ is not an independent variable, but its equilibrium value is determined as the solution of the equilibrium condition (5). This solution can be expressed as a function of ΔT and $\Delta\rho$ and must satisfy the scaling relation

$$\lambda_{eq}(L^{\alpha_T}\Delta T, L^{\alpha_\rho}\Delta\rho) = L^{\alpha_\lambda}\lambda_{eq}(\Delta T, \Delta\rho). \tag{14}$$

From Eqs. (13) and (14) we then get on the critical isochore

$$m[\Delta T, 0, \lambda_{eq}(\Delta T)] = |\Delta T|^{(d-2\alpha_\lambda/\alpha_T)}m[1, 0, \lambda_{eq}(1, 0)] \tag{15}$$

and therefore the new relation between exponents

$$\frac{d - 2\alpha_\lambda}{\alpha_T} = -\eta\nu. \tag{16}$$

Substituting the value of α_T from (11) we finally get

$$\alpha_\lambda = \frac{d + \eta}{2}. \tag{17}$$

In three dimensions η has been calculated to be of the order of a few hundredths, so the exponent α_λ will differ very little from its classical value. In two dimensions however one has $\eta = 1/4$ and therefore $\alpha_\lambda = 9/8$ instead of the classical value $\alpha_\lambda = 1$.

We must now develop an expression for the surface tension. At first sight, having lost most of the simplifying features of the square gradient theory, this might appear as a rather difficult task. It is at this point however that our previous definition of the stress tensor σ_{iJ}, as the derivative of the total free energy F with respect to the metric tensor g^{iJ}, turns out to be an essential tool. We need an explicit expression of the stress components in order to calculate the surface tension according to the mechanical definition

$$\sigma = \int_{-\infty}^{+\infty}(p_N - p_T)dz. \tag{18}$$

where $p_N = -\sigma_{zz}$ is the normal (to the interface) component of the pressure tensor $(-\sigma_{iJ})$ and $p_T = -\sigma_{xx} = -\sigma_{yy}$ its tangential component. On the other hand, according to our definition[11], when F is given by Eq. (3), the stress components are given by

$$\sigma_{iJ} = -2\frac{\partial f}{\partial g^{iJ}} + (f - \mu\rho)g_{iJ}. \tag{19}$$

It is not difficult to see that this tensor necessarily satisfies the condition of mechanical equilibrium.

The explicit evaluation of Eq. (19) is obtained observing that

$$\frac{\partial f}{\partial g_{iJ}} = \frac{\partial f}{\partial \lambda} \frac{\partial \lambda}{\partial g_{iJ}} = \frac{\nu'}{2\lambda} \rho_i \rho_J \qquad (20)$$

and therefore

$$\sigma_{iJ} = -\frac{\nu'}{\lambda} \rho_i \rho_J + (f - \mu\rho) g_{iJ}. \qquad (21)$$

The expression for the surface tension of the plane interface (normal to the z axis) is simplified by the fact that we can take $\rho_x = \rho_y = 0$, $g^{iJ} = \delta_{iJ}$ and therefore $\lambda = \left| \frac{dp}{dz} \right|$. We thus obtain

$$p_T = \mu\rho - f, \quad p_N = \nu'\lambda + \mu\rho - f \qquad (22)$$

and, from Eq. (18)

$$\sigma = \int_{-\infty}^{+\infty} \nu'\lambda dz. \qquad (23)$$

From the mechanical equilibrium condition it follows that p_N is constant and it can be easily verified that it is a first integral of the Euler-Lagrange Eq. (5), which now simplifies as

$$\frac{d\nu'}{dz} = \mu' - \mu. \qquad (24)$$

All these equations reduce to the corresponding classical one if $\nu' = m\lambda$; they are, however, more generally valid.

Equation (23) can also be written as an integral over ρ

$$\sigma = \int_{\rho_G}^{\rho_L} \nu' d\rho \qquad (25)$$

and from this one can evaluate the critical exponent μ of σ. Indeed, the scaling law for ν' is

$$\nu'(L^{\alpha_T}\Delta T, L^{\alpha_\rho}\Delta\rho, L^{\alpha_\lambda}\lambda) = L^{d-\alpha_\lambda}\nu'(\Delta T, \Delta\rho, \lambda). \qquad (26)$$

On the other hand in Eq. (25) we must take the values of ν' along the equilibrium profile. As a consequence we find

$$\nu(\Delta T, \Delta\rho, \lambda_{eq}) = |\Delta T|^{(d-\alpha_\lambda/\alpha_T)}\nu'[1, \frac{\Delta\rho}{|\Delta T|^{(\alpha_\rho/\alpha_T)}}, \times$$
$$\times \lambda_{eq}(1, \frac{\Delta\rho}{|\Delta T|^{(\alpha_\rho/\alpha_T)}})]. \qquad (27)$$

Substituting this in Eq. (25), putting $y = \frac{\Delta\rho}{|\Delta T|^{(\alpha_\rho/\alpha_T)}}$, we get

$$\sigma = 2|\Delta T|^{(d-\alpha_\lambda/\alpha_T) + (\alpha_0/\alpha_T)} \int_0^B \nu'(y)dy, \qquad (28)$$

where on the coexistence curve we have taken $\Delta\rho = \pm B|\Delta T|^{(\alpha_\rho/\alpha_T)}$. We therefore find

$$\mu = \frac{d - \alpha_\lambda}{\alpha_T} + \frac{\alpha_\rho}{\alpha_T} = \frac{d - \eta}{2d}(\delta + 1)\beta + \beta. \qquad (29)$$

This result is significantly different from that one would obtain according to the square gradient theory, namely

$$\mu = \frac{(\delta + 1)\beta}{2} + \beta. \qquad (30)$$

For the two dimensional Ising model we get

$$\mu \text{ (scaling law)} = 1; \quad \mu \text{ (square gradient)} = 9/8.$$

It is interesting to observe at this point that also the coefficient in front of the singularity of the surface tension is expected to be different in the two theories. This follows from the fact that in the square gradient theory

$$\nu'\lambda = m\lambda^2 = 2(p_N - \mu\rho + f_o)$$

while in the scaling theory, from Eq. (22)

$$\nu'\lambda = p_N - \mu\rho + f,$$

and there is no reason, outside the square gradient theory, to expect f to be such a function of its arguments that

$$f(T,\rho,\lambda) = p_N - \mu\rho + 2f(T,\rho,0).$$

REFERENCES

1. J. D. van der Waals, _Zeit. Phys. Chem._, 13:657 (1894); English translation, _J. Stat. Phys._, 20:197 (1979).
2. S. Ono and S. Kondo, Encyclopedia of Physics, S. Flugge, ed., vol. 10, p. 134, Springer, Berlin (1960).
3. R. Evans, _Adv. Phys._, 28:143 (1979).
4. H. T. Davis and L. E. Scriven, _Adv. Chem. Phys._, XLIX:357 (1982).
5. J. S. Rowlinson and B. Widom, "Molecular Theory of Capillarity", Clarendon Press, Oxford (1982).
6. S. Fisk and B. Widom, _J. Chem. Phys._, 50:3219 (1969).
7. B. Widom, _J. Chem. Phys._, 43:3892 (1965).
8. For the definition of the critical exponents and a recent review of critical thermodynamics in fluids we refer to J. V. Sengers and J. M. H. Levelt Sengers, _in_ "Progress in Liquid Physics", C. A. Croxton, ed., Ch. 4, p. 103, Wiley, Chichester (1978).
9. M. Moldover, "Interfacial Tension of Fluids Near Critical Points and Two-Scale Factor Universality", to be published in _Phys. Rev. A._
10. A similar point of view was adopted almost thirty years ago by E. W. Hart, _Phys. Rev._, 113:412 (1959).
11. L. Mistura, "Stress Tensor and Scalar Pressure in Non-Uniform Systems", to be published in _J. Chem. Phys._

THE EYRING SIGNIFICANT STRUCTURE THEORY APPLIED TO THE CALCULATION OF SURFACE TENSION OF SIMPLE LIQUIDS

G. K. Johri[*]

Department of Electronics
D. A. V. College
Kanpur 208001, India

INTRODUCTION

The study of surface tension is of immense importance and it has been intensively studied by experimental and theoretical scientists[1-4,5]. Several theories have been proposed for the surface tension of liquids. March and Tosi[1] have comprehensively reviewed various theoretical models. However, due to characteristics of the liquid state, i.e., the strong interaction of particles and their state of disorder, the theoretical analysis has lagged far behind theories of the gaseous and the crystalline states. The alternative procedure is the model approach, in which one visualizes a physical model of the liquid, translates the picture in the mathematical language, i.e., a partition function, and then calculates the properties of the liquid. Such a model is the Eyring's Significant Structure theory[6]. The theory is based on the idea that the vapor is mirrored in the liquid as vacancies which transform solid-like into gas-like degrees of freedom. The usefulness of the model cannot be doubted. However, criticisms have been raised against the theory. One of the disadvantages is that it has not been derived from an exact partition function by any mathematically well-defined approximation, but is a result of intuition. But when properly formulated, such a model should be a useful description of what actually happens, as in fact it is. In this paper, Eyring's significant structure theory with monolayer approximation [7] has been used to calculate the surface tension of simple liquids in the temperature range from triple point to the critical temperature. The calculations for neon, krypton and xenon have been done using Lennard-Jones potential and such results have not been reported earlier. A comparison of the calculated surface tension is made with the experimental and the reported values by Wu and Yan[5] using density functional theory.

THEORY

The X-ray diffraction data show no appreciable change in the nearest neighbor distance in spite of the increase in volume when argon melts and

---

*These calculations were done while working as a guest scientist at the International Centre for Theoretical Physics, Trieste, Italy during the period from September 7 to October 26, 1985.

the decrease of volume when ice melts. Also, the X-ray data indicate that
the nearest neighbors of a molecule are arranged in an orderly manner;
whereas the second and third nearest neighbors are more randomly distributed, and beyond the third there is complete disorder. The liquid
possesses short range order. It is assumed that the liquid has fluidized
vacancies on an average having molecular size. These holes and the
molecules move in random fashion. Such an assumption has two effects.
Firstly, it increases the volume simply by increasing the number of sites,
at the same time keeping their intermolecular distance constant. Secondly,
when two or more molecules share a vacancy, the lattice structure is
destroyed, so that we no longer have long range order. Therefore the
liquid has an excess volume, $(V - V_s)$, where V and V_s are the molar volume
of the liquid and of the solid, respectively. The liquid phase differs
from a solid phase in that the kinetic energy of molecules in the liquid
has become large enough that it can balance the potential energy tending
to make molecules collapse into the holes that are present. The entropy
gained from the appearance of the holes is negligible at low concentrations
of the holes but increases cooperatively with these holes becoming
fluidized so that the Helmholtz free energy at any temperature goes through
a minimum with volume near the melting point. The distribution of holes
in a liquid may be assumed to have the average volume of a vacancy, are
mobile, and are called fluidized vacancies. These vacancies in a liquid
mirror the molecules in the vapor in concentration and in behavior.

In a mole of liquid $(V - V_s/V_s)$ moles confer gas-like properties on
(V_s/V) molecules. The quantity (V_s/V) is the fraction of positions next
to a vacancy which are occupied by molecules. Therefore a fraction
$(V - V_s/V_s) \times V_s/V$, i.e., $(V - V_s/V)$, has degrees of freedom acting like
gas molecules and the remainder (V_s/V) are solid-like.

The mean value of a liquid property X, is accordingly given as

$$ X = X_s \left(\frac{V_s}{V} \right) + X_g \left(\frac{V - V_s}{V} \right) \tag{1} $$

with X_s and X_g as values of the property in the solid and vapor states,
respectively.

According to the significant structure theory of liquids, there are
two typical approaches for developing a theory of surface tension. One is
an iteration procedure which calculates the sum of the contributions of
successive molecular layers; the other is a monolayer approximation which
yields a simplified approximate calculation. The iteration procedure for
calculating surface tension of a liquid is very useful in its application
to various liquids ranging from inert gases to polar liquids. However,
this approach fails to provide a simple equation for calculating the
surface tension since, after all, an iteration procedure is needed in any
satisfactory calculation of the contribution to the Gibbs free energy of
successive layers. Therefore, monolayer approximation is useful due to
its simplicity and to interpret the observed surface tension satisfactorily.

According to the monolayer approximation the dividing surface between
a liquid and its vapor phase is a monolayer, in which a molecule has a
free volume larger than for an interior molecule and a potential energy
less than for the latter. There is some evidence that the boundary
between a liquid and its vapor is sharp and that the boundary extends over
only about one molecular layer.

Using Lennard-Jones potential and Devonshire cell model the partition
function is given as

$$f = \{ \frac{(2\pi mkT)^{3/2}}{h^3} \, v'_f [\exp - \frac{z\psi'(o)}{2kT}] \times$$

$$\times [1 + m \frac{(V - V_s)}{V_s} \exp \frac{a'\psi'(o)V_s}{(V - V_s)kT}]J(T)\}^{N'(V_s/V)} \times$$

$$\times \{ \frac{(2\pi mkT)^{3/2}}{h^3} \times v_f [\exp - \frac{z\psi(o)}{2kT}] \times \qquad (2)$$

$$\times [1 + m \frac{(V - V_s)}{V_s} \exp \frac{a'\psi(o)V_s}{(V - V_s)kT}]J(T)\}^{N''/V_s/V)} \times$$

$$\times \{ \frac{(2\pi mkT)^{3/2}}{h^3} \frac{eV}{N} J(T)\}^{(V - V_s/V)N}.$$

In this equation the assumption has been made that the solid-like and gas-like molecules are randomly distributed in the surface as well as in the bulk liquid. Here $J(T)$ is the partition function for the internal degrees of freedom of a molecule, m is the mass of the molecule, h is the Planck's constant, k is the Boltzmann's constant, and v_f the free volume of a molecule, $\psi(o)$ is the potential energy possessed by a molecule as it vibrates about its equilibrium position, z is the coordination number of the molecule, the single prime (except for parameter a) indicates surface quantities, and $N' + N'' = N$ where N' and N'' are the number of molecules in the surface and in the bulk liquid, respectively.

To calculate surface tension 'γ' we use Helmholtz free energy, A, as follows:

$$\gamma = (\frac{\partial A}{\partial \Omega})_{N,V,T} = \omega^{-1}(\frac{\partial A}{\partial N_c})_{N,V,T} \qquad (3)$$

where Ω is the surface area and N_c, the total number of sites available on the surface, is related to N by the relation $N' = (V_s/V)N_c$, since a random distribution is assumed. The symbol ω represents the area occupied by one molecule and is equal to $(V3/2)a^2$ for close packing, where a is the nearest neighbor distance, and is given by $(V_2 V_s/N)^{1/3}$. By substituting 'f' in the following relation

$$A = - kT \ln f \qquad (4)$$

and using an expression for the surface tension Eq. (4) we get the desired expression to calculate surface tension for simple liquids using monolayer approximations, i.e.,

$$\gamma = \frac{2}{\sqrt{3}} (\frac{N}{\sqrt{2}V_s})^{2/3}(\frac{V_s}{V})^2[\frac{3}{2} U(o) - k_B T \ln \frac{1 - 0.875(\sigma/a)}{1 - (\sigma/a)}] \qquad (5)$$

where $U(o)$ is the Lennard-Jones potential which is expressed as follows:

$$U(o) = \varepsilon[2.4090(\frac{N\sigma^3}{V_s})^2 - 1.0109(\frac{N\sigma^3}{V_s})^4], \qquad (6)$$

"a" is related to the molar volume V_s as follows:

$$a = (\frac{\sqrt{2}V_s}{N})^{1/3}, \qquad (7)$$

σ and ε are the distance and the energy characteristics of the system. N is the Avogadro's number and k_B is the Boltzmann's constant.

Table 1. Potential Parameters[5,13,14]

System	Ne	Ar	Kr	Xe
ε/k_B (K)	39.6	119.8	197.5	221.0
σ (A)	2.749	3.405	3.597	4.10
N_s (cc/mole)	13.98	24.98	29.60	36.50
Molecular weight	20.18	39.944	83.80	131.30

DISCUSSION

The surface tension of neon, argon, krypton and xenon are calculated using Eq. (5). The potential parameters used in the calculation are given in Table 1. The values of ε/k_B and σ obtained from second virial coefficients data fits the observed values of surface tension better than those from viscosity data. It is because the surface tension is a thermodynamic quantity. Using the calculated values of molar volume at various temperatures from the Guggenheim's method[8], the surface tension is then computed as tabulated in Table 2. A comparison with the observed data quoted from References (Ne)[9], (Ne and Ar)[8], (Ar)[10], (Xe)[11] and (Kr)[12] and those calculated by earlier workers[5] using Lennard-Jones potential shows that there is a close agreement with the experimental values. Johri and Saxena[15] also calculated the surface tension of water using the same approximation as applied here and their results are given in Table 3 for illustration. An examination of the Table 3 shows that the calculated[15] values of surface tension are about 10 - 24% higher than

Table 2. Calculated Surface Tension (dyne/cm) of
Neon, Argon, Krypton and Xenon

| | Neon | | |
| | Ref.[8,9] | Ref.[5] | |
T(K)	(Experimental)	(Calculated)	(Present)
25	5.50	6.52	5.58
26	5.17	6.10	5.41
27	4.80	5.74	4.23
28	4.45	5.31	4.05
29	–	4.92	3.88
30	–	4.54	3.71
31	–	4.13	3.53
32	–	3.75	3.36
33	–	3.38	3.19
34	–	3.03	3.03
35	–	2.66	2.86
36	–	2.31	2.69
37	–	1.97	2.53

Table 2. (Continued)

	Argon		
T(K)	Ref.[8,10] (Experimental)	Ref.[5] (Calculated)	(Present)
83.82	13.45	16.48	15.88
85	13.06	16.06	15.70
87.29	12.42	15.27	15.14
90	11.91	14.48	14.71
95	10.15	12.89	13.54
100	9.32	11.50	12.51
105	8.33	10.21	11.48
110	7.08	8.93	10.47
120	4.93	6.24	8.35
130	2.91	3.57	6.19
140	1.32	1.61	3.83
	Ref.[12]	Krypton	
120	15.49	19.33	13.10
130	–	16.69	11.85
140	–	14.10	10.63
150	–	11.60	9.39
160	–	9.20	8.10
170	–	6.93	6.83
180	–	4.79	5.52
190	–	2.86	4.08
200	–	1.17	2.44
	Ref.[11]	Xenon	
170	17.46	22.80	17.31
180	15.66	20.40	16.05
190	13.84	18.12	14.56
200	12.09	15.89	12.89
210	10.38	13.78	11.65
220	8.74	11.70	10.26
230	7.16	9.67	8.90
240	5.66	7.72	7.55
250	4.24	5.84	6.22
260	2.91	4.09	4.90
270	1.73	2.47	3.55
280	0.61	1.02	2.10

Table 3. The Surface Tension of Water

T(K)	Ref.[16] (Experimental)	Ref.[15] (Calculated)
273.15	75.60	83.20
283.15	73.49	82.81
293.15	72.75	78.60
313.15	69.56	81.63
353.15	62.6	75.82
373.15	58.9	73.07

the experimental[16] values but the temperature dependence is almost similar to the observed surface tension. The deviation from the experimental values increases with temperature. These discrepancies may be partly explained by considering the orientational effect of polar molecules and the contribution of the other layers within a few molecular diameters of the interface. But for an H-bonded associated liquid-like water, orientation effect arising due to the concentration gradient between liquid and gas and to the dipole moment of surrounding molecules should be included to interpret the observed values of the surface tension. It is found that if an appropriate partition involved in obtaining Eq. (5) is obtained for larger molecular systems with known values of V_s and potential parameters the Eyring significant structure theory is satisfactory to calculate surface tension.

REFERENCES

1. N. H. March and M. P. Tosi, Atomic dynamics of liquids, Chapter 10, pp 256–278, McMillan, London (1976).
2. F. F. Abraham, Phys. Rep., 53C:93 (1979).
3. R. Evans, Adv. Phys., 28:143 (1979).
4. C. Ebner, W. F. Saam and D. Stroud, Phys. Rev., A14:2264 (1976).
5. S. T. Wu and G. S. Yau, J. Chem. Phys., 77:5799 (1982).
6. H. Eyring, D. Henderson and W. Jost, Physical Chemistry, An Advanced Treatise, Vol. II, Academic Press, New York, London (1967).
7. T. S. Ree, T. Ree and H. Eyring, J. Chem. Phys., 41:524 (1964).
8. E. A. Guggenheim, J. Chem. Phys., 13:253 (1945).
9. N. B. Vargftik, Tables on the Thermophysical Properties of Liquids and Gases, 2nd ed., Wiley, New York (1974).
10. F. B. Sprow and J. M. Prauswitz, Trans. Faraday Soc., 62:1097 (1966).
11. B. L. Smith, P. R. Gardner and E. H. C. Parker, J. Chem. Phys., 47:1148 (1967).
12. S. Fuks and A. Bellemans, Physica, 32:594 (1966).
13. J. O. Hirschfelder, C. F. Curtiss and R. B. Bird, Molecular Theory of Gases and Liquids, Wiley, New York (1964).
14. E. F. Washborn et al., International Critical Tables, McGraw-Hill Brook Company Inc., New York (1926).
15. G. K. Johri and D. Saxena, Indian J. Pure Appl. Phys., 22:124 (1984).
16. H. Eyring and M. S. Jhon, Significant Liquid Structures, Wiley and Sons, New York.

EXPERIMENTAL STUDIES OF MICROEMULSION SYSTEMS

A GLOBAL DESCRIPTION OF PHASE EQUILIBRIA IN

THE QUATERNARY MICROEMULSION SYSTEM:

WATER-DODECANE-PENTANOL-SODIUM DODECYLSULFATE

A.M. Bellocq and D. Roux

Centre de Recherche Paul Pascal (CNRS)
GRECO "Microemulsions" du CNRS-Domaine Universitaire
33405 Talence Cedex, France

INTRODUCTION

The knowledge of the phase diagram of a multicomponent system is a
fundamental necessary step for the understanding of the physics of the
system. It has been clearly and intensively shown by Ekwall and co-
workers[1] that phase equilibria in multicomponent aqueous mixtures of
amphiphilic molecules can be richly diverse and intricate. Due to their
considerable potential for aggregation the surfactant solutions show a
multiplicity of structures (bilayers, cylinders, spherical micelles) which
can organize and produce a great variety of isotropic and mesomorphic
phases. The phase diagrams of several water-surfactant-alcohol and water-
surfactant-oil mixtures have been investigated in detail [1-6]. They
display a rich variety of phases and also complex multiphase regions where
two or three liquid isotropic and mesomorphic phases are in equilibrium.
As we will see later in this paper, the phase diagrams of the quaternary
mixtures which give rise to microemulsion phases are not qualitatively
different from those of ternary mixtures described by P. Ekwall. One still
encounters ordered phases but their extent is considerably reduced. Generally mesomorphic regions are replaced over a broad domain of water and
oil concentrations by an isotropic liquid microemulsion phase where no long
range order occurs.

Up to now only very restricted portions of the phase diagrams of
quaternary mixtures water-oil-surfactant-alcohol have been explored. The
extent and the shape of the region of existence of microemulsion in quaternary systems were determined but the multiphase regions occurring in
these mixtures are little known[7-9]. In contrast the multiphase regions
of quinary mixtures containing salt have received much interest in con-
nection with their potential use in oil recovery[10,11]. A great number of
studies have focused attention on the so-called Winsor III three-phase
region where a microemulsion is in equilibrium with both an organic and an
aqueous phase[12]. Previous studies of mixtures of aqueous solutions of
surfactant in presence of alcohol, salt and hydrocarbon have also shown
that a multiplicity of equilibria between isotropic liquid phases can be
produced by these systems[13]. Then, for example, in addition to the
Winsor III equilibria two three isotropic phase equilibria have been
observed in the system water-dodecane-pentanol-NaCl-sodium octylbenzene-
sulfonate[13]. These studies have also established that microemulsion

systems can give rise to critical points and critical end points[12,13].
Besides, recent results suggest that surfactant multicomponent systems can
exhibit tricritical points[14,25]. The location of the critical points or
the lines of critical points and their evolution with temperature or other
variables are of fundamental importance for the study of critical phenomena. In multicomponent mixtures, this study is made difficult by the large
number of independent variables required to describe the system. In 1970,
Griffiths and Wheeler[16] have shown that it is useful in discussions of
phase transitions in multicomponent mixtures to distinguish two classes of
intensive thermodynamic variables: fields and densities. The variables
such as pressure, temperature, chemical potential which are always identical in the coexisting phases are called "fields" and those which are
generally different in the coexisting phases such as concentration, refractive index are called "densities". In addition to being useful in considering the relative magnitudes of the divergences of critical points,
changing the representation can also simplify the appearance of the diagram
and make easier the understanding of the complicate evolution of phase
equilibria occurring in multicomponent systems. In practice representation
of the diagram in a field space encounters difficulties related to the
nature itself of these variables and to their experimental measurements.
The main difficulty is to find appropriate field variables different from
pressure or temperature which may be experimentally controlled. In a
recent paper we have presented experimental evidence that in oil rich
microemulsions the water over surfactant ratio (termed X in the following)
behaves as a field variable[17,18]. In the following we have taken advantage of this finding to propose an entire description of the phase diagram
of the quaternary system which consists of water-dodecane-pentanol-sodium
dodecylsulfate (SDS). Prior to describe the phase diagram of the quaternary mixture we have investigated that of the ternary system water-
pentanol-SDS. In the description of both systems we will emphasize the
details of the evolution of the phase equilibria as the composition or
temperature are varied. We have focused our attention not only on the
characterization and the location of the boundaries of the various phases
but also on the equilibria between the phases i.e. we have determined the
changes in direction of the tie-lines and tie-triangles as X is varied.

An outline of the paper is as follows: We will first briefly discuss
the experimental procedure then we describe the experimental diagrams of
the ternary and quaternary mixtures. Finally in the last section we give a
representation of these diagrams in a field space. The most important
features of these diagrams are summarized in the conclusion.

EXPERIMENTAL PROCEDURE

Experimental procedure is described in detail in reference 19. Tubes
were allowed to equilibrate in a chamber at constant temperature (T was
controlled within 0.1°C). The time for equilibrium to be attained varied
and was dependent on composition, viscosity, structure and vicinity of
critical points. In the oil rich part of the diagram phase equilibrium was
reached after a few hours except near critical points. Tubes were
maintained at constant temperature for several months and no change was
detected. For the parts of the phase diagram where a great complexity is
found, several hundreds of tubes have been prepared. Consistency in the
data has been achieved by showing continuity of the multiphase regions
which surround a given region.

The rich water plus surfactant part of the sections defined by X below
3 was not investigated precisely. Due to the high viscosity of the phases
occurring in these regions, phase separation was very difficult and the
boundaries marking the phase transitions were not determined accurately.

Identifications of mesomorphic phases were made by observation of optical textures in a polarized microscope. In the following, we have used the Ekwall's system of nomenclature to represent the different phase structures observed.

Compositions in the diagrams are expressed in weight fractions. The determination of the compositions of the co-existing phases in multiphase equilibria was performed by gas chromatography and gravimetric analysis. Prior to analysis, the samples were allowed to equilibrate in cells for several days or weeks in a chamber at 21°C. T was controlled to 0.1°C. Water, dodecane, and pentanol concentrations were determined by gas chromatography. Heptanol-1 and propanol-1 have been used as an internal reference. Their separation was effected on a one meter long Porapack P column at 190°C. The surfactant was weighed after evaporation under vacuum of all liquids. The relative precision obtained on the concentrations depends on the component. It is about 1.5% for water, 1% for pentanol and dodecane and 2% for SDS.

Alcohol and dodecane were Aldrich products ("purum" reagents) and SDS was purchased from Touzart et Matignon ("pur" reagent).

PHASE DIAGRAM OF THE TERNARY SYSTEM: WATER-PENTANOL-SDS

Diagram at 25°C

The diagram at 25°C of the ternary system consisting of water-pentanol-SDS is represented in Figure 1. Four one-phase regions are observed; in three of them (D, E, R) the mixtures are birefringent, in the last one L they are isotropic. At 25°C, the ternary mixtures water-pentanol-SDS display a continuous band of isotropic solution L between pure water and pure pentanol. This property is characteristic of microemulsion systems. Provided that the surfactant concentration is adjusted, it is possible to go continuously from the water corner to the pentanol one without any phase separation occurs. The main result of this continuity is the occurrence of a critical point P_C^A. The phase E corresponds to the hexagonal phase observed in the binary system H_2O-SDS. The new evidenced phases D and R are respectively lamellar and rectangular phases. Phases E and R are only obtained at high surfactant content ($\sim$30%). The boundaries of the regions of existence of these phases are not easy to define precisely because the difficulty in separating them. The lamellar mesophase D occurs over a wide range of surfactant concentration. Its region of existence extends down toward the water corner, up to 79% of water. The isotropic-lamellar phase transition is a first order transition. As a result, the phases L and D are separated by a two-phase region whose extent is found to decrease as a surfactant content decreases. At the point M, the boundaries of the isotropic and lamellar phases are tangential. The phase situation in this point resembles that at the azeotropic point in the liquid-gas equilibria. Analysis of several two-phase equilibria located on both sides of the lamellar phase shows that the length of the tie-lines vanishes at the point M. Then in this point the two phases D and L coexist with exactly the same composition. While at a critical point, certain thermodynamic properties have an universal behavior and follow power laws, at an azeotropic like point no particular behavior is expected.

Effect of temperature

Significant changes in the phase equilibria occur on lowering the temperature. Below 20°C, the isotropic domain L is no longer continuous; it is separated into two one-phase regions L_1 and L_2 which respectively expands from the water and pentanol corners (Figure 2). These two iso-

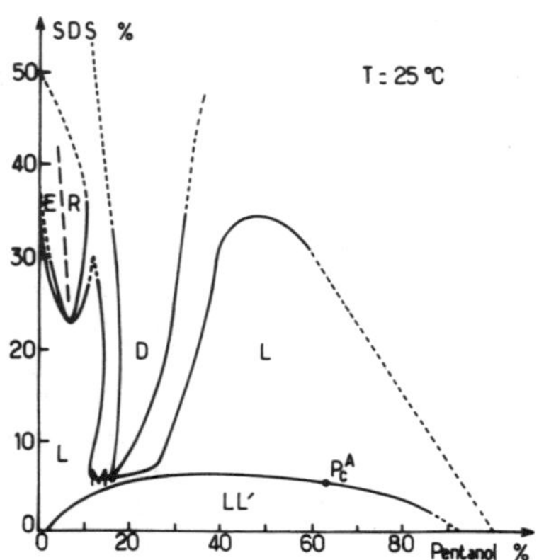

Fig. 1. Phase diagram at 25°C of the ternary water-pentanol-SDS system. L
is an isotropic phase, E, R and D are mesomorphic phases. The
dashed boundaries have not been determined accurately. The point M
is an azeotrope-like point; P_C^A is a critical point.

tropic regions are separated by a complex multiphase regions which involves
two three-phase domains t_1 and t_2. In these equilibria, two of the phases
are isotropic, the third one is the lamellar phase D. Composition analysis
of the coexisting phases in the equilibria t_1 and t_2 at different temperatures shows that these equilibria originate in an indifferent state at
20°C[20] (Figure 2b). Experimental data are reported on Figure 3. Two
major consequences of increasing temperature are found:

(1) the concentration of each component (water-pentanol-SDS) in the
corresponding phases of the equilibria t_1 and t_2 tends toward the
same value;

(2) the tie-triangles become thinner.

At 20°C, the points ABC representative of the three coexisting phases
of both equilibria are located on the same straight line. The system is
then in an indifferent state. Let us note that here too, as for the azeotropic state, the system does not show a particular behavior. A schematic
picture of the temperature dependence of the t_1 and t_2 regions is given in
Figure 4. At 20°C, the regions L_1 and L_2 have one common point and they
merge above this temperature. To our knowledge, these results are the
first experimental evidence for the formation of a three-phase region
through an indifferent state.

Shape of the coexistence curve LL'

The coexistence curve which separates the regions L and LL' was determined by visual examination of a large number of samples. This observation
was made after storage at 25°C of the samples for several days in order to
achieve a complete phase separation. Precise determination of tie-lines in
the region LL' around the critical point P_C^A was obtained from density
measurements. In a first step, densities of several one-phase mixtures
located along the coexisting curve of the LL' region were measured in using
a high precision picnometer (relative densities are measured with a precision better than 10^{-5}). The experimental variation of the density d
along the coexisting curve was found to be a linear function of the alcohol
concentration X_a:

$$d = d_w + aX_a \qquad (1)$$

where d_w = 1.0012 (at 25°C) and a = 0.19689.

In a second step two-phase equilibria were prepared in the LL' region and the density of both coexisting phases was measured. From the equation (1) and the knowledge of the coexisting curve, we were able to locate precisely the tie-lines. Figure 5 gives the positions of these tie-lines around the critical point. It appears that around the critical point the tie-lines seems to be parallel, this holds also for mixtures far from the critical point. This remark was made several years ago for another system by Zollweg[21]. In addition, the midpoints of the tie-lines are on a straight line perpendicular to the tie-lines. This direction is defined by a relation between two concentrations. Let us choice, for example, X_s (surfactant concentration and X_a (the alcohol concentration). From experimental behavior, we have noticed that a linear relation exists between δX_s and δX_a (Figure 5):

$$\delta X_s = -A \cdot \delta X_a \text{ with } A = + 0.07 \text{ and } \delta X_s = X_s^1 - X_s^2, \ \delta X_a = X_a^1 - X_a^2,$$

where X_a^j and X_s^j are the alcohol and surfactant concentrations in the upper phase (j = 1) and lower phase (j = 2).

Consequently for each two-phase equilibrium the function $Y(X_a,X_s)$ is conserved in each phase in equilibrium:

$$Y(X_a^1,X_s^1) = Y(X_a^2,X_s^2) \text{ with } Y(X_a,X_s) = A \cdot X_a + X_s.$$

The direction parallel to the tie-lines corresponds to the density conjugated to the field Y, i.e. to the order parameter O of the phase transition; it is given by the function

$$O(X_a,X_s) = -X_a/A + X_s.$$

It is possible to redraw the phase diagram as a function of the field Y and the order parameter O at constant temperature. Figure 6 shows this phase diagram representation. The critical point P_C^A is obviously an extremum of the coexisting curve. However it is more surprising that the points M and A are also extrema; this indicates that the variable Y seems to behave as a field even far from the critical point P_C^A. We may also remark that the indifferent state which appears at lower temperature is also nearly perpendicular to this direction. In fact a precise analysis shows that Y is not exactly constant on each tie-line, specially far from the critical point, it would be better to consider that $Y(X_a,X_s)$ is an approximate expression for a field variable and that a more complex expression exists. The determination of a more accurate and consequently a more complex formula for Y has probably not a great interest and must require more precise analysis.

This knowledge of an approximate expression for a chemical potential has allowed us to approach the critical point P_C^A along two different paths; results are published elsewhere[22].

DIAGRAM OF THE QUATERNARY SYSTEM: WATER-DODECANE-SDS-PENTANOL

At fixed pressure and temperature, the phase diagram of a quaternary mixture may be represented in a tetrahedron. Experimentally, we have

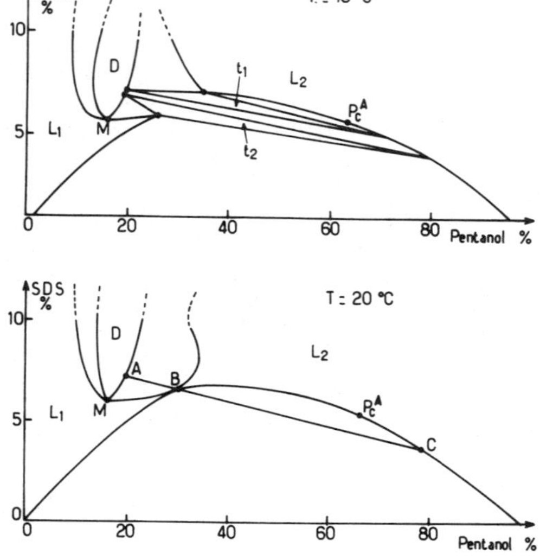

Fig. 2. Phase diagrams of the ternary water–pentanol–SDS system at 18°C and 20°C. L_1 and L_2 are isotropic phases, t_1 and t_2 are two three-phase regions. The line ABC is an indifferent three-phase state.

investigated sections of the tetrahedron defined by a fixed value of the water over surfactant ratio X. The choice of this variable allows a good description of the oil rich part of the diagram.

In the planes X below X = 0.76, in the oil rich region, only one wide isotropic region termed L_2 is detected. This region corresponds to the microemulsion region. For X = 0.76 a birefringent phase appears (Figure 7). This phase termed D has a lamellar structure, it can contain up to 98 percent of dodecane and alcohol in volume. In this plane, the polyphasic region corresponds to two-phase equilibria between the isotropic micro-emulsion and the liquid crystalline phase. In the plane X = 0.76, for the high oil content, the two-phase region becomes very thin. It seems prob-able that the microemulsion L_2 phase and the lamellar phase D exhibit an azeotropic like point (Figure 7). As one can see on the Figures 8-13 both regions L_2 and D exist in all the diagrams X > 0.76. Their extent obvi-ously depends on X. In the section corresponding to X = 1.034 (Figure 8) in addition to the two one-phase domains occurring at low X, there are two phases L_1 and L_3. While motionless these phases are isotropic, they exhibit flow transient birefringence as soon as any disturbance is created. In particular flow birefringence is easily generated by shaking the sample tube. Besides both phases L_1 and L_3 scatter light.

The diagram found for X =1.207 is quite similar to that observed for X = 1.034 (Figure 9). Whereas in both planes, X = 1.034 and X = 1.207, the regions L_1 and L_3 are separated, as X increases above X = 1.38 they form a single region which is referred in the following as the phase L_1 (Figure 10). The latter, displays shear birefringence which rapidly decreases as the oil content decreases and as X increases. For X = 1.89 the mixture water–SDS becomes homogeneous; it can dissolve a small amount of dodecane and alcohol. For X = 2.586, the region L_1 extends up to water-SDS corner (Figure 11). Finally for X = 3.017, the regions L_1 and L_2 have merged and the diagram presents a very large isotropic one-phase region named L (Figure 12). This domain remains continuous up to X = 5.3 (Figure 13). For this last value, the amount of surfactant with respect to water is no

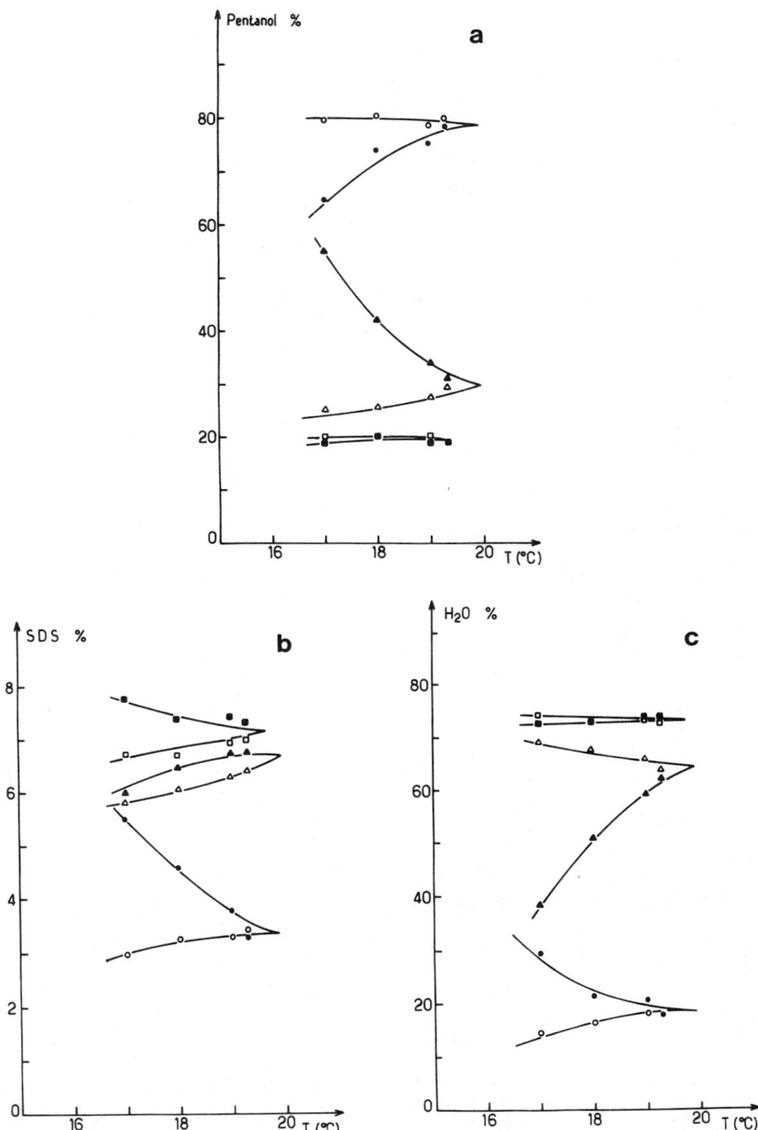

Fig. 3. Temperature dependence of the pentanol (a), SDS (b) and water (c) concentrations in each phase of the equilibria t_1 and t_2, ●, ▲, ■, correspond respectively to the upper, middle and lower phases of the equilibria t_1, the open symbols ○, △, □, correspond to the same phases for the equilibria t_2.

longer sufficient to achieve the continuity of the single phase domain. For X above 5.3, the latter has splitted into two regions, one rich in oil and alcohol and one rich in water and surfactant.

As mentioned earlier, the lamellar phase D occurs in all the X planes above X = 0.76. As X increases its region of existence becomes smaller and is progressively shifted towards the alcohol-water plus surfactant side of the diagram. This mesophase which appears as an extension of that observed

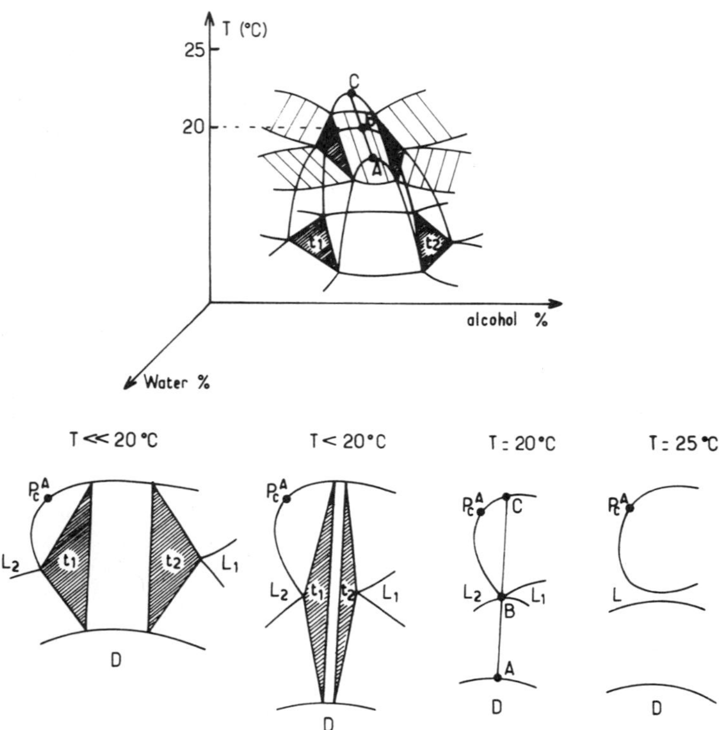

Fig. 4. Di and three-dimensional schematic representations of the
tie-triangles t_1 and t_2 at different temperatures.

in the ternary system water-pentanol-SDS can contain up to 98% of oil and
alcohol. A second liquid crystalline phase occurs in the planes X below
2.8. It is probably the extension of the E or R phases found for the
ternary system water-SDS-pentanol. The limits of this mesophase have not
been determined accurately.

In each X plane above X = 0.95 a critical point P_C exists on the
coexistence curve of the microemulsion region L_2. The direct consequence
of this critical point is the existence of a two-phase region ($L_2L'_2$) where
two isotropic microemulsions are in equilibrium. The richest microemulsion
in oil is termed L'_2. The extent of this two-phase domain sharply increases
with X. The set of the critical points generates a critical line P_C^l which
extends from X = 0.95 up to X = 6.6 However we are not sure that this last
point (P_C^A) which is located on the face of the tetrahedron water pentanol
SDS belongs to the critical line P_C^l . Indeed all the points of the line
defined by the values of X ranging from 0.95 to 6.47 are lower critical
points whereas the point P_C^A is an upper critical point. A detailed study
of the diagrams for X ranging between 6.47 and 6.6 is currently under
progress in order to clarify this point.

In the planes X above X = 0.95 appearance of new phases (L_1 and L_3) on
one hand, and that of the critical point P on another hand generate several multiphase regions. In particular in the plane X = 1.034 a four-phase
region (T) is seen where the four phases L_2, L'_2, D and L_1 coexist simultaneously. Because of their density, the sequence of phases in a tube is
L'_2-L_1-D-L_2 ; L'_2 is the upper phase, L_2 the lower phase. In a four component

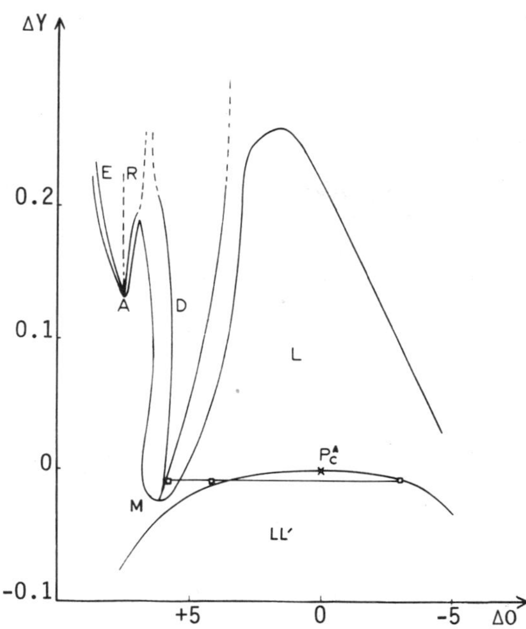

Fig. 5. Representation of the water-pentanol-SDS diagram (at T = 25°C) in the mixed field density space Y, O.

$$\Delta Y = Y (X_s, X_a) - Y (X_s^c, X_a^c)$$

$$\Delta O = O (X_s, X_a) - O (X_s^c, X_a^c)$$

The subscript c corresponds to the critical point,

$$[X_s^c = 0.0539; \quad X_a^c = 0.604]$$

The squares represent the positions of the three phases ABC of the indifferent state evidenced at T = 20°C.

system, at constant pressure and temperature such a region is invariant. In the density space, this region is a tetrahedron; it is surrounded by the four following three-phase regions: $L_2^!L_1D$ (t_1), L_1DL_2(t_2), $L_2^!L_1L_2$(t_3) and $L_2^!DL_2$ (t_4). These domains are themselves separated by the six two- phase regions: $L_2^!L_2$(d_1), L_2L_1(d_2), L_2D(d_3), $L_2^!D$(d_4), L_1D(d_5), $L_2^!L_1$(d_6). Each of these six regions starts on the sides of the tetrahedron. in the plane X = 1.034, the three-phase regions t_1, t_Z and t_3 are observed. In t_3 the three coexisting phases are isotropic, in t_1 and t_2 one of them is birefringent. At low alcohol content, two three-phase regions t_5 and t_6 are seen, they connect respectively the phases $L_2^!$, L_1, D (t_5) and $L_2^!$, L_3, D (t_6) (Figure 8). Due to the very small extension of most of the multiphase regions observed in the plane X = 1.034, it has been difficult to determine accurately their boundaries. Figure 14 shows the sequence of equilibria found along the AA' line of Figure 8.

The multiphase region found in the plane X = 1.034 undergoes significant changes as X is varied. In the plane X = 1.55, the four-phase region T and the three- phase regions t_1, t_5, t_6 do not exist. The equilibria t_2 has disappeared in the plane X = 3.017 where the phases L_1 and L_2 have merged. In this plane, the equilibria t_3 is still seen as well as another three-phase region (t_7) where the coexisting phases are all isotropic. The regions t_3 and t_7 have both disappeared for X = 3.5. So far we do not know for which value of X, the t_7 equilibrium appears.

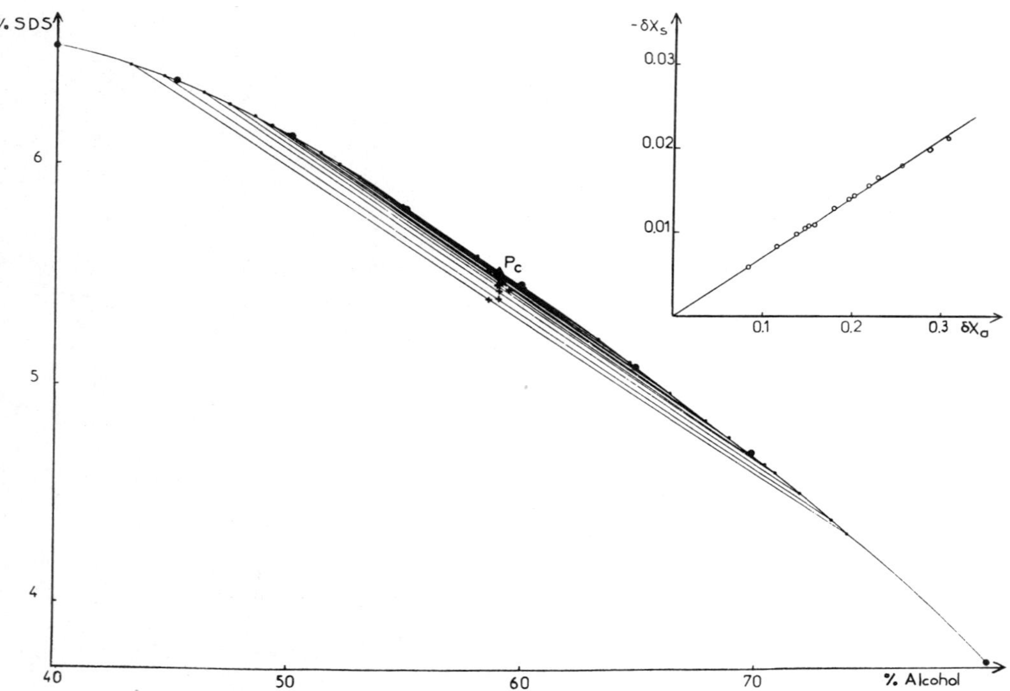

Fig. 6. Magnification of the two-phase region LL' around the critical point P_C^A. The segments represent the tie-lines, the full points correspond to the visual determination of the coexistence curve. The crosses are the positions of the middle of the tie-lines and the triangle position of the ciritical point.
Plot of the difference of the surfactant concentration between the two phases in equilibrium, δX_a in the same two-phase equilibria.

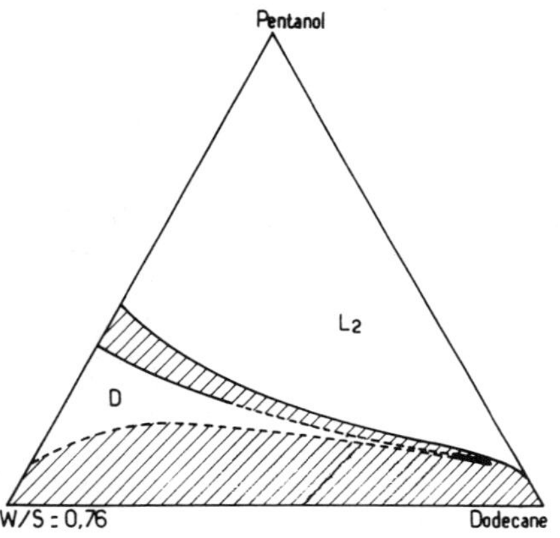

Fig. 7. Section at constant water over surfactant ratio X = 0.76 of the phase diagram of the water–dodecane–pentanol–SDS system at 21°C, L_2 and D are respectively liquid isotropic and lamellar liquid crystalline regions. The hatched domain is the multiphase region.

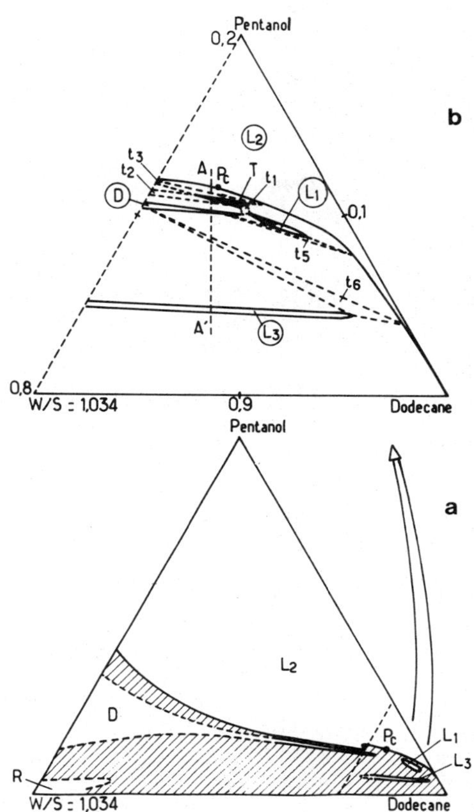

Fig. 8. (a) Section X = 1.034 of the water-dodecane-pentanol-SDS system at
21°C. R is a rectangular liquid crystalline phase, L_1 and L_3 are
two flow birefringent microemulsion phases. P is a critical point.
Other abreviations are the same as in Figure 7.
(b) Magnification of the lower right corner of Figure 9a.
t_1, t_2, t_3, t_5, t_6 are three-phase regions, T is a four-phase
region.

Compositions of the coexisting phases in multiphase equilibria

Determination of the composition of the coexisting phases in multiphase equilibria gives access to the shape of the coexistence phase surfaces and to the directions of tie-lines and tie-triangles. We have measured the concentrations of the four components in each phase of one four-phase sample T and of several two- and three-phase samples.

In Table 1 are reported the data for the equilibrium T. The compositions of the four phases are found to be very close and explains the very small extent of the four-phase tetrahedron in the phase diagram at 21°C. It should be interesting to investigate the temperature dependence of this volume and examine how it disappears.

We have seen earlier that in the plane X = 1.55 the phases L_2, L_1 and D are separated by a complex multiphase region which includes in particular the three-phase regions t_2 and t_3 and the two-phase region d_1 where two microemulsions are in equilibrium (Figure 10). Measurements were made on ten samples denoted D_1-D_{10} lying in region d_1, one sample in region t_2 and

169

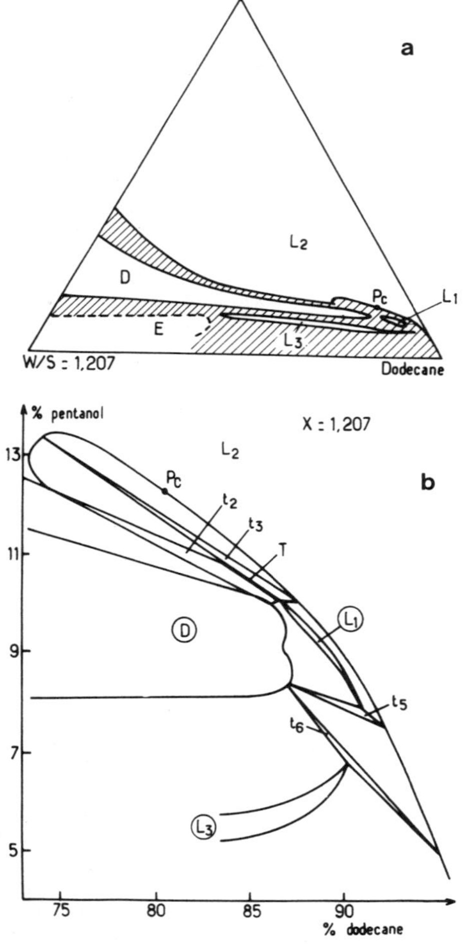

Fig. 9. (a) Section X = 1.207 of the phase diagram of the water-dodecane-
pentanol-SDS system at 21°C. E is an hexagonal phase.
(b) Magnification of the lower right corner of Figure 9a.
Abreviations are as in Figures 7 and 8.

one in region t_3. The samples D_1-D_{10} were located along the line BB' drawn
in Figure 10b. In all these samples the global water to surfactant ratio
is 1.55. The water over surfactant ratio in each phase was determined from
the data of the phase analysis [18]. Figure 15 gives the value of X
measured (denoted X_i) in each phase (i) in equilibrium for the three-phase
equilibria t_2 and t_3 and for the ten two-phase equilibria (denoted D_1-D_{10}).
In every case, the value of X_i is close to 1.55 which is the value of the
global preparation. This result is the indication that the two triangles,
which represent the two three-phase equilibria, and the tie-lines associated with the 10 two-phase equilibria all lie on the plane X = 1.55.

In view to examine whether this property holds for other values of X
we have prepared and analyzed further three-phase samples t_2[17, 18] and
two-phase mixtures D_1. The data of the analysis of all the components in
each phase for the whole X range where t_2 exists are plotted in Figure 16.
The value of X measured in each phase (X_i) is compared in Figure 17 with

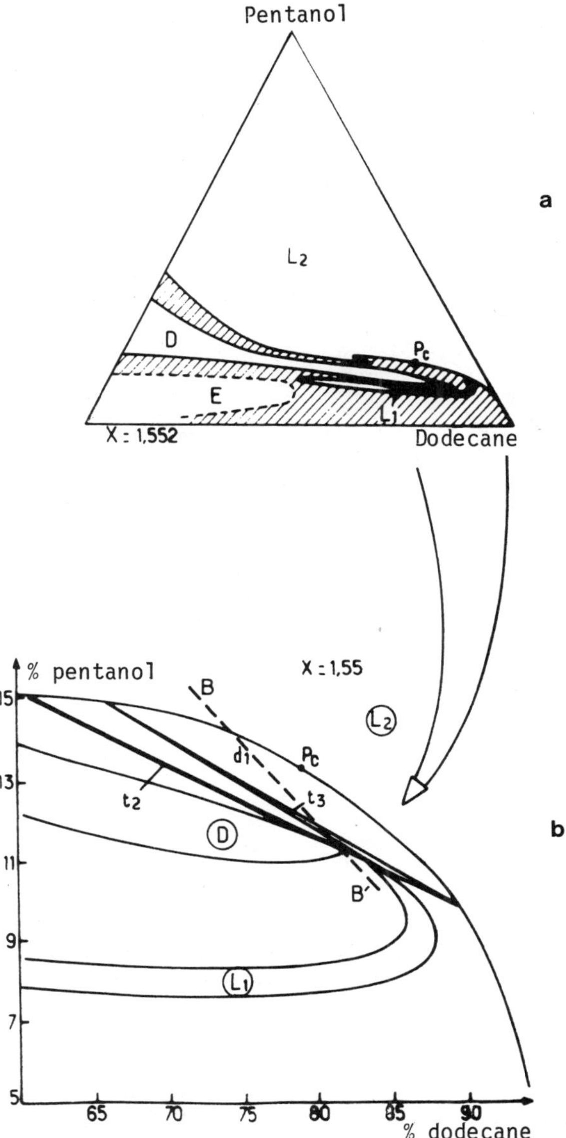

Fig. 10. (a) Section X = 1.55 of the phase diagram of the water–dodecane-
pentanol–SDS at 21°C. (b) Magnification of the lower right corner
of Figure 13. Abreviations are as in Figures 7–9.

the value of the overall mixture (X). The conservation of X in each phase
is better than 1% which is within the experimental accuracy. This remarkable result is the indication that the X ratio has the characteristics of a
field variable. The major consequence of this property is that the phase
diagrams experimentally determined in which X is maintained constant are
true pseudoternary diagrams. The three-phase volume t_2 consists then of a
stack of tie-triangles each of them lying in a X plane. As already mentioned, analysis of several three-phase samples allows us to determine the
boundaries of the volume in the three-dimensional phase diagram. Data

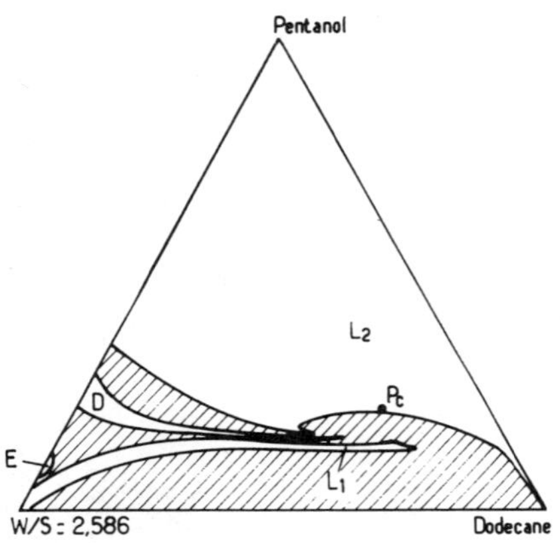

Fig. 11. Section X = 2.586 of the phase diagram of the water-dodecane-pentanol-SDS at 21°C. Abreviations are as in Figures 7-9.

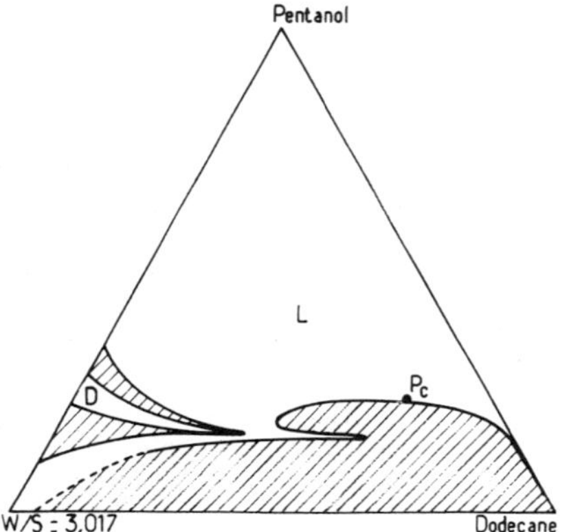

Fig. 12. Section X = 3.017 of the phase diagram of the water-dodecane-pentanol-SDS at 21°C. Abreviations are as in Figures 7-9.

reported in Figure 16 shows that the volume t_2 extends from the four-phase region T at X=1.05 to a critical end point (CEP) at X = 2.98. At low X, each of the three phases has the same composition as the corresponding phase of the four-phase equilibrium T (Cf. Table 1). The volume t_2 is situated in the oil rich part of the phase diagram (the water content never exceeds 25% in weight). Between X = 1.05 and X = 2.80 the two isotropic phases are the upper and the lower phases, the middle phase is the liquid crystalline phase D. For low values of X, the isotropic upper phase (L_1) exhibits flow birefringence. Above X = 2.80, due to the increasing proximity of the two isotropic phases density, the liquid crystalline phase becomes the lower phase and the two isotropic phases are adjacent. At X = 2.98 these two phases merge at a critical end point (CEP). A schematic representation of the volume t_2 in the space alcohol concentration, oil concentration-X ratio is given in Figure 18. At the point P, the lamellar phase and the isotropic L_2 phase have the same composition for the four components. This point corresponds to an azeotrope-like point.

Measurements of compositions of conjugate phases of 10 samples occurring in region d_1 have been carried out in order to determine in which region of the phase diagram the field character of X is altered or lost.

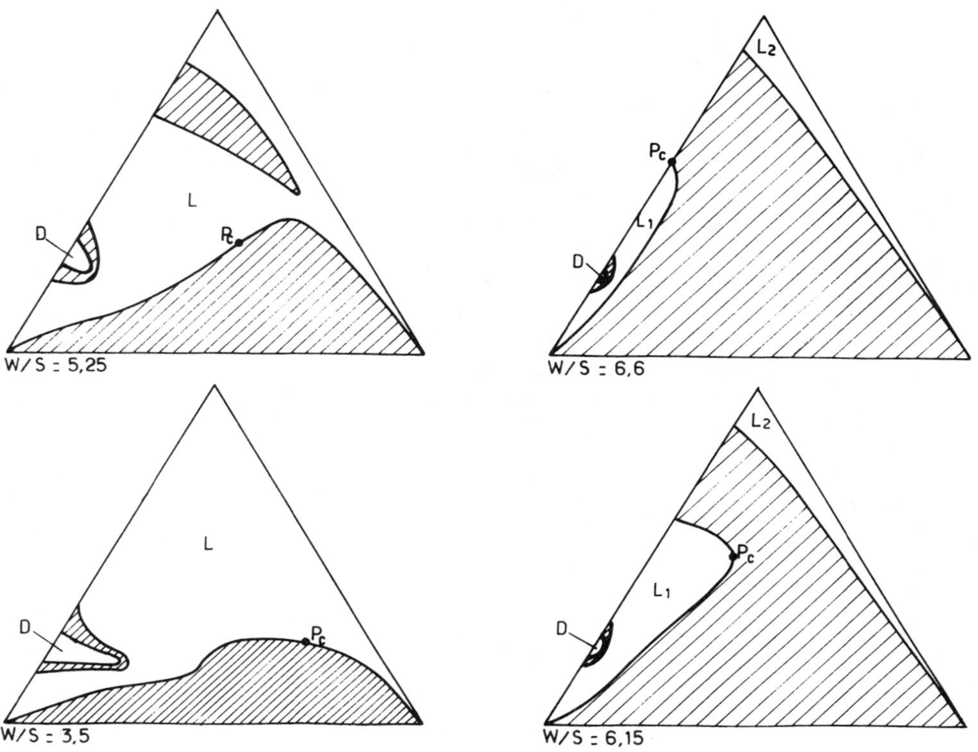

Fig. 13. Boundaries of the isotropic L and lamellar D regions in the sections of the quaternary phase diagram defined by X above 3. The hatched domains are multiphase regions.

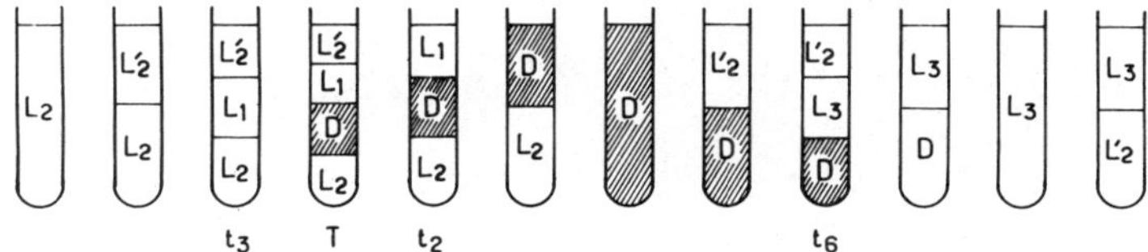

Fig. 14. Pattern of phase behavior observed in the section X = 1.034 along the line AA' drawn in Figure 8b.

Table 1.

	L'_2	L_1	D	L_2	Unc.
Water	2.0	3.4	4.6	6.9	± 0.4
Pentanol	10.7	10.9	10.9	12.5	± 0.5
Dodecane	85.1	82.8	80.4	74.5	± 1
S.D.S.	2.2	3.0	4.1	6.1	± 0.4

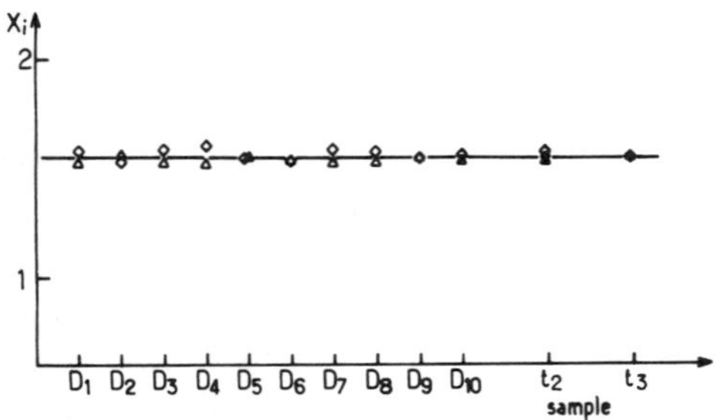

Fig. 15. Value of the water over surfactant ratio (X_i) measured in each
phase (i) for ten two-phase equilibria (termed D_1–D_{10}) lying along
the line BB' in region d_1 and for two three-phase equilibria of
the regions t_2 and t_3 (Figure 10b). For all these samples, the
global value of X is equal to 1.55 (T = 21°C).

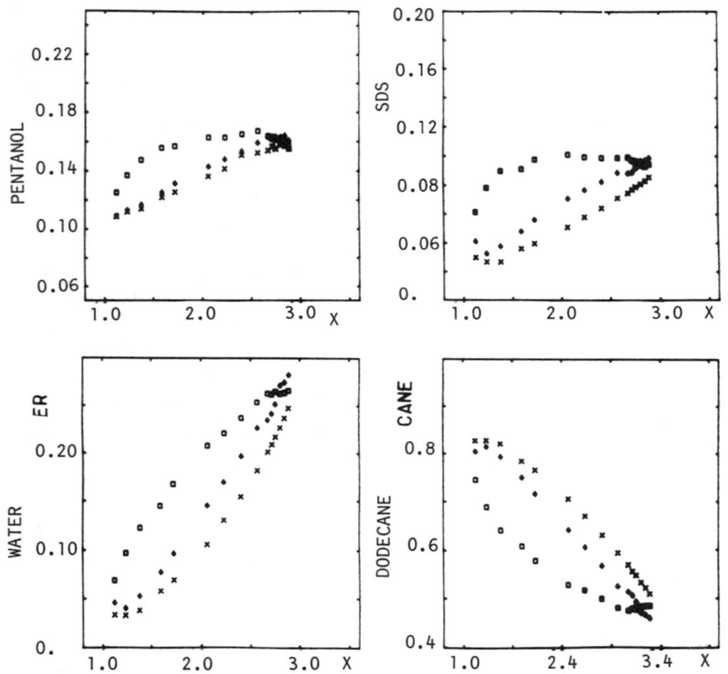

Fig. 16. Variation as a function of X of the water-pentanol-SDS and
dodecane concentrations in the three coexisting phases of several
t_2 equilibria (T = 21°C). x , ◇ and □ correspond respectively to
the upper phase (L_1), the middle phase (D) and the lower phase
(L_2).

In the mixtures investigated, X varies between 2 and 5.6, the global percentage of oil between 71 and 40.35, the global percentage of water between
8.17 and 21.65 and the global percentage of pentenol between 16.89 and
34.14. The ratio X_i measured in each phase is reported versus the value X
of the overall mixture (Figure 19). For X < 4 the conservation of X in
each phase is within the experimental accuracy. For X > 4, the volume of
the upper phase of the samples investigated is very small in comparison to

174

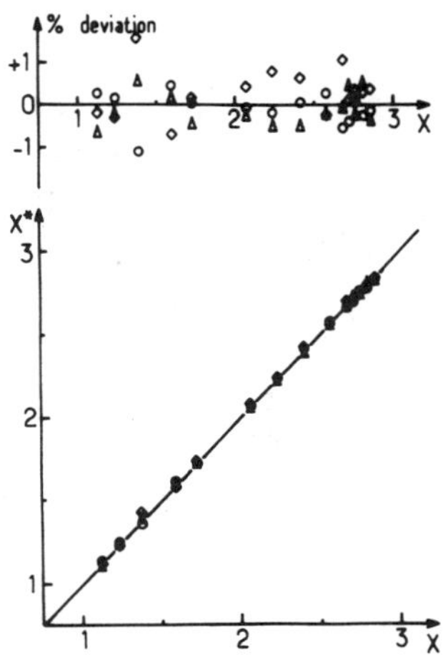

Fig. 17. Value of the water over surfactant ratio (X*) measured in each
phase of the three coexisting phases of the equilibria t_2 as a
function of the value X of the global mixture. The standard
deviation is also reported (T = 21°C). $\diamond$, $\triangle$ and $\square$ respectively
represent the value of X in the lower, middle and upper phases.

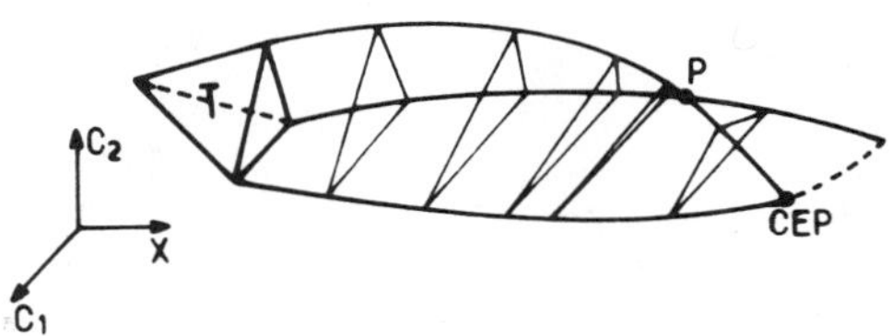

Fig. 18. Schematic representations of the three-phase volume t_2 in the C_1,
C_2 - X space. C_1 and C_2 correspond respectively to the alcohol
and oil percentages. The volume t_2 is limited at low X by the
four-phase tetrahedron T and at high X by the critical end point
CEP. At the point P, the lamellar phase and the isotropic L_2
phase have the same composition (azeotropic point).

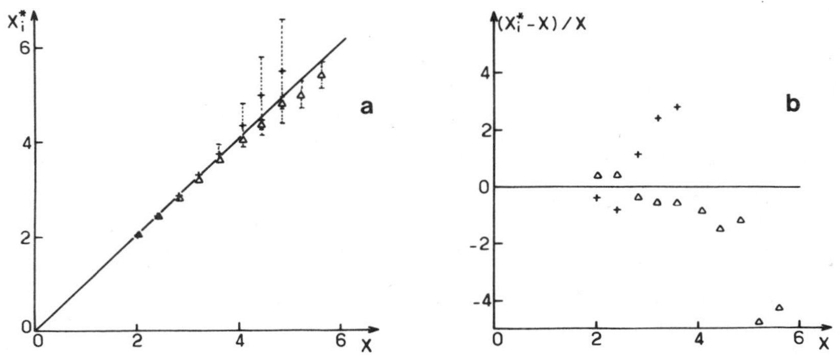

Fig. 19. (a) Value of the water over surfactant ratio (X*) measured in both
coexisting phases of the equilibria d_1 observed in various X
sections as a function of the value X of the global mixture (T =
21°C). (b) Deviation between X_i^* and X.

175

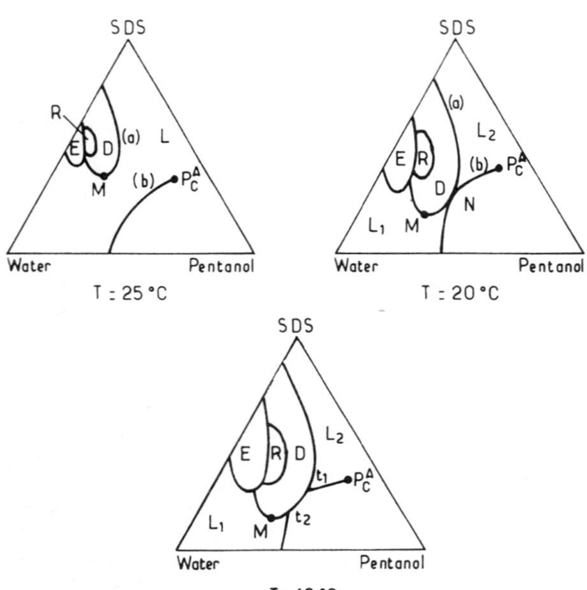

Fig. 20. Schematic phase diagrams of the water-pentanol-SDS system at 18°C, 20°C and 25°C in the field space consisting of the chemical potentials of the components. This figure is to be compared with Figures 1 and 2.

that of the lower phase; besides, the upper phase contains very small amounts of water and surfactant. These two effects show that our data do not have the precision required to conclude whether X behaves as a field variable when its value is above 4.

In summary, phase analysis of a large number of two- and three- phase equilibria which involve both isotropic and mesomorphic phases has provided evidence that the X ratio is field-like in the oil-rich mixtures.

REPRESENTATION OF THE PHASE DIAGRAMS IN A FIELD SPACE

As pointed out by Griffiths and Wheeler[16], one way to simplify the representations of the phase diagrams of multicomponent mixtures is to use field variables. The replacement of a density by a field variable reduces the dimensionality of the coexistence phase regions. Because fields take the same value in the coexisting phases, multiphase regions are reduced to the common boundaries of the regions of existence of the concerned phases. In a field representation, the dimensionality of each region is equal to its number of degrees of freedom. Besides the diagram of a section obtained by holding n field variables constant displays the same qualitative features as a system with n less degrees of freedom. Consequently the phase diagram of a quaternary mixture obtained at fixed temperature, pressure and constant value of the chemical potential of one of the components has features similar to those of a ternary mixture at fixed P and T.

One of the essential consequence of the simplification of the appearance of the diagrams is to make easier the understanding of the complicate evolution of phase equilibria occurring in multicomponent systems. A convenient choice of fields might be temperature, pressure and the chemical

176

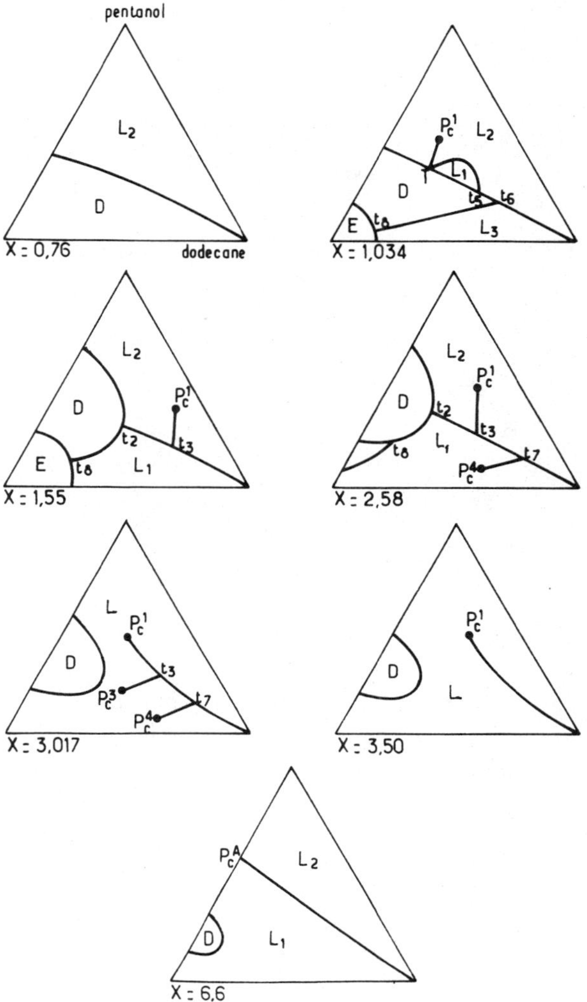

Fig. 21. Schematic representations in a field space of the diagrams shown
in Figures 7-13. The field variables consist of X and the
chemical potentials of pentanol and dodecane.

potentials of the components. The measurement of the latter requires
additional difficult investigations. However in most cases it is not
necessary to draw quantitatively the phase diagram. A schematic representation is often sufficient and can serve as a guide to follow the evolution
of phase equilibria. The experimental observation that in the quaternary
mixtures water-dodecane-SDS-pentanol, the water to surfactant ratio X
behaves as a field is of fundamental importance for the transcription of
experimental density diagrams in a field space. As mentioned earlier, in
this last representation the X sections should present the same features as
a ternary system at constant pressure and temperature. Therefore as an
example we will first focus attention on the phase diagrams of the ternary
water-pentanol-SDS system experimentally found at 25°C, 20°C and 18°C.
These diagrams shown in Figure 20 contain some illustrative features occurring in the quaternary mixture. Each diagram obtained at constant P and T
is drawn schematically in a triangle. In this representation, phases are

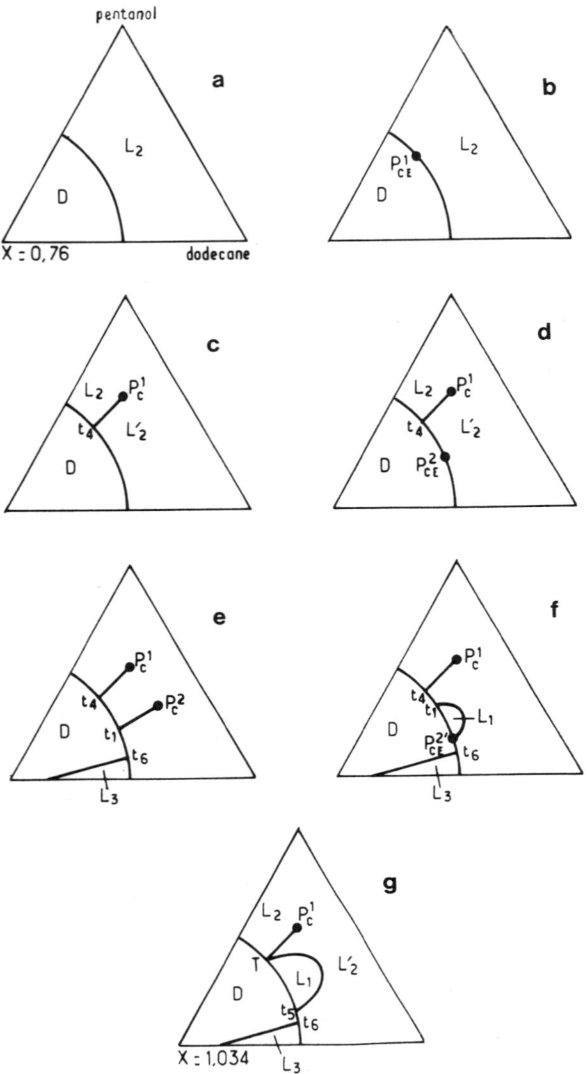

Fig. 22. Schematic representations in a field space of two sequences of
diagrams which account for the development of phase equilibria
between X = 0.76 and X = 1.034. In sequence (a) (Fig. 22a) three
critical end points are involved. In sequence (b) (Fig. 22b) the
creation of the phase L_1 results from a bitriple point (diagram
d). In each triangle, the field variables are X (lower left
corner), the chemical potentials of dodecane (lower right corner)
and pentanol (upper corner).

represented by surfaces, two-phase coexistence by lines and three-phase
coexistence by points. For the three temperatures, the coexistence line
(a) between region L and D shows a minimum which corresponds to an azeotropic like point (point M) and the coexistence line (b) terminates at a
critical point (P_C^A). As temperature decreases the coexistences lines (a)
and (b) move closer together until they become tangential at the point N (T
= 20°C). As temperature is further decreased, the lamellar phase inter-
sects the coexistence line (a). This intersection gives rise to the two-

178

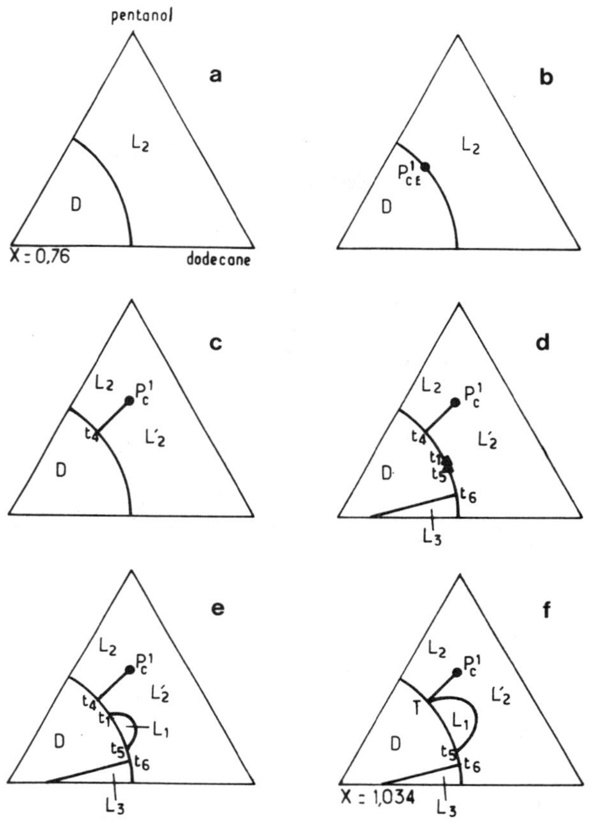

Fig. 22 (b)

three phase regions t_1 and t_2 and also to the two single phase regions L_1 and L_2. The quadruple point N represents the indifferent state. At this point the regions L_1 and L_2 coalesce and the three-phase regions t_1 and t_2 are degenerated.

On Figure 21 are drawn in a field space the diagrams of the X sections of the quaternary mixture experimentally investigated. In the following we present an overall view of the phase diagram showing its qualitative behavior in the entire space considered. For that, we have attempted to imagine a simple gradual conversion of the equilibria as X varies between zero and infinite consistent with the experimental diagrams. In Figures 21-23 triple points are denoted by t, critical points by P^i_C and critical end points by P^i_{CE}.

The increase of X from 0.76 to 1.034 produces strong changes in the diagram. The two different sequences of diagrams shown in Figure 22 yield the correct qualitative features of the diagram X = 1.034. In the first one, the creation of the island of the phase L_1 results from the existence of a line of critical points P^2_C ended by two critical end points P^2_{CE} and $P^{2'}_{CE}$. In the second scheme of evolution the creation of the phase L_1 in the interior of the triangle is explained by the occurrence of an indifferent state involving the equilibria t_1 and t_5. Both solutions have in common some salient features. In particular, the line of critical points P^1_C and the critical end point P^1_{CE} occur in both sequences. They also contain a point of four-phase coexistence labeled T. This point appears at the

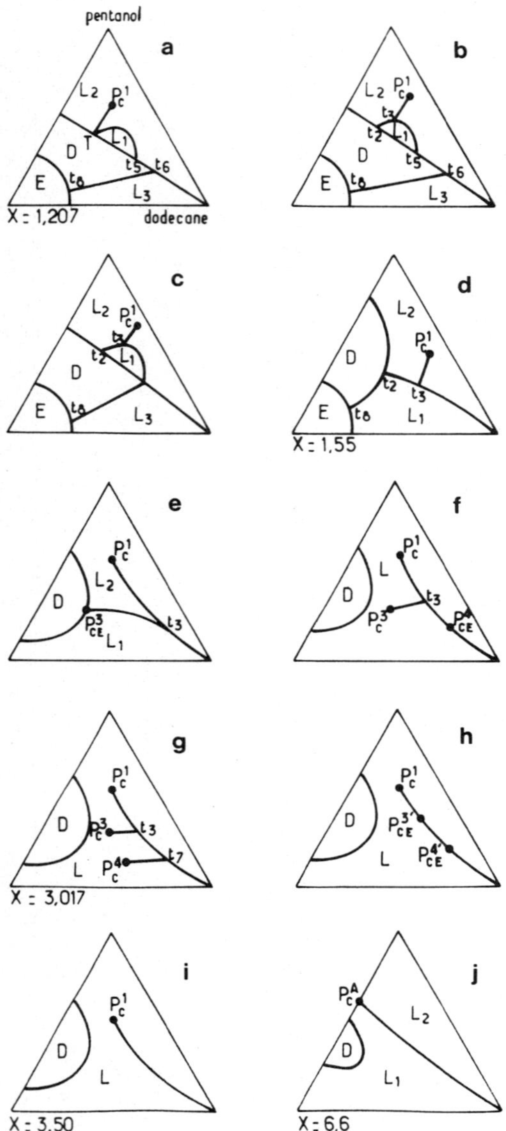

Fig. 23. Schematic representation of a sequence of diagrams which accounts
for the development of phase-equilibria between X = 1.207 and X =
6.6. The field variables are the same as in Figure 22.

merging of the four triple points t_1, t_2, t_3, t_4. Correlation length
measurements in the phase L_2 of the equilibrium t_1 support the first
sequence of diagrams[18].

The essential feature found between the planes X = 1.207 and 1.552 is
the connectivity of the phases L_1 and L_3. In the proposed development of
equilibria (Figure 23), as X increases, the boundary between the phases D
and L_2^1 contracts until the two triple points t_5 and t_6 coalesce into a
quadruple point where the phases L_1 and L_3 merge.

180

Between the planes X = 1.55 and X = 3.017, the coalescence of the phases L_1 and L_2 is achieved by the shrinking of the three-phase region t_2 at the critical end point P_{CE}^3 (Figure 23) the critical line P_C^3 appearing in this last point goes to the critical end point $P_{CE}^{3'}$ where the line of triple points t_3 ends. The last line of triple points t_7 might extend between the two critical end points P_{CE}^4 and $P_{CE}^{4'}$.

In conclusion, two descriptions of the overall diagram of the quaternary mixture water-dodecane-pentanol-SDS at 21°C are proposed. Each of them includes a four-phase region and eight lines of triple points. Figure 24 summarizes the evolution of the lines of triple points and of the lines of critical points as a function of X. In the first model four lines of critical points and seven critical end points occur. In the second model, only three lines of critical points and five critical end points are required to explain the development of equilibria. Let us note that both descriptions imply the occurrence of degenerate triple points. Such points have been found to exist in the ternary mixture water-SDS-pentanol. Their existence has also been suggested by T. Lang[14,23] to interpret the formation of the anomalous phase observed in the diagrams of certain water non ionic surfactant mixtures.

Most of the features predicted in the above scheme have been observed. The three-phase equilibria experimentally observed are given in Table 2. Apart the equilibria t_4 and t_8, all the three-phase equilibria have been observed. Concerning critical points, only the line P_C^1 and the critical end points P_{CE}^1 and P_{CE}^3 have been accurately located. The other lines have still to be researched but it is likely that their extent in the compositional diagram is very short. The choices of phase equilibria that we propose, in particular between the sections X=0.76 and X=1.034 are not the only possibilities which exist, but they are likely one of the simplest. One might expect that further accurate investigations of certain sections ranging between X = 0.9 and 1.034 should allow to identify features anticipated in the various possible progressions of equilibria and to distinguish the correct one. However, such a study in the density space presents practical difficulties related to the large number of equilibria expected, and also to the very small extension of most of the multiphase regions found in this part of the diagram.

CONCLUDING REMARKS ON THE PHASE DIAGRAM

The results presented above give for the first time a global picture of the phase diagram of a quaternary mixture. In addition to the very wide microemulsion region, we have identified two mesomorphic phases in the diagram of the quaternary mixture water-dodecane-pentanol-SDS. These phases appear as an extension of those found for the ternary mixture water-pentanol-SDS. The lamellar phase D possesses the exceptional property to dissolve considerable amounts of alcohol and oil up to 98%. A structural study of similar phases has revealed the existence of a large number of structural defects[24]. In certain regions of the phase diagram the microemulsion present particular characteristics; then some rich in oil mixtures are flow birefringent. The phases are connected by a complex multiphase region where two, three, or four phases coexist. These equilibria are different from the so-called Winsor equilibria. It is worth mentioning that the three-phase regions observed in these mixtures do not necessarily emerge from critical end points[26,27]. Indeed our data clearly show that in the ternary system water-pentanol-SDS, the formation of the three-phase regions t_1 and t_2 arises from an indifferent state. Likewise in the four-component system, it is likely that the merging of the phases L_1 and L_3 and the disappearance of the regions t_5 and t_6 as X increases result from an indifferent state. Recent results suggest that the occurrence of such

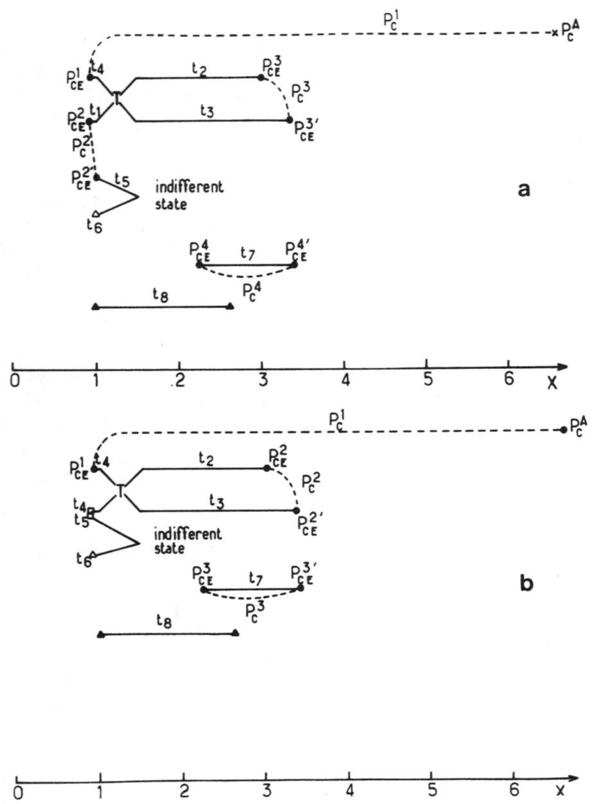

Fig. 24. Evolution of the lines of triple points (t_1-t_8) and of the lines of critical points P_c as a function of X. Model (a) and (b) correspond respectively to the sequences (a) and (b) given in Figures 22a and 22b.

Table 2.

equilibria	X = 1.034	1.207	1.37	1.55	2.558	3.017	3.5
T — $L_2'\,L_1\,D\,L_2$	x	x	—	—	—	—	—
t_1 — $L_2'\,L_1\,D$	x	x	x	—	—	—	—
t_2 — $L_1\,D\,L_2$	x	x	x	x	x	—	—
t_3 — $L_2'\,D\,L_2$	x	x	x	x	x	x	—
t_4 — $L_2'\,D\,L_2$	—	—	—	—	—	—	—
t_5 — $L_2'\,L_1\,D$	x	x	x	—	—	—	—
t_6 — $L_2'\,L_3\,D$	x	x	x	—	—	—	—
t_7 — $L\,L\,L$	—	—	—	—	x	x	—
t_8 — $D\,E\,L_1$	—	—	—	—	—	—	—

x : equilibria observed   —— : equilibria not observed

states is quite frequent in microemulsion systems[28]. Very significant
changes in the phase equilibria occur as pentanol is replaced by
hexanol[29]. In particular the line of critical points P_C^1 disappears.
This effect of alcohol may be simply understood as a decrease of attractive
interactions between micelles. Indeed recent experimental studies have
shown the importance of interactions between water in oil micelles on the
stability of oil rich microemulsions[29].

One of the prominent feature common to the ternary and quaternary
systems studied above is the existence of field variables related in a very
simple way to concentrations. In the quaternary mixture, the water to
surfactant ratio (X) behaves as a chemical potential in the oil rich part
of the phase diagram. The relation between the ratio X and the variable Y
evidenced in the ternary mixture is not straightforward. However we may
remark that another way to express the variable Y is to use the water
concentration (X_w). We may define the variable Z, strictly equivalent
to Y.

$$Z(X_a, X_s, X_w) = \frac{X_w + (1 - A) \cdot X_a}{X_x + AX_a}$$

$Z = 1/Y-1$. This new expression of the field emphasizes the role of
the water over surfactant ratio. In microemulsion systems it is possible
to relate the field character of this ratio to a characteristic length.
Indeed, the microscopic structure of oily microemulsions can be described
as a dispersion of water droplets in an oily continuous phase; the water
over surfactant ratio fixes the size of the particles[27]. In the ternary
mixture studied here, the structure is probably more complex. Besides the
partitioning of the alcohol between the continuous phase, the interfacial
film and the water region could be the origin of the introduction of the
parameter A.

It is worth mentioning that the complete description of the quaternary
diagram that we proposed, which accounts for the continuity of the multiphase regions, has only be possible because the field variable X has been
evidenced. Another major consequence of the field character of X has been
to permit the study of the critical behavior of several critical points at
fixed temperature by using the variables X or Y as fields[21,30,31].

<u>Acknowledgements</u>

It is a pleasure to acknowledge M. Maugey and O. Babagbeto for their
technical assistance.

REFERENCES

1. P. Ekwall, in Advances in Liquid Crystals, Vol. 1, G.H. Brown, Ed.
    Acad. Press New York, 1975, p. 1.
2. P. A. Winsor, <u>Chem. Rev.</u>, <u>68</u>, 1 (1968).
3. K. Fontell, in Liquid Crystals and Plastic Crystals, Vol. 2, G. W.
    Gray, and P. A. Winsor, Eds. Ellis Howood, Chichester, 1984, p.80.
4. G. J. T., Tiddy, <u>Physics Reports</u>, <u>57</u>, 1 (1980).
5. G. J. T. Tiddy and M. F. Walsh, in Aggregation Process in Solution,
    Vol. 28, E. Wyn Jones and J. Gornally, Ed. Elsevier Sci. Pub. Co.,
    1983, p. 151.
6. J. Charvolin, in Colloides et Interfaces, Editions de Physique, 1983,
    p. 33.
7. G. Gillberg, H. Lehtinen, and S. Friberg, <u>J. Coll. Interface Sci.</u>, 33,
    40 (1970).

8. S. E. Ahmad, K. Shinoda, S. Friberg, J. Coll. Interface Sci., 47, 32 (1974).

9. S. Friberg and I. Burasczenska in Micellization, Solubilization and Microemulsions, Vol. 2, K. L. Mittal, Plenum Press, 1977, 791.

10. R. N. Healy, R. L. Reed, D. G. Stenmark, S.P.E. Journal, 16, 147 (1976).

11. "Improved oil recovery by surfactant and polymer flooding," Eds. D. O. Shah and R. S. Schechter, Acad. Press, N.Y., 1977.

12. A. M. Bellocq, J. Biais, P. Bothorel, B. Clin, G. Fourche, P. Lalanne, B. Lemaire, B. Lemanceau, D. Roux. Adv. Coll. Interface Sci., 20, 167 (1984).

13. A. M. Bellocq, D. Bourbon, and B. Lemanceau, L. Coll. Int. Sci., 79, 419 (1981).

14. J. C. Lang, in Physics of Amphiphiles, Micelles, Vesicles and Micro-emulsions, V. Degiorgio and M. Corti, Ed. North Holland, Amsterdam, 1985, p. 336.

15. M. Kahlweit, E. Lessner and R. Strey, J. Phys. Chem., 89, 163 (1985).

16. R. B. Griffiths and J. C. Wheeler, Phys. Rev. A2, 1047 (1970).

17. D. Roux and A. M. Bellocq, Phys. Rev. Lett., 52, 1895 (1984).

18. D. Roux and A. M. Bellocq, in Surfactants in Solution, K. L. Mittal, P. Bothorel, Eds. Plenum Press, New York, (1987), 6, p. 1247.

19. A. M. Bellocq and D. Roux, in Microemulsions, Ed. by S. Friberg, and P. Bothorel, CRC Press, in press.

20. I. Prigogine, and R. Defay, Thermodynamique Chimique, Liége, 1950.

21. J. A. Zollweg, J. Chem. Phys., 55, 1430 (1971).

22. D. Roux and A. M. Bellocq, in Statistical Thermodynamics of Micelles and Microemulsions, Ed. by S. H. Chen, Springer-Verlag, in press.

23. J. C. Lang and R. D. Morgan, J. Chem. Phys., 73, 5849 (1980).

24. M. Dvolaitzky, J. M. Dimiglio, R. Ober and C. Taupin, J. Phys. Lett., 44, L229 (1983).

25. M. Kahlweit, J. Coll. Interface Sci., 90, 197 (1982).

26. P. K. Kilpatrick, L. E. Scriven, and H. T. Davis, SPE J. 11209 (1982).

27. D. Roux, A. M. Bellocq and P. Bothorel, in Surfactants in Solution, Ed. by K. L. Mittal and B. Lindman, Plenum Press, 3, 1984, 1843.

28. G. Guerin, Thesis Bordeaux 1985.

29. D. Roux and A. M. Bellocq, in Physics of Amphiphiles. Micelles, Vesicles and Microemulsions, V. Degiorgio and M. Corti, Ed. North Holland, Amsterdam, 1985, p. 842.

30. A. M. Bellocq, P. Honorat and D. Roux, in Surfactants in Solution, K. L. Mittal, P. Bothorel, Eds. Plenum Press, New York, (1987), 6, p. 126.

31. A. M. Bellocq, P. Honorat and D. Roux, J. Phys., 46, 743 (1985).

STATIC SMALL ANGLE NEUTRON SCATTERING ON CONCENTRATED WATER-

IN-OIL MICROEMULSION : THE PARTICULAR CASE OF POTASSIUM OLEATE

E. Caponetti and L. J. Magid*

Istituto di Chimica Fisica, 26 V. Archirafi
90123 Palermo, Italy
*Department of Chemistry, University of Tennessee
Knoxville, Tennessee 37996-1600, USA

1. INTRODUCTION

It is well-known that surfactants can solubilize hydrocarbons in aqueous phases and water in hydrocarbon phases. When a cosurfactant, commonly an alcohol, is added, the solubilization can become much greater. In the presence of surfactants and alcohols, stable, clear dispersion of hydrocarbons in water called microemulsions may be formed. These microemulsions can be either of an oil-in-water (o/w) or water-in-oil (w/o) category, just as the unstable, coarser emulsions can be. Under certain conditions, bicontinuous arrangements of oil, amphiphile, and water regions have been proposed. Many techniques have been used to get information on these kinds of systems.

In this lecture, we will present the application of the Small Angle Neutron Scattering (SANS) technique coupled with isotopic substitution on a concentrated w/o microemulsion: potassium oleate/hexadecane/water/1-pentanol or 1-hexanol. Section 2 summarizes the results of other studies on the same microemulsion made using different techniques. Section 3 gives a brief description of a static SANS apparatus. Section 4 presents a short review of basic neutron scattering theory. Section 5 elucidates what sort of information can be obtained from SANS measurements on colloidal dispersions and presents a comparison between SANS and other scattering techniques such as Light Scattering (LS) and Small Angle X-ray Scattering (SAXS). Section 6 describes some SANS studies accomplished for several microemulsions along with the models for particle structures and interactions used to fit the experimental results. Section 7 presents our experimental data on the potassium oleate microemulsion and describes the procedure used to fit the experimental data and our conclusions.

2. POTASSIUM OLEATE/HEXADECANE/WATER/1-PENTANOL OR 1-HEXANOL MICROEMULSION

The potassium oleate/hexadecane/water/hexanol or pentanol microemulsion has been studied by several authors.

D. O. Shah et al.[1-3] using several different techniques such as NMR, electrical conductivity, viscosimetry and turbidimetry, investigated

the system comprised of five ml of hexadecane and two ml of hexanol or
pentanol per gram of potassium oleate, as a function of added water.
They concluded that when hexanol is present in the system, in going
from a water-in-oil microemulsion to an oil-in-water biphasic
dispersion, there is a transition from spheres to cylinders to
lamellae. In contrast to this behavior, in the presence of pentanol,
they concluded that amphiphiles and water were molecularly dispersed
(from 0.1 to 0.6 volume ratio of water to oil).

E. Sjoblom and S. T. Friberg[4] investigated the water-in-oil domain
of the potassium oleate system containing pentanol and decane instead of
hexadecane, and concluded that inverse micelle formation began when the
mole ratio of water to oleate exceeded eight; A. M. Bellocq and G. F.
Fourche[5] reached a similar conclusion for the hexanol/dodecane pair.

A. Boussaha et al.[6] interpreted positron annihilation rates as
indicating that aggregation occurs at a 0.5 water/oil volume ratio in the
presence of pentanol and at a 0.2 ratio in the presence of hexanol for
hexadecane as the hydrocarbon.

S. S. Atik and J. K. Thomas[7] found differences in quenching by
potassium ferricyanide of fluorescence in hexanol- and pentanol-containing
microemulsions. Their results for the pentanol system with the same
components Shah used (water oil volume ratio > 0.2) could have been
interpreted to indicate a molecularly dispersed system. However, on the
basis of other measurements in which the hydrocarbon was dodecane rather
than hexadecane and in which the pentanol-to-surfactant ratio was varied
to give a single phase, they concluded that water pools exist under all
the conditions they investigated.

C. Boned et al.[8] measured conductivity and permittivity over
several compositional paths of the w/o region of this system with hexanol,
the surfactant/alcohol ratio being the same as used by Shah et al. They
divided the apparently isotropic region into three subregions: a
dispersion of hydrated soap aggregates, inverse micelles, and coalesced or
clustering micelles.

The differences in properties of visually clear microemulsions
arising from a difference of one carbon in the alcohol is interesting, as
are the sources of the major changes in properties with water content.
SANS can provide information on structural changes accompanying these
differences. The size of the aggregates is in the range accessible to the
technique and the differences in scattering by hydrogen and deuterium
allow inferences concerning the location of protiated and deuterated
components.

3.   SANS EXPERIMENTAL APPARATUS

Figure 1 depicts the 30-m (source to detector) SANS spectrometer of
the National Center for Small-Angle Scattering Research (NCSASR) located
at the Oak Ridge National Laboratory (ORNL), USA, where our scattering
experiments on the potassium oleate microemulsions were carried out.

A thermal neutron beam emerges from the core of the High Flux Isotope
Reactor (HFIR), passing first through a beryllium filter and then through
two sets of graphite monochromator crystals which may be set to select a
wavelength either 4.75 Å or 2.37 Å with a spread, $\Delta\lambda/\lambda$, of 6%. The
neutron beam goes through an evacuated beam flight path of 7.6 m
before impinging upon the sample.

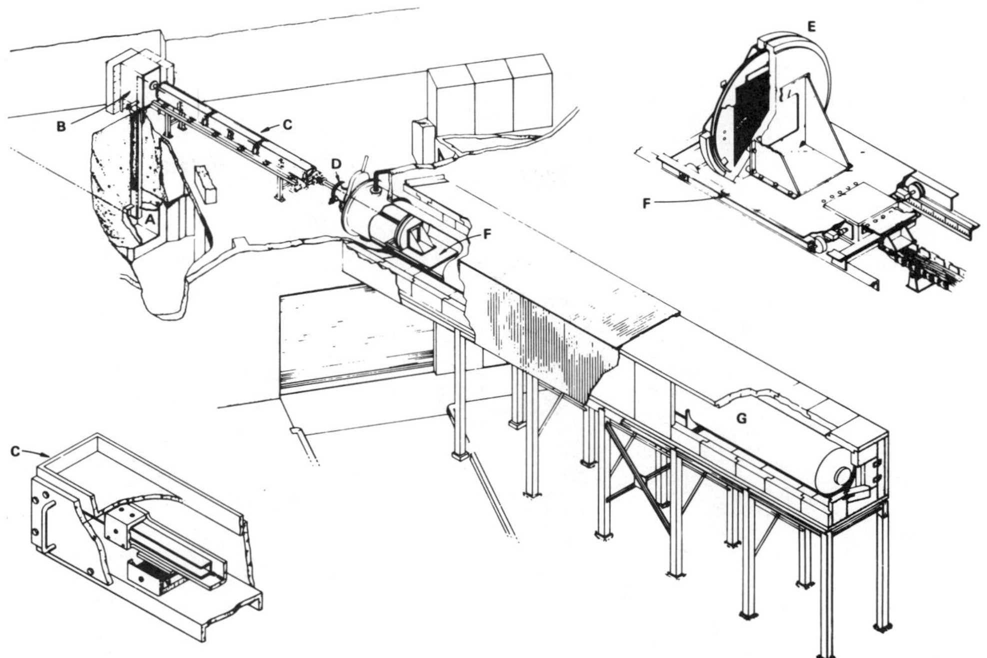

Fig. 1. Thirty meter SANS spectrometer at the National Center for Small-
Angle Scattering Research, Oak Ridge National Laboratory.
A: lower monochromator housing; B: upper monochromator housing;
C: collimator and neutron beam guide; D: sample chamber;
E: 112 cm diameter two-dimensional position-sensitive detector;
F: detector carriage; G: 56 cm gate valve.

Solid samples may be mounted directly in the evacuated sample chamber
on a 12-position sample wheel whose motor is under computer control.
Solutions are typically studied in quartz spectrophotometric cells having
path-lengths of 1 or 2 mm; these are mounted on a six sample position
motor-driven slide.

Figure 2 shows a schematic diagram of the scattering experiment. The
scattered neutron intensity, I_S, is detected for many scattering angles at
once by a helium-filled Kopp-Borokowski two-dimensional multidetector
having an active area of 64 x 64 cm^2. The unscattered beam is intercepted
by a cadmium beamstop. The sample-to-detector distance can be varied from
1.4 to 19 m so that the total accessible Q range is 0.002 - 0.3 Å^{-1}.
Figure 3 depicts a typical intensity map (contour plot) for an ionic
micellar solution.

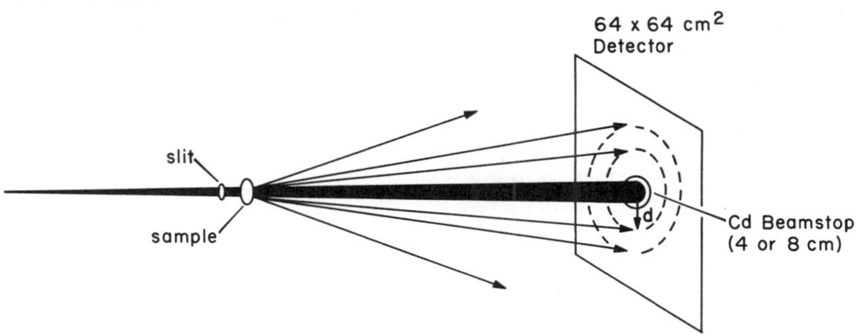

Fig. 2. A schematic diagram of the scattering experiment.

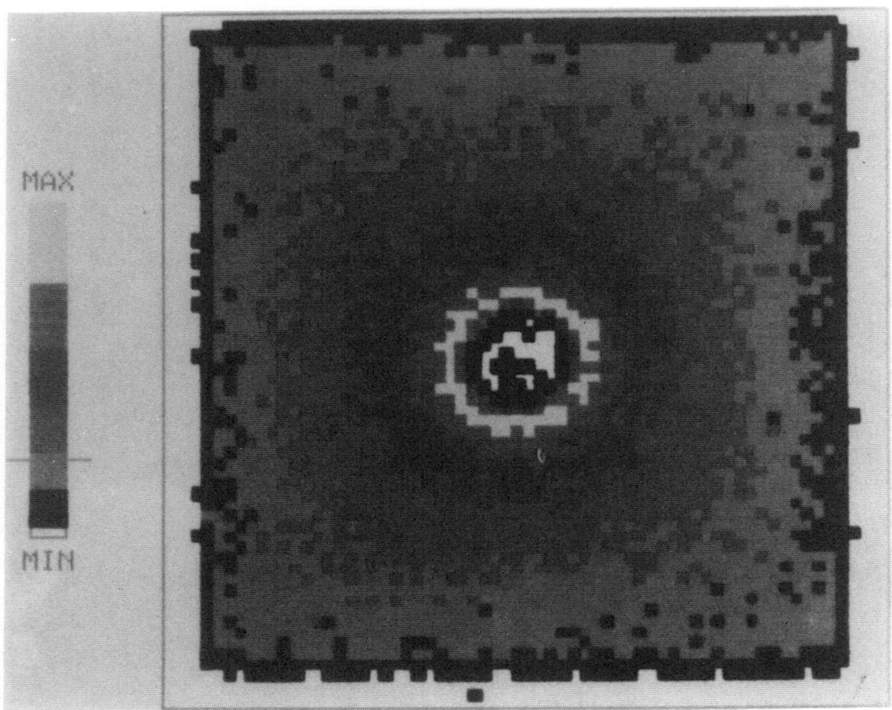

Fig. 3. Contour plots and one-dimensional radial average for a D_2O solution
of 0.01 M octylhexadecyldimethylammonium bromide in 0.02 M NaCl
at 45°C. The radial average depicted involves overlap of data
taken at two different sample-to-detector distances.

     The total scattered intensity for a solution includes contributions
from the background, the empty cell and the solvent, in addition to the
solute. Programs developed by the NCSASR staff are used to correct first
the raw data (both solution and solvent) for background, empty cell
scattering and transmission coefficient. The corrected scattered
intensities are converted into absolute differential cross sections by
using a standard scatterer to calibrate the detector. The one-dimensional
radial average is the conventional scattering curve (intensities vs.
modulus of the wave vector) generally presented in the literature.

     Programs developed by the NCSASR staff are also available to subtract
the computed incoherent scattering level for the solution and the
solvent's contribution to the coherent scattered intensity, leaving the
coherent scattering cross section. All of the data-correction routines at
the NCSASR also take into account variations in sensitivity among the
detector elements.

4. SANS THEORY

     The scattering of thermal neutrons by the atoms in a liquid sample
arises essentially from interaction with the nuclei[9]. Furthermore, we
are concerned with quasi-elastic scattering, in which the energy ($E_i - E_f$)
transferred in the scattering process is small compared to the incident
energy (E_i). The momentum transferred in the scattering event is $\underline{Q} = \underline{k}_i - \underline{k}_f$; $|\underline{Q}| = (4\pi/\lambda) \sin\theta$ (2θ being the scattering angle and λ the neutron
wavelength).

188

The scattered waves interfere. The superposition of the coherent scattered waves is a function of nuclear distribution in the sample and hence contains some information about intraparticle and interparticle correlations. Interferences associated with the incoherent cross section are averaged and give rise to an isotropic scattering, which generates a background contribution.

For coherent scattering, a nucleus' scattering cross section, σ_{coh}, is given by $4\pi|b_{coh}|^2$, where b_{coh} is the scattering length. In what follows, we use b to represent b_{coh}.

The partial coherent cross section is defined as

$$\left(\frac{d\sigma}{d\Omega}\right)_{coh} = \left\langle \sum_{i,j} b_i b_j \exp[i\underline{Q}\cdot(\underline{r}_i - \underline{r}_j)]\right\rangle, \tag{1}$$

which is a sum of amplitude-weighted phase shifts over all atoms in the sample; $\underline{r}_i$ and $\underline{r}_j$ are nuclei positions relative to an arbitrary origin in the sample. We now rewrite (1) as

$$\left(\frac{d\sigma}{d\Omega}\right)_{coh} = \left\langle \sum_{p,q} \sum_{i(p)}^{n_p} \sum_{j(q)}^{n_p} b_i b_j \exp(i\underline{Q}\cdot\underline{r}_{i,j})\right\rangle, \tag{2}$$

where $\underline{r}_{i,j} = \underline{R}_j - \underline{R}_i$ is the relative separations of two atoms i and j that may be within the same particle or in two different particles (whose centers of mass are at R_p and R_q respectively).

Equation (2) can be separated into intraparticle ($p = q$) and interparticle terms ($p \neq q$), giving

$$\left(\frac{d\sigma}{d\Omega}\right)_{coh} = \left\langle \sum_{i(p)}^{n_p} \sum_{j(p)}^{n_p} b_i b_j \exp[i\underline{Q}\cdot(\underline{r}_j - \underline{r}_i)]\right\rangle + \left\langle \sum_{p\neq q=1}^{N} \times \right.$$
$$\left. \times \exp[i\underline{Q}\cdot(\underline{R}_p - \underline{R}_q)] \sum_{i(p)}^{n_p} \sum_{j(q)}^{n_q} b_i b_j \exp[i\underline{Q}\cdot(\underline{r}_j - \underline{r}_i)]\right\rangle. \tag{3}$$

Here $\underline{R}_p - \underline{R}_q$ is the center-to-center distance for a pair p,q of particles; N is the number of particles. The single-particle form factor is defined as

$$F(Q) = \sum_{j(q)} b_j \exp(i\underline{Q}\cdot\underline{r}_j), \tag{4}$$

with the sum being over all atoms in the particle.

The intraparticle term in Eq. (3) is just N times the square of the single particle form factor averaged over all particle orientations relative to Q, $(N\langle|F(\underline{Q})|^2\rangle_Q)$. The quantity $\langle|F(q)|^2\rangle$ is the Fourier transform of the distribution of distances $P(r)$ within one particle.

The evaluation of second term of Eq. (3) involves the knowledge of both the position and the orientational correlation between the scattering particles. Recently, Vrij[57] has evaluated numerically the above mentioned term for a Percus Yevick fluid.

Only very recently[69] the analytical solution of Eq. (3) for uncharged polydisperse spheres with Schulz distributed diameter has been obtained in the Mean Spherical Approximation (MSA)[22].

For an arbitrary mixture of particles the formidable problems encountered in giving explicit analytical expression for Eq. (3) can be by-passed if one assumes that there is little correlation between particle size and orientation and interparticle distance. In this last case, using the decoupling approximation, the average of the product (second term in Eq. (3)) may be set equal to the product of the two terms averaged separately; Eq. (3) becomes

$$
\left(\frac{d\sigma}{d\Omega}\right) = N\langle|F(Q)|^2\rangle_Q + \langle\sum_{p\neq q}\exp[iQ\cdot(\underline{R}_p - \underline{R}_q)]\rangle \cdot
$$
$$
\cdot \langle\sum_{i(p)}^{n_p}\sum_{j(q)}^{n_q} b_i b_j\exp[iQ\cdot(\underline{r}_j - \underline{r}_i)]\rangle .
\tag{5}
$$

For charged particles, the decoupling approximation works well, since contact among neighboring particles is unlikely; however, it is particularly poor in the case of uncharged polydisperse hard spheres.

The average structure factor, $S(Q)$, for the correlations between the particles, averaged over all orientations and positions, is

$$
S(Q) = N^{-1}\langle\sum_{p,q}\exp[i\underline{Q}\cdot(\underline{R}_p - \underline{R}_q)]\rangle = 1 + N^{-1}\langle\sum_{p\neq q}\exp[i\underline{Q}\cdot(\underline{R}_p - \underline{R}_q)]\rangle ,
\tag{6}
$$

so that

$$
\langle\sum_{p\neq q}\exp[i\underline{Q}\cdot(\underline{R}_p - \underline{R}_q)]\rangle = N[S(Q) - 1].
\tag{7}
$$

Provided that $\underline{r}_{i(p)}$ and $\underline{r}_{j(q)}$ are statistically independent,

$$
\langle\sum_{i(p)}^{n_p}\sum_{j(q)}^{n_q} b_i b_j\exp[iQ\cdot(\underline{r}_i - \underline{r}_j)]\rangle = \langle\sum_{j(q)}^{n_q} b_j\exp(i\underline{Q}\cdot\underline{r}_j)\rangle^2_Q =
$$
$$
= |\langle F(\underline{Q})\rangle|^2_Q.
\tag{8}
$$

Using Relation (7) and (8), Eq. (5) becomes

$$
\left(\frac{d\sigma}{d\Omega}\right)_{coh} = N[\langle|F(\underline{Q})|^2\rangle_Q - |\langle F(\underline{Q})\rangle|^2_Q] + N\cdot S(Q)\cdot|\langle F(\underline{Q})\rangle|^2_Q.
\tag{9}
$$

The quantity $\langle F(\underline{Q})\rangle$ is the Fourier transform of the radial distribution of scattering length within one particle. From now on, for simplicity we shall omit the subscript Q.

4.1 Single Particle Form Factor

For particles in solution, it is actually the scattering length density in excess of ρ_{solv} in which we are interested. We can thus write Eq. (4) as

$$
F_p(\underline{Q}) = \int_{particle\ p} [\rho_p(\underline{r}) - \rho_{solv}]\exp[iQ\cdot\underline{r}]^3 dr,
\tag{10}
$$

where the integral is over the particle volume only. Let us consider the simplest case.

Monodisperse spheres. The form factor for a homogeneous sphere was first derived by Rayleigh[13]; Eq. (10) becomes

$$
F_p(\underline{Q}) = \int_0^R \frac{\sin(Qr)}{Qr} 4\pi r^2 dr = \Delta\rho\cdot(4\pi/3)R^3\cdot\Phi(QR),
\tag{11}
$$

190

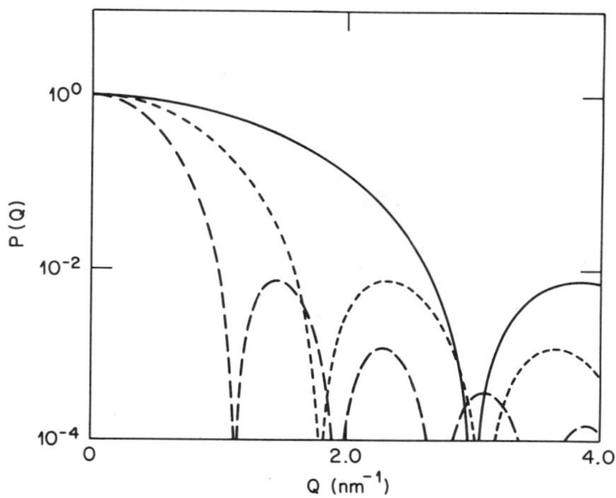

Fig. 4. Form factors for homogeneous spheres: R = 15 Å (——); 25 Å (———);
40 Å (·····).

with $\Phi(QR) = 3\ (\sin(QR)\ \cos(QR))\cdot(QR)^{-3}$. Here $\Delta\rho$ is equal to $\bar{\rho} - \rho_{solv}$ where $\bar{\rho}$ is the mean scattering length density of the solvent-excluded particle, and $(4\pi/3)R^3$ refers to the solvent-excluded particle as well. For monodisperse spheres, we find further that $<|F(Q)|^2> = |<F(Q)>|^2$; we shall designate $<|F(Q)|^2>$ as P(Q). Figure 4 shows normalized P(Q)'s for homogeneous spheres of various radii. The function has pseudo-oscillations of period $QR = 2\pi$ superimposed on a Q^{-4} law. Using the designation N_p for number particle density, Eq. (9) can thus be rewritten as

$$\left(\frac{d\sigma}{d\Omega}\right)_{coh} = I(Q) = N_p \cdot S(Q) \cdot P(Q). \tag{12}$$

We shall be particularly interested in the scattering patterns for particles with scattering length density inhomogeneities; a centrosymmetric distribution of scattering length density may be modelled by a set of n concentric shells of radii $R_1, R_2, \ldots, R_n$ (Figure 5 shows the ease where n = 3); one may then write

$$F_p(\underline{Q}) = (\rho_1 - \rho_2)V_1\phi(QR_1) + (\rho_2 - \rho_3)V_2\phi(QR_2) + \ldots$$
$$+ (\rho_n - \rho_{solv})V_n\phi(QR_n), \tag{13}$$

where $\rho_1, \rho_2, \ldots, \rho_n$ are the scattering length densities of the n shells; $V_1, V_2, \ldots, V_n$ are the volumes corresponding to the radii $R_1, R_2, \ldots, R_n$.

For departure from sphericity and for polydispersity the inequality $<|F(Q)|^2> \neq |<F(Q)>|^2$ holds; for this reason, one must average over the particle orientation or over the particle size distribution. Form factors have been calculated[14] for many different cases. We will consider only the analytical expression for polydisperse spheres and monodisperse ellipsoids (prolate and oblate) that we will use to discuss our data on potassium oleate microemulsion.

Polydisperse spheres. The size-averaged scattering functions for polydisperse spheres (which may be homogeneous or core-plus-shell(s)) are

$$<|F(\underline{Q})|^2> = \int_0^\infty |F_p(Q,R)|^2 f(R)\,dR \tag{14}$$

$$|<F(\underline{Q})>|^2 = |\int_0^\infty F_p(Q,R)f(R)\,dR|^2. \tag{15}$$

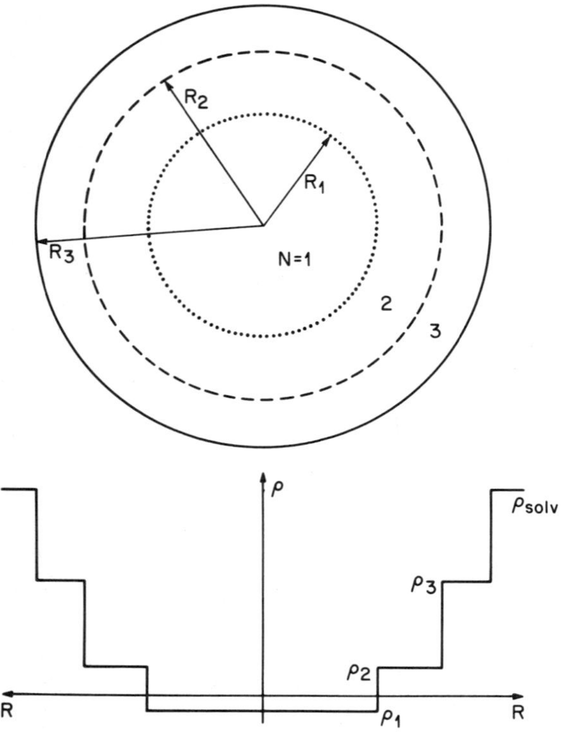

Fig. 5.   A sphere having three distinct regions of scattering length
          density, ρ.

Analytical solutions for the integrals in Eqs. (14) and (15) have been
presented[10,12] for a Schulz distribution of radii:

$$f(R) = (\frac{Z+1}{\bar{R}})^{Z+1} R^Z \exp[- (\frac{Z+1}{\bar{R}})R]/\Gamma(Z+1), \quad Z > -1, \tag{16}$$

where $\bar{R}$ is the number-average particle radius and Z is the width parameter
of the distribution[15]. Gaussian distributions[16], zero-order log
normal distributions[17], rectangular distributions[12] and simple
histograms[18] have also been used for f(R).

   Monodisperse ellipsoids. In this case, the orientational averages
are computed according to Eq. (12)

$$<|F(\underline{Q})|^2> = \int_0^1 |F(Q,a,ECC,t,\mu)|^2 d\mu \tag{17}$$

$$|<F(\underline{Q})>|^2 = |\int_0^1 F(Q,a,ECC,t,\mu)d\mu|^2. \tag{18}$$

The function F is given by

$$F(Q,a,ECC,a',ECC',t,\mu) = V_C \cdot (\rho_1 - \rho_2) \cdot \phi(u') + V_T \cdot (\rho_2 - \rho_{solv}) \cdot \phi(u), \tag{19}$$

where the ellipsoid's core (of volume V_C) has axes a', a' and ECC'·a'
(semi-major for prolate, semi-minor for oblate) and the overall ellipsoid
(of volume V_T) has axes a, a and ECC·a. The thickness of the shell region
is given by t; ECC is the eccentricity (axial ratio) of the ellipsoid.
The quantity u(a,ECC) is Q[ECC·a·μ)^2 + a^2(1 - μ^2)]$^{1/2}$.

192

It is worth noting that polydispersity in particle size and deviations from spherical shape both have the effect of damping the oscillations observed in the form factor for monodisperse spheres (Figure 4).

4.2 Structure Factor

$S(Q)$ specifies the correlation between the centers of different particles, produced by particle-particle interaction in the dispersion.

$S(Q)$ is the Fourier transform of the radial distribution function $g(r)$ for the centers of mass of the particles

$$S(Q) = 1 + 4\pi N_p \int_o^\infty r^2 [g(r) - 1][(\sin Qr)/Qr] dr, \tag{20}$$

$g(r)$ measures the probability of finding another particle at a distance r from a reference particle centered at the origin; $g(r)$ is zero when r is less than the distance of closest approach of two particles; and $\lim_{r\to\infty} g(rO = 1$.

The total correlation between two particles separated by a distance r is given by the total correlation function, $h(r) = g(r) = 1$.

This function is composed of two contributions, given by the Ornstein-Zernike (OZ) equation[21],

$$h(r_{12}) = c(r_{12}) + N_p \int c(\underline{r}_{13}) h(\underline{r}_{32}) d\underline{r}_3. \tag{21}$$

The direct influence of particle 1 on particle 2 is expressed as $c(r_{12})$; the indirect influence of particle 1 on particle 2, transmitted via a third particle (situated at $\underline{r}_{13}$ from the first and $\underline{r}_{32}$ from the second) is integrated over all possible positions of the third particle. To solve Eq. (21) one needs closure relations. For an interaction potential with a hard core, in the MSA, these take the form

$$g(r) = 0 \quad r \leqq 2R_{HS} \tag{22}$$

$$C(r) = -U(r)/k_B T, \ r > 2R_{HS} \tag{23}$$

where $U(r)$ is the pair potential for interaction between the two particles and R_{HS} is the hard-sphere radius, and hence $2R_{HS}$ is the distance of closest approach between particles. At low volume fractions, a rescaled version of the MSA must be used[23]. For non-interacting hard spheres, $U(r) = 0$ for $r \geqq 2R_{HS}$ and ∞ for $r < 2R_{HS}$; and Eqs. (22) and (23) reduce to the Percus-Yevick approximation[24,25]. In fact, the actual $C(r)$ for a hard sphere system is not equal to zero for $r > 2R_{HS}$. Approximations which relate the actual $g(r)$ to $g_{HS}(r)$ have been developed[26-28]. Using the hard sphere model, Ashcroft and Lekner[25] obtained an expression for $S(Q)$ in the form

$$S(Q) = \frac{1}{[1 - N_p C(2QR_{HS})]} \tag{24}$$

$$C(2QR_{HS}) = 32\pi R_{HS}^3 \int_o^1 \left(\frac{\sin(2sQR_{HS})}{2sQR_{HS}} \right) (\alpha_o + \beta s + \gamma s^3) s^2 ds \tag{25}$$

with the coefficients α_o, β, and γ defined by

$$\alpha_o = (1 + 2\phi_{HS})^2 / (1 - \phi_{HS})^4 \tag{26}$$

$$\beta = -6\phi_{HS}(1 + 0.5\phi_{HS})^2/(1 - \phi_{HS})^4 \tag{27}$$

$$\gamma = 0.5\phi_{HS}(1 + 2\phi_{HS})^2/(1 - \phi_{HS})^4. \tag{28}$$

The hard sphere volume fraction, ϕ_{HS}, is simply:

$$\phi_{HS} = N_p \, (4\pi/3)R_{HS}^3. \tag{29}$$

Note that in general R_{HS} may be greater than, roughly equal to, or even less than the total particle radius, R_{tot}.

In the case of ionic dispersions, Hayter and Penfold[29] have obtained a closed analytic form for S(Q) in the MSA, for a one component macroion system with a total interaction potential containing a hard-sphere term plus a Debye term.

For polydisperse uncharged hard spheres, a rigorous solution for S(Q) has been obtained recently by Griffith, Triolo and Compere[70].

The attractive energy of interaction for two spherical particles of equal radius (R_{tot}), due to van der Waals attractions may be expressed according to Hamaker as

$$U(y) = -(A/12)\,\frac{1}{y^2 + 2y} + \frac{1}{y^2 + 2y + 1} + 2\ln\frac{y^2 + 2y}{y^2 + 2y + 1}, \tag{30}$$

with $y = x - 1$ and A as the composite Hamaker constant for the particles in the medium, given by

$$A = (A_{11}^{1/2} - A_{22}^{1/2})^2. \tag{31}$$

Here A_{11} = the Hamaker constant of the particles and A_{22} that of the medium.

Attractive interactions in aqueous micellar solutions of surfactants and in water-in-oil microemulsions (where the surfactants used are ionic but the microemulsion particles are essentially uncharged) are short-ranged, and can be expressed in the Debye form, usually called a Yukawa tail

$$U(r) = U_o\,\frac{\exp[-k(x-1)]}{x}, \quad x > 1 \tag{32}$$

with k^{-1} characterizing the range of the potential and x equalling r/σ, where σ is the effective particle diameter (i.e., the point at which the hard sphere portion of the potential becomes operational).

Hayter and Zulauf[30], in studying a non-ionic micellar system have calculated S(Q) using the Eq. (32). The values of U_o needed to fit the data were comparable to those calculated from Eq. (30) at a surface-to-surface separation of 3 Å.

For several water-in-oil microemulsions, it has been established that the surface layer of the opposing particle surfaces may inter-penetrate to a considerable extent[42,43,64-69] so that R_{HS} is less than R_{tot}.

Long range van der Waals attractions between the aqueous cores lead to values of A which are too large, so it is believed that this inter-penetration gives rise to short-ranged attractive interactions.

A square-well potential[31] with λ on the order of 1.1 and a well depth which depends on the extent of inter-penetration has been used to describe these systems[32.33].

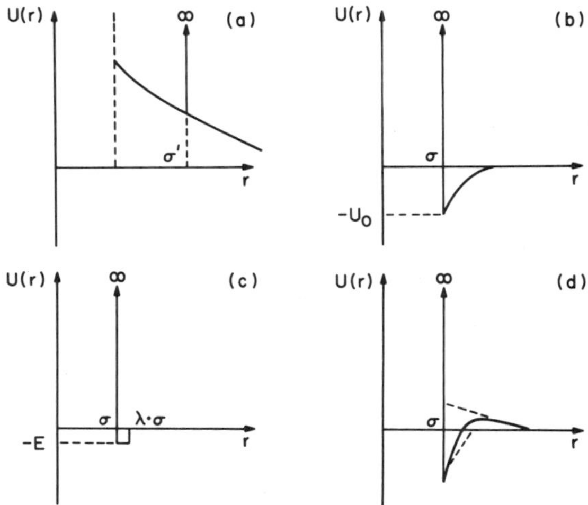

Fig. 6. Various interaction potentials for interacting spherical
particles: (a) repulsive interaction between charged spheres
with effective HS diameter σ' (see ref. 23); (b) short-ranged
attractive interaction of the Yukawa form between uncharged
spheres according to Equations 22–32; (c) short-ranged squarewell
attractive potential of depth ε and width $(\lambda-1)\cdot\sigma$ for uncharged
spheres; (d) interaction potential for nonionic micelles
according to Reatto and Tau (see ref. 65).

Reatto and Tau[65] have expressed the full interaction potential for
a non-ionic micellar system adding to the hard sphere potential two Yukawa
tails, taking into account the van der Waals interactions and the
hydration forces.

Figure 6 summarizes the various interaction potentials.

In the limit $Q \to 0$, $S(Q)$ is related to the osmotic pressure π by the
compressibility relation

$$S(0) = (k_B T/V_p)(\partial\pi/\partial\phi_p)^{-1} \tag{33}$$

with V_p and ϕ_p referring to the total particle (of radius R_{tot}). The
hard-sphere component of the osmotic pressure may be given by the semi-
empirical Eq. (34)

$$\pi_{HS} = (k_B T/V_{HS})\phi_{HS}(1 + \phi_{HS} + \phi_{HS}^2 - \phi_{HS}^3)/(1 - \phi_{HS})^3. \tag{34}$$

For systems with attractive interparticle interactions, the total osmotic
pressure is $\pi = \pi_{HS} + \pi_A$, with the attractive perturbation equal to

$$\pi_A = (k_B T/V_p)(A^*/2)\phi_p^2 \tag{35}$$

where

$$A^* = (4\pi/k_B T V_p) \int_{2R_{HS}}^{\infty} U_A(r)r^2 dr. \tag{36}$$

If the attraction is sufficiently strong, phase separation is
observed and $(\partial\pi/\partial\phi_p)^{-1}$, or in other words, $S(0)$ diverges as $T \to T_c$.

At very low volume fractions, the total osmotic pressure can be
approximated by

$$\pi = (k_B T/V_p)\phi_p(1 + B\phi_p); \tag{37}$$

here B is

$$B = 8(\phi_{HS}/\phi_p) + \overset{*}{A}, \text{ with } \phi_{HS}/\phi_p = (R_{HS}/R_{tot})^3. \tag{38}$$

5. POTENTIAL OF SANS

Certain characteristics of SANS make it a powerful complement to conventional methods of obtaining information on the structure of micro-emulsions and micellar solutions. In comparison with LS, the short wavelength (λ) of neutrons (or X-rays) makes accessible much larger values of the modulus of the wave vector Q. This makes it possible to obtain information about particle size and (especially) particle shape through P(Q) for particles as small as ca. 10 Å radius.

Some SAXS studies of non-ionic micellar solutions have appeared [35,36]. It is a fair question to ask why small-angle X-ray scattering (SAXS), which uses a far less expensive and more readily available radiation source than a neutron beam and probes essentially the same range of momentum transfer, is not used to the exclusion of SANS in the study of microemulsions and micellar solutions. There are two commanding advantages of neutrons over X-rays for these systems. The scattering length in SAXS is proportional to atomic number, usually low for the elements making up micellar solutions and microemulsions, whereas neutron scattering by hydrogen isotopes is high. The second advantage arises from the fact that scattering of neutrons by protons is largely incoherent and that by deuterons, coherent. Consequently, scattering by protiated components is distinguishable from that by deuterated components; this makes possible an array of interesting contrast matching experiments.

The external contrast variation is obtained by changing ρ_{solv}; that means by varying the C_nD_{2n+2}/C_nH_{2n+2} ratio in a aliphatic hydrocarbon medium and the D_2O/H_2O ratio in aqueous solution. Internal contrast variation involves changing $\bar{\rho}$ (or ρ_1, ρ_2, etc.) by appropriate isotopic substitution in the particles themselves.

Figure 7 presents three possible contrast conditions for a core-plus-shell particle. When $\rho_1 = \rho_{solv}$, one observes scattering only from the cores, whose size can therefore be obtained from analysis of the full scattering curve, since Eq. (13) becomes

$$F_p(\underline{Q}) = (\rho_1 - \rho_{solv})V_1\phi(QR_1). \tag{39}$$

Similarly, when $\rho_1 = \rho_{solv}$, one sees scattering only from the shell, since Eq. (13) is now

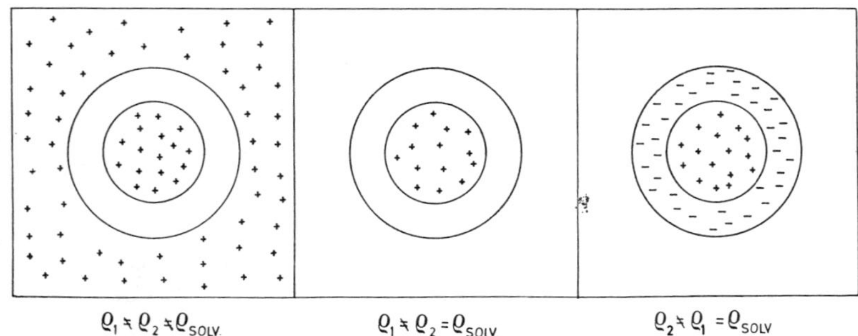

Fig. 7. Contrast matching conditions for a spherical core-plus-shell particle.

$$F_p(Q) = (\rho_2 - \rho_{solv})V_2\phi(QR_2).\tag{40}$$

Such contrast-matching experiments have been carried out for micellar[37-40] and microemulsion[41-46] systems, in addition to other types of colloidal dispersions[21,47-49]. We used such a technique in our study on potassium oleate microemulsions.

6. APPLICATION OF THE SANS TECHNIQUE TO MICROEMULSIONS

SANS has been applied to a number of w/o microemulsions. For example, water/cyclohexane/sodium dodecylsulfate/1-pentanol were investigated with D_2O, with C_6D_{12}, and with D_2O and various ratios of C_6H_{12} to C_6D_{12} [12,50,51]. The volume ratios were low enough in some cases[42] for the effect of the interparticle structure factor on scattering to be small, and interpretation was feasible in terms of Guinier radii[14]. Values of the radius of the water core and of R_G for the overall particle (affected by penetration of hydrocarbon in the shell) were consistent with results of ultracentrifugation. At high water fractions[50], onset of polydispersity and inversion to an o/w system was inferred. Ober and Taupin[52] determined osmotic compressibilities by extrapolation of scattered intensity curves and inferred hard-sphere radii of the particles. They found that attraction was too strong to be explained by van der Waals potentials, a conclusion earlier reached by Agterof et al.[53] from light scattering of w/o microemulsions. For systems without added salt, Ober and Taupin found it necessary to postulate prolate ellipsoids to fit their results.

Recent SANS studies[44,45] on the Winsor III and Winsor II domain for water/toluene/1-butanol microemulsion have identified regions where discrete water droplets as well as regions where the droplet model fails, and the microemulsion is considered to be bicontinuous.

Cabos and Delord[54] investigated a ternary w/o system, sodium bis-(2-ethylhexy)sulfosuccinnate (AOT)/water/n-heptane. The solutions were dilute enough for the results not to be affected seriously by interparticle interactions. Surfactant molecules per aggregate estimated from variation of contrast, C_7H_{16}/C_7D_{16} and H_2O/D_2O, agreed well; aggregation numbers increased sharply with the ratio of water/AOT. B. H. Robinson and co-workers[18,55,56] have also made measurements on the water/AOT/n-alkane systems; they and Huang and Chen[67,68] have investigated AOT microemulsions close to phase separation.

With respect to interpretation of experimental results, two investigations are more pertinent to the approach used here than those previously cited. Cebula et al.[16,43] applied several techniques to the system sodium dodecyl benzene sulfonate/xylene/water/hexanol, the SANS measurements covering a volume fraction of water from about 0.15 to 0.5. At these concentrations, interparticle interactions are expected to be important, and this was evidenced by peaks in the scattering intensity as a function of angle. In the fit to the results, they used a hard-sphere structural factor[25] and a model involving a water core plus a shell of somewhat different scattering density than the xylene continuous phase. The varied parameter was the radius of the core, the hard-shell radius being taken to be this value plus a constant thickness. Computations were carried out for various values of thickness, scattering density of the shell, and a standard deviation of size for a polydisperse distribution. They were able to reproduce the features of the curves with plausible values of these parameters. The shell thickness, also inferred[43] from experiments with deuterated hydrocarbon coupled with

those with D_2O, was less than one would compute from the total
stoichiometric particle less the water core, an indication of inter-
penetration of surfactant hydrocarbon in the closest approach of the
particles. The value was approximately the length of the alcohol.

Huang et al.[32] also carried to high volume fractions measurements
of AOT micelles in deuterated octane and to a moderate concentration
measurements of an AOT/D_2O microemulsion in decane. They also used a
hard-sphere model to calculate structural factors, but added a
perturbation from a short-range square-well attractive potential[31].

7. POTASSIUM OLEATE MICROEMULSION RESULTS AND DISCUSSION

7.1 Experimental

Data have been obtained for one composition (C) with hexanol as
cosurfactant and D_2O the only deuterated component, and for a set of
solutions with pentanol as cosurfactant:

(1) A, B and D with different amounts of D_2O that was the only deuterated
component.
(2) E, F and G at the same molar concentration, but with water,
hexadecane or both as deuterated component. In such a way we have used
the isotopic substitution technique, realizing the experimental
situations presented in Figure 7.

All the compositions are given in Table 1.

7.2 Analysis of Experimental Data

Scattering patterns for composition A-D and E-G are presented in
Figures 8 and 9 respectively: the strong patterns obtained in all cases
unequivocally demonstrate that aggregates are present. The minimum
dimension within an aggregate may be estimated from the decay of intensity
in reciprocal space, without regard to a model. This gives a dimension of

Table 1. Microemulsion Compositions

Code	K Oleate	$C_{16}H_{34}$	$C_{16}D_{34}$	1-Pentanol	1-Hexanol	D_2O	H_2O	Vol. Frac.
A	1 gm	5 ml		2 ml		1 ml		0.41
Moles	1	5.54		5.96		17.8		
B	1 gm	5 ml		2 ml		2.5 ml		0.52
Moles	1	5.52		5.97		44.7		
C	1 gm	5 ml			2 ml	2.5 ml		0.52
Moles	1	5.54			5.18	44.6		
D	1 gm	5 ml		2 ml		3.27 ml		0.55
Moles	1	5.52		5.97		57.0		
E	1 gm	4.42 ml		1.99 ml		2.55 ml		0.55
Moles	1	4.85		5.89		45.3		
F	1 gm		4.42 ml	2.02 ml		2.49 ml		0.55
Moles	1		4.84	5.99		44.2		
G	1 gm		4.36 ml	2.06 ml			2.51 ml	0.55
Moles	1		4.79	6.11			44.4	

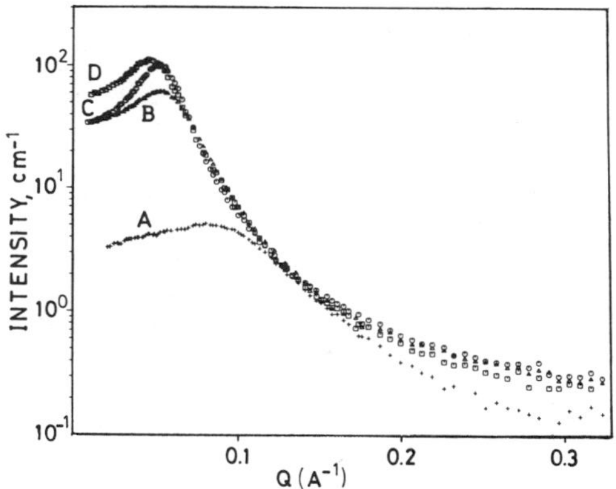

Fig. 8.   Scattering pattern for composition A, B, C, and D.

the order of $\pi/0.1$ or 30 Å, in contradiction to inferences from other techniques in some of the references cited that solutes are molecularly dispersed at some of the stoichiometries investigated.

A number of models were tested in reproducing the scattering patterns. All postulated a water core, with inclusion of the carboxylate head groups and potassium counterions of the oleate and OH group of the alcohol. The core is surrounded by a shell made up of the hydrocarbon parts of the surfactant and of the alcohol in the initial model. Approximately half of the stoichiometric alcohols per oleate were arbitrarily assigned to the aggregates; tests of other plausible assumptions indicated little effect on fits or parameters.

<u>Monodisperse spheres, polydisperse spheres and monodisperse ellipsoids</u>. In the simplest model, monodisperse spheres, the intensities are given by Eq. (12) to which was added a constant contribution of residual incoherent scattering (B)

$$I(\underline{Q}) = N_p \cdot S(Q) \cdot P(Q) + B. \tag{41}$$

The single particle form factor was calculated using Eq. (13), that for a core plus shell model becomes

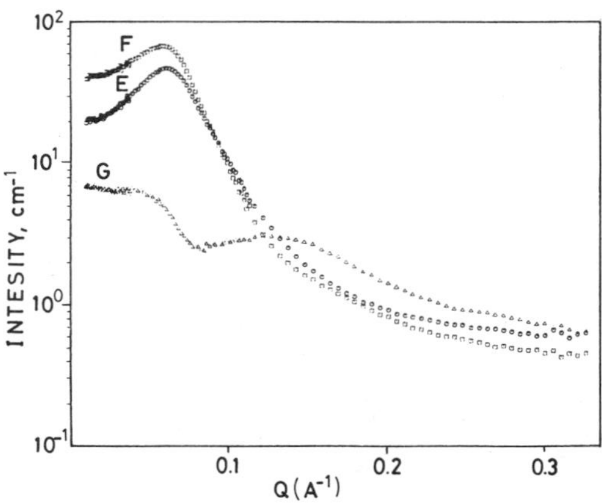

Fig. 9.   Scattering pattern for composition E, F, and G.

$$F_p(\underline{Q}) = (\rho_1 - \rho_2)V_1\phi(QR_1) + (\rho_2 - \rho_{solv})V_2\phi(QR_2). \tag{42}$$

A hard-sphere structure factor, calculated by the program of J. B. Hayter and J. Penfold[11,29] was used for S(Q). Adequate fits were not obtained by postulation of monodisperse spheres. The absence of any evidence in our scattering curves for the secondary minima and maxima in P(Q) which occur for monodisperse spheres suggests that the micellar population is polydisperse, anisometric or both. In fact, polydispersity and anisotropy would be expected to damp such oscillations. For departure from sphericity, polydisperse spheres and monodisperse (prolate and oblate) ellipsoids were tested. Here the term B was added to Eq. (9)

$$I(Q) = N_p \cdot [\, <|F(Q)|^2> - |<F(\underline{Q})>|^2\,] + N_p \cdot S(Q) \cdot |<F(\underline{Q})>|^2 + B. \tag{43}$$

In the case of polydisperse spheres, the size-averaged scattering function was computed according to Eqs. (14) and (15), using a Schulz distribution of radii (Eq. (16)). In the case of monodisperse ellipsoids the orientational averages were computed according to Eqs. (17) and (18).

In both the monodisperse ellipsoids and the polydisperse sphere case, the structure factor we shall use is that computed (in the same way discussed above for monodisperse spheres model) for a sphere of size equivalent to the ellipsoid or the (weighted) average particle size. This is, of course, an approximation, and for polydisperse hard-sphere systems, not a very good one[57].

It is worth noting that the volume fraction used in these fits was obtained from stoichiometry and the fraction of alcohols assigned to the disperse phase.

Although general features could be reproduced, we could not compute adequate fits to scattering patterns by fits with polydisperse spheres, nor monodisperse ellipsoids by adjusting dimensions in a variety of core plus shell models.

We have already referred to some other aspects of models we have tested, with worse fits or insignificant improvement. With respect to structural functions for polydisperse systems, we tried the alternative phenomenological procedure suggested in Ref. (66) without success. Inclusion of an attraction between particles has been popular of late. The attempts we have made have not indicated it would be very helpful.

<u>Penetrable particles</u>. The peaks in intensity indicate interparticle interference in the scattering. Because coulombic repulsions would not be expected to be important with electrically neutral reversed micelles, these interactions presumably arise from excluded volume effects.

It seemed possible that penetration of the shell by hydrocarbon solvent or by the shells of other particles might not be accounted for by the models used. In the cases for which D_2O provided the dominant contrast, the effect of shell penetration would be mainly on the structure factor, because the scattering length density of shell and the continuous phase are very nearly the same. Analyses in the literature[10,58] suggest that allowance can be made by multiplying the volume fraction by a factor varying from one for no penetration, to zero for the case of complete penetration. The theory is not strictly applicable, because it implies the densities in the inter-penetrating regions are the sum of the components, but it should give a better approximation, if inter-penetration occurs.

Using the adjustable "penetration factor" (which in view of its
approximate relation to theory might better be termed "permeability
factor"), the fits for all patterns and for all models have been greatly
improved. In the case of core plus invisible shell model ($\rho_2 \simeq \rho_{solv}$),
the values of core radii are not tremendously different for the four
geometries: the best fits have been obtained by monodisperse core +
invisible oblate ellipsoids model.

Figure 10 shows a comparison with the other three geometries tested
for composition E. Monodisperse spheres (I) fail most clearly; the
computed curves for polydisperse spheres (II) and monodisperse prolate
ellipsoids (III) are also less like the experimental intensities than for
monodisperse oblate ellipsoids (IV).

The model parameters varied in these fits, R_{core} (the radius of the
core of the sphere of volume equivalent to the ellipsoid or to the
weighted average sphere), the penetration factor and distribution breadth
Z (when polydispersity was assumed), or axial ratio (ECC) (when
ellipsoidal configurations were tested). The number of potassium oleates
per micelle and the aggregation number, were computed from R_{core}; the hard
sphere radius (R_{HS}) was set equal to R_{core} plus the alcohol extended chain
length.

<u>Contrast</u>. Varying the visibility to neutrons of different regions of
the aggregates (see Figure 7) can be expected to illuminate further
aspects of their structure. It is clear in such a case that an invisible
shell model is not appropriate, and the fits to be presented of the
systems E-G having different components deuterated are carried out with a
core + shell model. For computation of the particle scattering function,
the overall radius (R_{tot}) and the shell scattering length density
consistent with R_{tot}, are computed by assigning to the shell the
hydrocarbon parts of the surfactant and half of stoichiometric micelle
cosurfactant; the value of AGG is obtained as before from the least-
squares derived core radius (the same result is obtained, as expected,

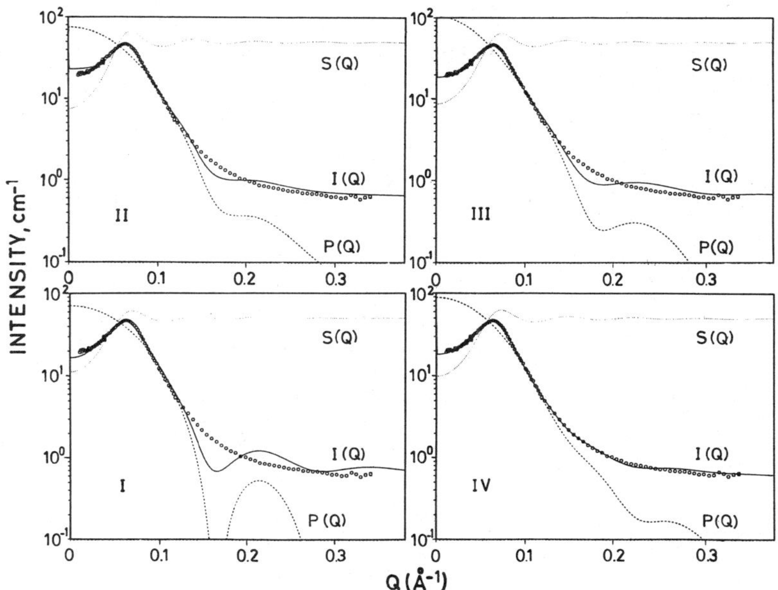

Fig. 10. Fits of experimental data for composition E: core + invisible
shell model. I: monodisperse spheres; II: polydisperse spheres;
III: prolate ellipsoids; IV: oblate ellipsoids.

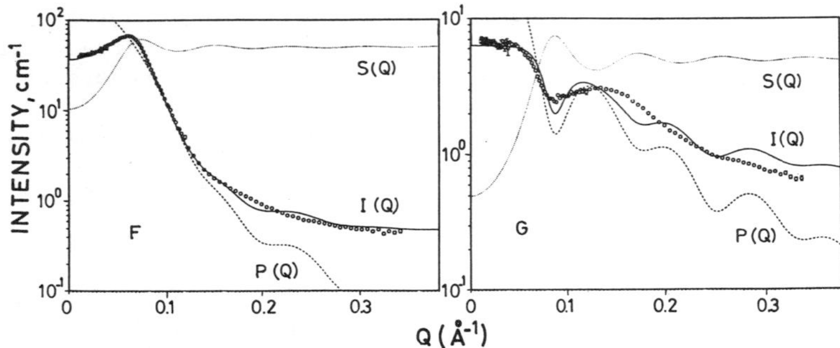

Fig. 11. Fits of experimental data for composition F and G: core + shell monodisperse oblate ellipsoid model.

from varying R_{tot} to determine AGG and deducing R_{core}); R_{HS} was set equal to R_{tot}. This model gave an essentially identical fit and the same parameters as the invisible shell for E, which has only water deuterated. Figure 11 shows the fits for composition F and G.

The fit to the case of deuterated hexadecane (Figure 11,F), in which dominant contrast is between the total aggregate and the continuous medium, is good, though not so good as for D_2O. The magnitudes of the parameters in the two fits are very close. The agreement increases confidence in the basic features of the model.

The fit (Figure 11,G) to the case in which there is strong contrast between the shell and both the core and the medium reproduces only general aspects of the experiment. Part of the problem here may arise from instrumental smearing, to which this case would be particularly vulnerable, although it seems unlikely that loss of resolution could be this large. It may be that details of the structure to which the other two are relatively indifferent may be important here. Results for the fits of compositions E and F are reported in Table 2.

The assignment of half stoichiometric alcohol to the disperse phase perhaps merits comment. The selection was made partly on the basis of getting some sort of agreement of fit and experiment with solution G, which was particularly sensitive to this parameter. The number used had little effect on the D_2O-only microemulsions. For example, with solution E, monodisperse oblate ellipsoid/invisible shell model, putting all the alcohols in the shell instead of 3 affected only the penetration factor appreciably.

Table 2.   Parameters from Least Squares Fits to Scattering Patterns for Composition E and F: Core + Shell Monodisperse Oblate Ellipsoids Model

Composition	E	F
R_{core}	30 ± 4	30 ± 10
$R_{tot.}$	40	40
Axial Ratio	0.5 ± 0.3	0.5 ± 0.3
Penet. Fac.	0.4 ± 0.2	0.4 ± 0.5
AGG	81	81

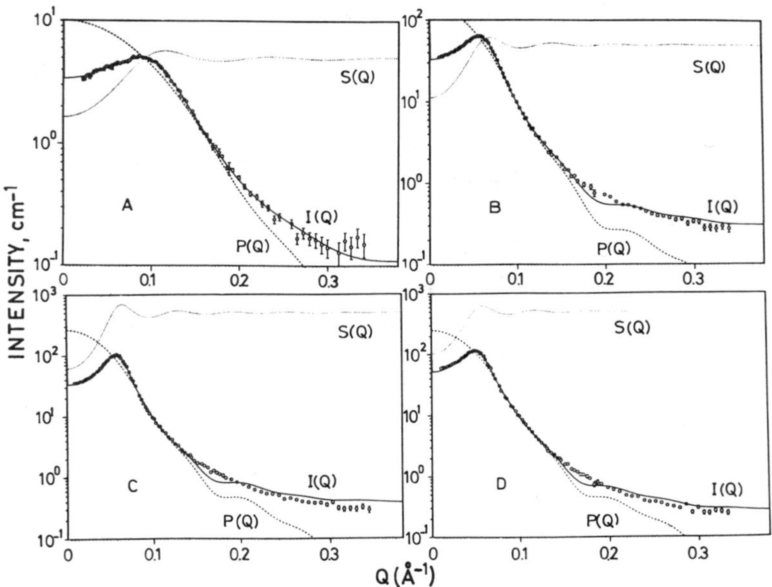

Fig. 12. Fits of experimental data for composition A, B, C, D: core +
invisible shell monodisperse oblate ellipsoids model.

Other compositions. Monodisperse oblate ellipsoids core plus
invisible shell model gives reasonable fits to the other compositions
reported, all having only water deuterated. Figure 12 shows the fits for
composition A, B, C and D, obtained by monodisperse core plus invisible
shell oblate ellipsoids model; results for these fits are reported in
Table 3.

Once occurrence of aggregation at all compositions is accepted,
variations in core radius are not surprising. Core radius and aggregation
increase as water/amphiphile ratio increases, for constant amphiphile/
hydrocarbon ratio (A, B and D).

Decrease in aggregate size (B vs E) is not quite as expected for
increase in disperse phase/hydrocarbon ratio, at constant water/amphiphile
ratio; possibly the distribution of alcohol into the aggregates is greater
in this case. Although the molar concentrations are not quite the same,
it appears that substitution of hexanol for pentanol increases aggregate
size (B vs C). Axial ratios are similar for the different concentrations.

Table 3. Parameters from Least Squares Fits to Scattering
          Patterns: Core + Invisible Shell Monodisperse
          Oblate Ellipsoids Model

Composition	A	B	C	D
R_{core}	18.2 ± 0.1	35 ± 3	40 ± 5	41 ± 20
$R_{tot.}$	26.1	43	49	49
Axial Ratio	0.46 ± 0.02	0.4 ± 0.1	0.5 ± 0.2	0.4 ± 0.5
Penet. Fac.	0.49 ± 0.01	0.5 ± 0.1	0.6 ± 0.2	0.5 ± 0.8
AGG	41.8	124	184	169

7.3 Conclusion

The scattering patterns establish, without recourse to models, that
aggregates having at least one dimension of approximately 30 Å are formed
in potassium oleate microemulsions (at least in the range of composition
analyzed), when either hexanol or pentanol is present. Although there
remains room for doubt, we believe the present observations and their
analyses argue the essential validity of a distribution of permeable
particles of a slightly anisotropic shape. Oblate ellipsoids gave clearly
the best fits.

A much broader survey of the single-phase water-on-oil region of
these systems is clearly of interest. The most surprising aspect of the
rather limited part of the single-phase region reported here is the lack
of correlation between microscopic structure and the strange behavior
reported in the literature reviewed earlier. We have mentioned the
persistence of aggregation at compositions for which they have been said
not to exist. We see nothing so far to explain the profound reported
differences in properties between systems containing pentanol and
hexanol.

ACKNOWLEDGEMENT

The authors are grateful to R. Triolo, J. S. Johnson, Jr., and J. B.
Hayter for many discussions and helpful guidance. We acknowledge support
from the "Progetto Chimica Fine e Secondaria" of CNR (Rome) and the
National Science Foundation (CHE-8308362).

Small-angle neutron measurements were carried out at the National
Center for Small-Angle Scattering Research (NCSASR). The NCSASR is funded
by National Science Foundation Grant No. DMR-77-244-58 through Interagency
Agreement No. 40-637-77 with the Department of Energy (DOE) and is
operated by the U.S. Department of Energy under contract DE-AC05-840R21400
with Martin Marietta Energy Systems, Inc.

REFERENCES

1. D. O. Shah and R. M. Hamlin, Science, 171:483 (1971).
2. J. W. Falco, R. D. Walker and D. O. Shah, Aiche J., 20:510 (1974).
3. D. O. Shah, V. K. Bansal, K. Chan and W. C. Hsieh, in: "Improved Oil
    Recovery by Surfactant and Polymer Flooding", p. 293, Academic
    Press (1977).
4. E. Sjoblom and S. Friberg, J. Coll. Interface Sci., 67:16 (1978).
5. A. M. Bellocq and G. Fourche, J. Coll. Interface Sci., 78:275 (1980).
6. A. Boussaha, B. Djermouni, L. A. Fucugauchi and H. J. Ache, J. Am.
    Chem. Soc., 102:4654 (1980).
7. S. S. Atik and J. K. Thomas, J. Phys. Chem., 85:3921 (1981).
8. C. Boned, M. Clausse, B. Lagourette, J. Peyrelasse, V. E. R. McClean
    and R. J. Sheppard, J. Phys. Chem., 84:1520 (1980).
9. G. Kostorz, ed., "Treatise on Material Science and Technology, Vol.
    15: Neutron Scattering", Academic Press, New York (1979).
10. J. B. Hayter, in: "Physics of Amphiphiles, Micelles, Vesicles and
    Microemulsions", V. Degiorgio and M. Corti, eds., North Holland,
    Amsterdam (1985).
11. J. B. Hayter and J. Penfold, Colloid Polym Sci., 261:1022 (1983).
12. M. Kotlarchyk and S.-H. Chen, J. Chem. Phys., 79:2461 (1983).
13. Lord Rayleigh, Proc. Roy. Soc. (London), A90:219 (1914).
14. A. Guinier and G. Fournet, "Small-Angle Scattering of X-Rays", Wiley,
    New York (1955).

15. S. R. Aragon and R. J. Pecora, J. Chem. Phys., 64:2395 (1976).
16. D. J. Cebula, R. H. Ottewill, J. Ralston and P. N. Pusey, J. Chem. Soc., Faraday Trans. 1, 77:2585 (1981).
17. R. H. Ottewill, in: Proc. 8th Scandinavian Symp. on Surface Chemistry, Lund (1984), to be published.
18. B. H. Robinson, C. Toprakcioglu, J. C. Dore and P. Chieux, J. Chem. Soc., Faraday Trans. 1, 80,13:413 (1984).
19. A. L. Patterson, Phys. Rev., 56:972 (1939).
20. H. Hoffman, J. Kalus, H. Thurn and K. Ibel, Ber. Bunsenges. Phys. Chem., 87:1120 (1983).
21. R. H. Ottewill, in: "Colloidal Dispersions", J. W. Goodwin, ed., Royal Society of Chemistry's Special Publication #43 (1983).
22. J. L. Lebowitz and J. K. Percus, Phys. Rev., 144:251 (1966).
23. J.-P. Hansen and J. B. Hayter, Mol. Phys., 46:651 (1982).
24. J. K. Percus and G. J. Yevic, Phys. Rev., 110:1 (1958).
25. N. W. Ashcroft and J. Lekner, Phys. Rev., 145:83 (1966).
26. H. C. Andersen and D. Chandler, J. Chem. Phys., 57:1918 (1972).
27. H. C. Andersen, D. Chandler and J. D. Weeks, in: "Advances in Chemical Physics", Vol. 34, p. 105, I. Prigogine and S. A. Rice, eds., Wiley, New York (1976).
28. L. Verlet and J.-J. Weis, Mol. Phys., 28:665 (1974).
29. J. B. Hayter and J. Penfold, Mol. Phys., 42:109 (1981).
30. J. B. Hayter and M. Zulauf, Coll. Polym. Sci., 260:1023 (1982).
31. R. V. Sharma and K. C. Sharma, Physica, 89A:213 (1977).
32. J. S. Huang, S. A. Saffran, M. W. Kim, G. S. Grest, M. Kotlarchyk and N. Quirke, Phys. Rev. Lett., 53:592 (1984).
33. J. S. Huang, J. Chem. Phys., 82:480 (1985).
34. N. F. Carnahan and K. E. Starling, J. Chem. Phys., 51:635 (1969).
35. H. H. Paradies, Colloids and Surfaces, 6:405 (1983).
36. H. H. Paradies, J. Phys. Chem., 84:599 (1980).
37. L. J. Magid, R. Triolo and J. S. Johnson, Jr., J. Phys. Chem., 88:5730 (1984).
38. R. M. Jones, S. B. Berr, E. Caponetti, L. J. Magid and J. S. Johnson, Jr., to be published.
39. D. Bendedouch, S.-H. Chen and W. C. Koehler, J. Phys. Chem., 87:153 (1983).
40. J. Tabony, Mol. Phys., 51:975 (1984).
41. L. J. Magid, R. Triolo and J. S. Johnson, Jr., J. Chem. Phys., 81:5161 (1984).
42. M. Dvolaitzky, M. Guyot, M. Laques, J. P. LePesant, R. Ober, C. Sauterey and C. Taupin, J. Chem. Phys., 69:3279 (1978).
43. D. J. Cebula, L. Harding, R. H. Ottewill and P. N. Pusey, Colloid Polym. Sci., 258:973 (1980).
44. L. Auvray, J.-P. Cotton, R. Ober and C. Taupin, J. Phys. Chem., 88:4586 (1984).
45. A. de Geyer and J. Tabony, Chem. Phys. Lett., 113:83 (1985).
46. P. D. I. Fletcher and B. H. Robinson, Ber. Bunsenges. Phys. Chem., 85:863 (1981).
47. D. J. Cebula, R. K. Thomas, N. M. Harris, J. Tabony and J. W. White, Disc. Faraday Soc., 65:76 (1978).
48. D. J. Cebula, R. K. Thomas and J. W. White, J. Chem. Soc., Faraday Trans. 1, 76:314 (1980).
49. D. J. Cebula, J. W. Goodwin, R. H. Ottewill, G. Jenkin and J. Tabony, Colloid Polym. Sci., 261:555 (1983).
50. M. Laques, R. Ober and C. Taupin, J. de Phys. (Lettres), 39:L-487 (1978).
51. M. Dvolaitzky, M. Laques, J. P. LePesant, R. Ober, C. Sauterey and C. Taupin, J. Phys. Chem., 84:1532 (1980).
52. R. Ober and C. Taupin, J. Phys. Chem., 84:2418 (1980).
53. W. G. M. Agterof, J. A. J. Van Zomeren and A. Vrij, Chem. Phys. Lett., 43:363 (1976).

54. C. Cabos and P. Delord, J. Appl. Cryst., 12:502 (1979).
55. C. Toprakcioglu, J. C. Dore, B. H. Robinson and A. Howe, J. Chem. Soc., Faraday Trans. 1, 80:413 (1984).
56. P. D. I. Fletcher, A. M. Howe, N. M. Perrins, B. H. Robinson, C. Toprakcioglu and J. C. Dore, in: "Surfactants in Solution", Vol. 3, pp 1745-1758, K. Mittal and B. Lindman, eds., Plenum Press (1984).
57. P. van Buerten and A. Vrij, J. Chem. Phys., 74:2744 (1981).
58. L. Blum and G. Stell, J. Chem. Phys., 71:42 (1979); 72:2212 (1980).
59. A. A. Calje, W. G. M. Agterof and A. Vrij, in: "Micellization, Solubilization and Microemulsions", Vol. 2, pp 779-790, K. L. Mittal, ed., Plenum Press, New York (1977).
60. A. Vrij, E. A. Nieuwenhuis, H. M. Fijnaut and W. G. M. Agterof, Faraday Soc., 65:101 (1978).
61. B. Lemaire, P. Bothorel and D. Roux, J. Phys. Chem., 87:1023 (1983).
62. S. Brunetti, D. Roux, A. M. Bellocq, G. Fourche and P. Bothorel, J. Phys. Chem., 87:1028 (1983).
63. J. S. Huang, S. A. Saffran, M. W. Kim, G. S. Grest, M. Kotlarchyk and N. Quirke, Phys. Rev. Lett., 53:592 (1984).
64. J. S. Huang, J. Chem. Phys., 82:480 (1985).
65. L. Reatto and M. Tau, Chem. Phys. Lett., 108:292 (1984).
66. S. H. Chen, T. L. Lin and M. Kotlarchyk, in: Proc. 5th Int. Symp. on Surfactant in Solution, to be published.
67. M. Kotlarchyk, S.-H. Chen and S. J. Huang, Phys. Rev., A28:508 (1983).
68. M. Kotlarchyk, S.-H. Chen, S. J. Huang and M. W. Kim, Phys. Rev., A29:2054 (1984).
69. R. Triolo, private communication.
70. W. L. Griffith, R. Triolo and A. Compere, Phys. Rev. Lett., submitted.

THERMAL ANALYSIS OF WATER-IN-OIL MICROEMULSIONS

Donatella Senatra, Gabriella Gabrielli*
and Giulio G.T. Guarini*

Department of Physics, University of Florence, Largo E.
Fermi, 2 Arcetri, Italy
*Department of Chemistry, University of Florence, Via G.
Capponi, 9 Italy

INTRODUCTION

In previous papers[1-4], we reported the study of the dielectric and
calorimetric properties of a water/oil (w/o) microemulsion at increasing
water concentration (C, mass fraction), by focusing attention mainly on the
roles of the water and the interphase of the system. In particular, by
Differential Scanning Calorimetry (DSC), the measurement of the enthalpy
change (ΔH) associated with well identified thermal events, showed that w/o
microemulsions are characterized by the presence of a "free water" fraction
($\Delta H_w \neq 0$ at 273 K), (Fig. 1). Furthermore, it was shown that, for very low
water concentrations, $(0.024 \leqslant C \leqslant 0.222)$, the enthalpy change due to the
melting of the dodecane present, (ΔH_d at 263 K), appeared as "more
endothermic" than expected; as if a larger amount of oil than that really
contained in the sample had melted! (Fig. 2). Finally, a careful analysis
carried out on several samples in the range $(0.222 < C < 0.4)$, allowed us to
detect two concentrations where time dependent phenomena were found to
occur upon ageing of the samples. (Fig. 2). As a result in these samples,
the free water fraction evaluated from the experimentally measured ΔH_w
tends to zero, paralleled by an increase of the heat of melting of the
dodecane up to a value much higher than that corresponding to the dodecane
content of the sample. Since the latter behaviour was observed every time
the enthalpic contribution ΔH_w was either absent or disappearing, the
larger amount of heat absorbed by the more endothermic dodecane transition
was interpreted in terms of the presence of another endothermic contribution occurring in the same temperature interval and ascribed to the
fusion of "interphasal water" at 263 K.

In order to verify the latter interpretation, a different microemulsion was studied with hexadecane as dispersing medium. In the
w/hexadecane system in fact, the endothermic contribution due to the
melting of the oil at 291 K does not interfere with the thermal events
associated with water in the temperature range of interest.

One of the purposes of this work is to show the effectiveness of w/o
microemulsions for the study of the properties of water both in the bulk
and in the interphasal configuration. The microemulsion droplets are
considered here as extremely small water containers whose "structural"

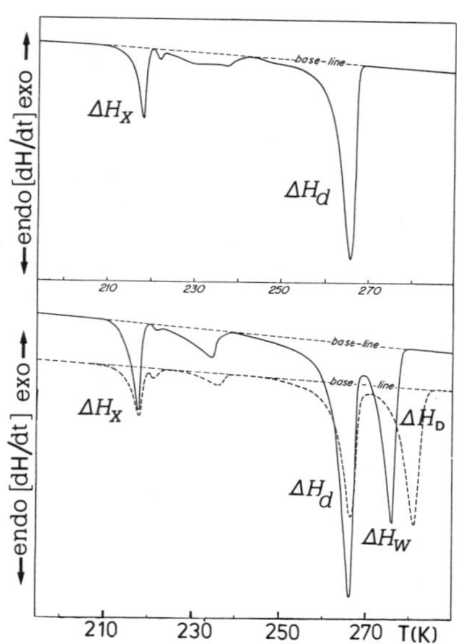

Fig. 1.   Typical DSC curves of w/dodecane microemulsion samples with
concentration in the intervals: $0.0245 \leq C \leq 0.222$ (top) and $0.222 < C < 0.4$ (bottom).   Continuous line: H_2O/oil microemulsion;
Dotted line: D_2O/oil microemulsion sample.   The endotherm due to
water melting at 273 K is clearly observable.

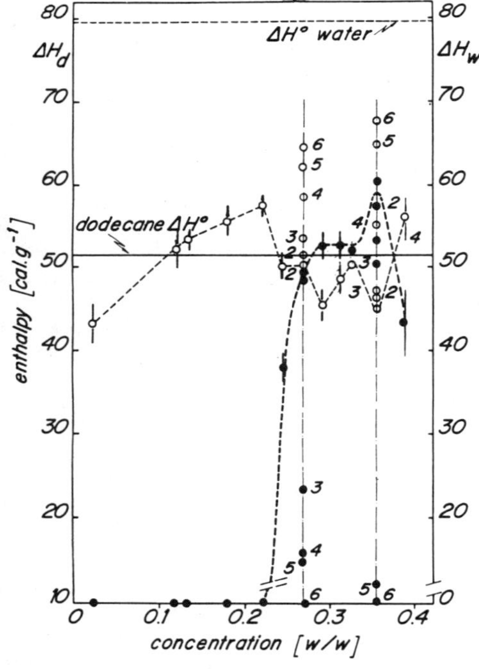

Fig. 2.   Behavior of the enthalpy changes due to the melting of dodecane
and water vs. concentration (w/dodecane microemulsion).   (ΔH
expressed in cal/g of component).   Note that, in the concentration
range 0.1-0.2, the experimentally measured ΔH_d values are
significantly higher than the theoretical ones ($\Delta H_d^o = 51.7$ cal/g).
Numbers: "Ageing Effect".

208

environment, for the kind of interphase considered, the size of the structured units and the thickness of the interfacial layers, can be compared to the one within which the water finds itself confined in many biosystems[5].

EXPERIMENTAL

Following the procedure described in Refs. [3 & 4], we measured, by means of a Perkin Elmer DSC 1B and a Mettler TA 3000 with a Low Temperature DSC Cell, the heat flow rate as a function of increasing temperature from 75 K to 313 K as well as the enthalpic changes of each recorded endothermic process.

The percentage weights of the components of the initial mixtures of the two microemulsions are: 57.22% Dodecane, 25.03% Hexanol, 15.30% K-Oleate and 2.45% Water for the water/dodecane system and, 57.73% Hexadecane, 24.35% Hexanol, 14.94% K-Oletae and 2.98% Water for the water/hexadecane system. A minimum amount of water was necessary in both cases in order to solubilize the potassium oleate. Microemulsion samples were produced by adding water to the above mixtures.

The variation of the thermal properties of both systems was followed by changing the water content, (C = weight ratio water/water + oil).

The concentration intervals investigated are: $0.0245 \leqslant C < 0.4$ and $0.0298 \leqslant C < 0.4$ for the w/dodecane and w/hexadecane system respectively.

As reference data for the enthalpies and the melting temperatures, the values measured on the components used to formulate the two microemulsions were adopted. (Table 1).

The phase diagrams and additional details on the above two systems can be found in Refs. [6-8].

RESULTS AND DISCUSSION

As previously reported in the case of the w/dodecane system[3,4], (Fig. 1), also in the w/hexadecane microemulsion, the samples can be distinguished depending on whether they exhibit an endothermic contribution due to the melting of water at 273 K, as shown in Figs. 3 and 4, where, for sake of comparison, the DSC curves of w/dodecane samples are also plotted.

Table 1. Reference Melting Temperatures and Enthalpy
          Values

Component	T_m(K)	Enthalpy (J/g)	Symbol
Dodecane	263	216.27	ΔH_d
Hexadecane	391	215.34	ΔH_h
Hexanol	225	150.52	ΔH_x
Water	273	333.42	ΔH_w
Interphasal Water	263	312.38	$(\Delta H_w)_{263}$
Heavy Water	277	313.54	ΔH_D

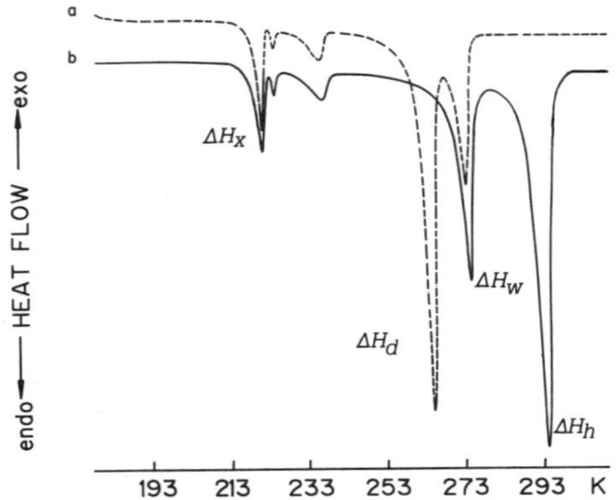

Fig. 3. DSC curves of microemulsion samples exhibiting a free water
endotherm at 273 K; Curve (a): w/dodecane system; Curve (b):
w/hexadecane system. The two termograms differ only in the
endotherms due to the melting of the hydrocarbon oils.

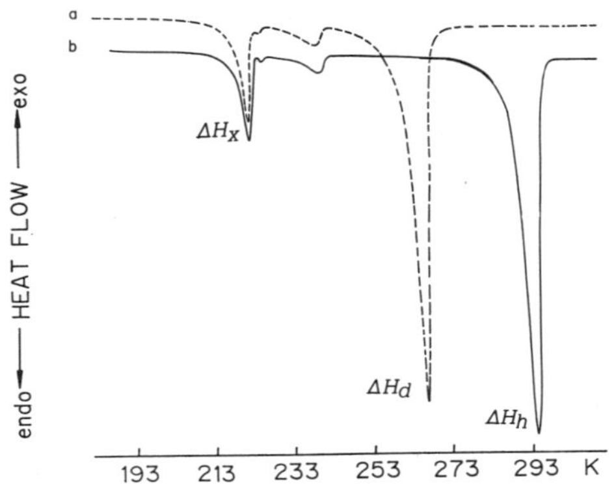

Fig. 4. DSC curves of microemulsion samples without a free water fraction
(ΔH_w = 0). Curve (a): w/dodecane system; Curve (b): w/hexadecane
system.

By means of the DSC analysis two concentration ranges have been identified for each system studied, namely:

w/dodecane system: 1) $0.0245 \leqslant C \leqslant 0.222$ (ΔH_w = 0)

                                         2) $0.222\ <\ C\ <\ 0.4$ ($\Delta H_w \neq 0$)

w/hexadecane system: 1) $0.0298 \leqslant C \leqslant 0.104$ (ΔH_w = 0)

                                         2) $0.18\ \ \ <\ C\ <\ 0.4$ ($\Delta H_w \neq 0$).

However, in the w/hexadecane microemulsion within the concentration
range (0.104-0.18) an additional thermal process can be observed around 263
K (Fig.5, bottom), in the same temperature interval where the melting of
the dodecane occurs in the first system studied (Fig. 5, top).

A test made by adding heavy water instead of normal water proved that,
besides the 273 K process (Fig. 6), the 263 K transition can also be
ascribed to water (Fig. 7). In fact, in both cases, the DSC endotherms

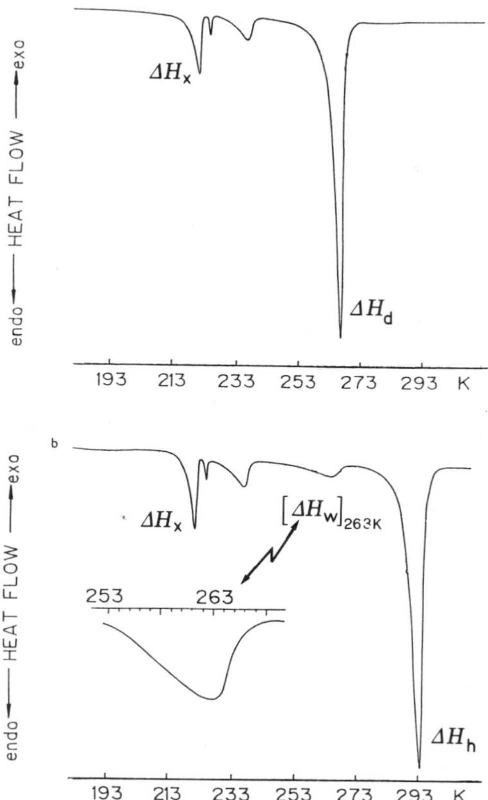

Fig. 5.   Interphasal water.   Top: w/dodecane microemulsion.   The presence
of the endothermic contribution due to the fusion of water at 263
K is sheltered by the melting of the dodecane oil.   Bottom:
experimental evidence of the endotherm due to the melting of
interphasal water at 263 K in the w/hexadecane system.   The water
content is the same in both samples.

shift their peak temperatures by about 4 K as expected.   Moreover, it is
worth noting that the substitution of the dispersing oil seems not to
affect the low temperature parts of the thermograms, as may be easily seen
in Fig. 3-6.

A synthesis of the thermal analysis of both systems is given in Fig.
8, where the ΔH experimental values of the hydrocarbon oils, water and
hexanol, are plotted as a function of increasing concentration.

A comparison between the trends of the enthalpy changes of the two
oils and the water vs. concentration shows that, in the case of the
w/hexadecane microemulsion, the relationships between ΔH_h and ΔH_w, as a
function of increasing water content, are linear and indeed consistent with
the dilution of the oil as well as the concentration of the water.   The
same does not apply to the w/dodecane microemulsion, where the presence of
the overimposed endotherm at 263 K conceals the decreasing behaviour of the
dodecane enthalpic contribution, whereas the melting of the hydrocarbon
shelters the endothermic process associated with the fusion of the interphasal water.

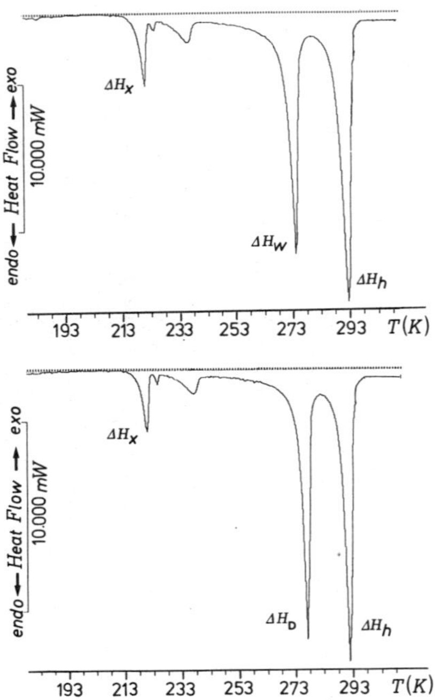

Fig. 6. Water-in-hexadecane microemulsion: thermal curves showing the difference between the H_2O/O (top) and the D_2O/O (bottom), sample with C = 0.31.

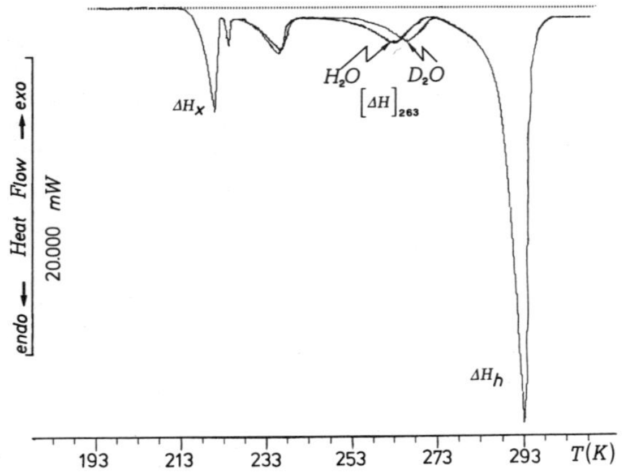

Fig. 7. Water-in-hexadecane microemulsion. The shift of the 263 K endotherm in the D_2O/O microemulsion sample with the same concentration; C = 0.104.

From Figure 8 it also follows that both systems exhibit a saturation trend at C∼0.3; moreover, the plot of the enthalpy-data, (ΔH), expressed in joules per gram of sample, offers a ruther immediate picture of the concentration region within which the microemulsion-state of the system in confined. In fact, when the linear relationship fails, the microemulsion, although macroscopically isotropic and transparent, is in a pretransitional state: further addition of water leads to a phase separation.

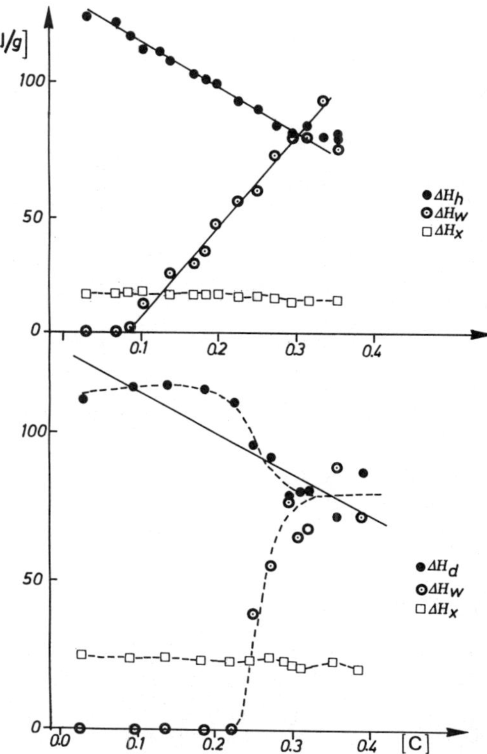

Fig. 8. Thermal analysis of the w/hexadecane system (top) and the
w/dodecane (bottom). The lines are least squares fit regression
lines on the experimental ΔH values of the w/hexadecane
microemulsion. While in the lower part of the figure, the line is
the computed regression line on the theroetical enthalpic changes
associated with the melting of the known amounts of dodecane in
the samples. Average uncertainty, less than 5%.

CONCLUSIONS

From the body of results the following conclusions can be drawn:

i)   Water-in-oil microemulsions are very useful systems for the study of
the properties of water, mostly with respect to the "different forms"
of water in highly disperse systems[9-10];

ii)   The results obtained for the w/hexadecane microemulsion offer a
further support to the interpretation of our previous findings on the
w/dodecane system as far as both the interphasal water and the ageing
effect is concerned;

iii) The experimental evidence of water melting at 263 K, for the first
time observed in w/o microemulsions, confirms also the behaviour
reported for the dielectric dispersion in the w/dodecane system (Fig.
9), if the latter is analyzed taking into account the fact that "bound
water" is supposed to relax at frequencies between those
characteristic of ice and normal water. Moreover, a broader spectrum
of time constants seems to be involved in the relaxation of bound

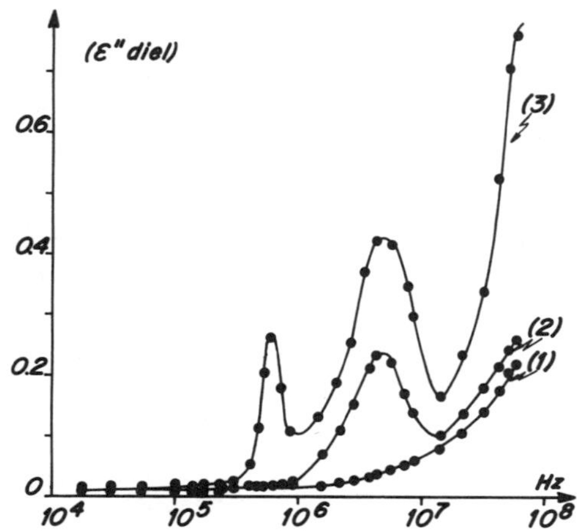

Fig. 9. Dielectric loss of the w/dodecane system as a function of
increasing frequency. (T = 293 K). Curve (1): ε''_{diel} for samples
without a free water fraction ($\Delta H_w = 0$); Curves (2-3): ε''_{diel} for
samples with a free water fraction ($\Delta H_w \neq 0$).

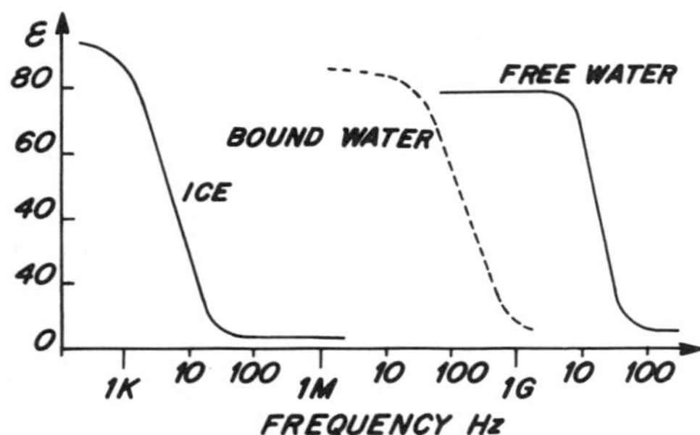

Fig. 10. Frequency dependence of the dielectric constant of ice, bound
water and free water.

water. In fact, the slope of the dispersion curve of the latter is
flatter than that of ice and also flatter than that of normal, massive
water. (Fig. 10). This suggests that there may be a variation in the
characteristics of water binding to which different activation energies correspond[11].

Acknowledgments

Financial support of this work by the "Centro Interuniversitario di
Struttura della Materia" (MPI) and the "Gruppo Nazionale di Struttura della
Materia" (CNR), is gratefully acknowledged.

REFERENCES

1. D. Senatra and G. Giubilaro, J. Colloid Interface Sci., 67 (1978), 448
    and ibidem, 457-464.
2. D. Senatra, G. G. T. Guarini, G. Gabrielli, and M. Zoppi, in "Physics
    of Amphiphiles. Micelles, Vesicles and Microemulsions", V.

214

Degiorgio and M. Corti Edrs., (North Holland, Amsterdam), Proc. Int. School of Physics "Enrico Fermi", Italian Physical Soc., Varenna July 1983, Italy, pp 802-826 (1985).

3. G. Senatra, G. G. T. Guarini, G. Gabrielli, and M. Zoppi, in "Macro-Microemulsions: Theory and Applications", (ACS Symposium Series No. 272), D. O. Shah Ed., 1985, pp 133-148.

4. D. Senatra, G. G. T. Guarini, G. Gabrielli, and M. J. Zoppi,J.Physique, 45, (1984) pp 1159-1174.

5. See, f.i., "Forms of Water in Biological Systems", Annals N.Y. Academy of Sci., Vol. 125, (1965).

6. G. Gillbert, H. Lehtinen and S. Friberg, J. Colloid Interface Sci., 33, (1970), 40-53.

7. D. O. Shah, and R. M. Hamlin, Science, 171, (1971), 483-485.

8. D. Senatra, J. Electrostatics, 12, (1982), 383-404.

9. V. Bassetti, L. Burlamacchi, and G. Martini. J. Am. Chem. Soc., 101, (1979) 5471-5477.

10a. F. Franks, "Water, A Comprehensive Treatise", Vol 4, (Plenum Press, N.Y., London), (1975).

10b. A. Banin, and D. M. Anderson, Nature, 255 (1975), 261-262.

11. H. P. Schwan, Annals N.Y. Academy of Sci., Vol. 125, pp 344-354, (1965)

STRUCTURAL PROPERTIES OF MICROEMULSION-DERIVED

Fe_2O_3 MICROPARTICLES

P. Ayyub

Tata Institute of Fundamental Research Homi Bhabha Road
Bombay 400 005, India

We propose to take an applications-oriented view of microemulsions
(ME); that is, we will not study the ME system per se, but use it to produce something else; then, by looking at the end product, we will try to
deduce something about the precurser ME. In particular, we have applied
the ME technique to synthesize microcrystals (or microparticles, since they
may sometimes be amorphous) of Fe_2O_3.

<u>What is so Interesting about Microcrystals?</u>

We can adopt the following two (not completely equivalent) approaches
for studying microcrystals (MC):

a. One can gradually build up a solid starting from a single atom or molecule and try to determine how many atomic units are necessary for it to
actually start showing bulk-like solid state properties. Such an approach
has been taken, for example, by Sattler's group[1] who have built up
metallic clusters (2-500 atoms) by inert gas condensation of metallic
vapours.

b. Alternatively one can attempt to synthesize smaller and smaller microparticles and observe just how the long-range cooperative properties (like
ferromagnetism, ferroelectricity and superconductivity) get affected by
decreasing the particle-size. We have adopted the second approach and have
been studying mainly oxidic ferromagnets and ferroelectrics.

<u>How Small is Small Enough?</u>

This really depends on the particular property that we wish to study.
For tightly bound electrons and many types of atomic oscillations, short-
range potentials are most significant. In such cases a MC of as small as
10 lattice units across would presumably show "bulk" behavior.

However, for the properties determined predominantly by long-range
Coulomb interactions or other collective excitations, a crystallite as
large as 1000 Å may show significantly different properties when compared
with the bulk solid.

<u>Which are the Solid State Properties Affected?</u>

1. Due to the significantly larger number of atoms lying on or close to the surface, various surface effects become predominant in MC. Since the surface in some sense forms an extended defect, this would also affect the core.

2. Due to the relatively smaller number of available electrons, the electronic energy level distribution may become discrete.

3. The phonon spectrum too shows significant changes on going over from bulk to MC due to: a) broken bonds at the surface, b) surface modes, c) change in magnitude of the long-range Coulomb forces and d) a low-frequency cut-off arising from the absence of phonons with a wavelength > D, the crystallite dimension.

4. The MC may be seen as a poor man's pressure apparatus. It can be predicted (qualitatively) from a first-principles theory using Lennard-Jones potentials, that the surface free energy term would lead to a lattice expansion (Negative Equivalent Pressure) in covalent MC, and a lattice contraction (PEP) in ionic MC. This has been experimentally verified in a few materials, e.g. a NEP = 180 kbars has been found in 50 Å Fe_2O_3 MC[2].

<u>How does one make Microcrystals?</u>

1. <u>Precipitation</u>: a finely precipitated material may be dried and heated at low temperatures. The resulting particle size (PS) distribution is rather broad.

2. <u>Hydrosol</u>: preparing a colloidal dispersion in water. This is mainly suitable for metallic MC of some types.

3. <u>Vacuum Evaporation</u>: may be used for some metals and simple compounds. Deposition is on a cooled substrate. Reproducibility is difficult to achieve.

4. <u>High-area Silicates</u>: Many types of MC may be produced by impregnating high-area silicates with a suitable solution and calcining at low temperatures. The PS is reporducible and controllable.

5. <u>Sol-gel</u>: a suspension of metallic oxides/hydroxides (sol) is converted to a semi-rigid mass (gel) by a dehydrating agent, which is heated in vacuum to get the MC.

And the latest technique...

<u>Microcrystals via Microemulsions</u>

Metallic MC have been recently produced by a similar method[3]. We have modified the method for metallic oxides as well. The process consists of:-

a) Preparing a thermodynamically stable, optically isotropic three component system consisting of water, hydrocarbon and surfactant. The aqueous phase contains a soluble salt of the metal concerned. The essential constituents are:

    - aqueous phase → ferric nitrate solution
    - oil/hydrocarbon phase → 2-ethyl hexanol
    - surfactant phase → sorbitol mono-oleate

b) Microemulsions (almost transparent) are formed on prolonged ultrasonic
agitation. Even though we have not attempted so far to establish the phase
diagram, the resulting system is probably oil-continuous where well-defined
closed nuclei (pools) of the aqueous phase are surrounded by the oil phase
and almost isolated from each other.

c) On passing liq.ammonia through the ME system - the nitrate is converted
to hydroxide - presumably without destroying the ME.

d) The hydroxide particles are allowed to settle slowly, separated and
dried under IR. TGA of this material shows a decomposition loss between
180 and 250°C [$Fe(OH)_3 \rightarrow Fe_2O_3$] So the material was calcined at 250 C to
give microcrystalline Fe_2O_3.

The true structure of the 3-component precursor phase (is it a microemulsion or not?) is under some doubt both for the present as well as the
previous (metallic ME) studies. When there is a high volume fraction of
the aqueous phase, we may not get well-defined closed islands in an oil-
continuous environment. However, such a system is unlikely to lead to the
almost spherical and nearly mono-disperse MC phase observed by us.

The Equivalent Spherical Diameter of the resulting MC have been determined by XRD line-broadening, and this tallies reasonably well with the PS
found from EM photographs. The latter study also confirms that the PS
distribution is quite narrow. It may be somewhat simplistic to assume that
each MC is formed from a single ME because this would imply the existence
of microemulsions large enough to produce 400 A microcrystals. Thus some
limited aggregation does probably take place during MC formation.

<u>Fe_2O_3 Microparticles: in Different Shapes and Sizes!</u>

Various forms of Fe_2O_3 could be produced by suitably changing the
initial concentration of the nitrate. The two major products are: a)
alpha-Fe_2O_3 (hematite) - which has a rhombohedral (corundum) structure and
is dominantly antiferromagnetic. It shows a small ferromagnetic moment due
to spin canting between 950 K (Neel temp.) and 260 K (Morin temp). b)
gamma- Fe_2O_3 (maghemite) - is a cation deficient form of Fe_3O_4 having an
inverse spinel structure with 8 occupied tetrahedral sites and 16 partially
occupied octahedral sites represented as

$$Fe^{3+}_8 (Fe^{3+}_{13.33} \quad []_{2.66}) O_{32} \quad , \text{ where } [ ] \text{ denotes a vacancy.}$$

It is ferromagnetic with an estimated Curie temp- 950 K, since it
actually transforms to the alpha form at 400C.

Only MC of the alpha phase is produced from comparitively higher
nitrate solution concentrations (C). As C is decreased from 0.2 to 0.312,
more and more of the gamma phase is nucleated in preference to the alpha;
ultimately, for the lowest C, the product becomes completely amorphous to
XRD. Also, with decreasing C, the PS decreases gradually in alpha and
rapidly in gamma. The unit cell volume also shows an interesting
dependence on C with an increasing PEP in alpha and an increasing NEP in
gamma.

The nitrate concentration is expected to determine the interfacial
(O/W) surface tension and hence the ME curvature. Thus the PS is
controlled by C. The alpha-gamma transition with decreasing C may arise
from : a) precursor effect- since at low concentrations FeOOH is formed in
preference to $Fe(OH)_3$ and at low temperatures the former is known to

decompose to the gamma phase; or b) a difference in the surface energy of
the two phases may lead to one phase being preferred at higher and the
other phase at lower particle sizes.

Mossbauer Study of Microcrystals

The Mossbauer effect offers four distinct windows for investigating
microscopic solid state properties: a) the isomer shift tells us about the
(s) electron density at the active nucleus, b) the quadrupole splitting
gives a measure of the EFG seen by the nucleus, c) the nuclear Zeeman
splitting allows us to determine the effective magnetic field at the
nucleus and d) the Lamb-Mossbauer factor provides information on the mean-
square vibrational amplitude of the emitting and absorbing nuclei. The
Mossbauer effect is particularly suitable for looking at MC because:

- it is not restricted to single crystals or well-crystallized samples
- it measures microscopic properties (magnetic, electronic, structural
  and chemical) and not gross effects,
- atoms with differing chemical and electronic environments give rise
  to separable contributions to the Mossbauer spectra, e.g. the atoms
  at the surface and within the bulk may exhibit distinct spectra.

What do our Mossbauer Spectra tell us?

We can summarize and discuss the novel features of the spectra as
follows:

1. The saturation magnetic field in both alpha and gamma phases are con-
sistently lower than the corresponding "bulk" values even at the lowest
available temperature (4.2 K). Further, this decrement is larger in
samples prepared from lower nitrate concentration (where the particle size
is smaller). This could have been due to the changes in the unit cell
dimensions (pressure effects) which are independently observed. However
this is unlikely simply because the lattice "contracts" in the alpha and
"expands" in the gamma phase and the hyperfine field should therefore have
changed in opposite directions in the two cases, unless the ionicity of the
Fe-O bond is drastically different in the two phases. A decrease in the
hyperfine field and broadening of the resonance lines is expected in
amorphous materials because of a distribution in the Fe-O bond lengths and
Fe-O-Fe bond angles, and a possible enhancement of the zero-point spin
deviation. Thus it appears that a substantial number of Fe ions (probably
at the MS surfaces) experience an amorphous-like environment.

2. The sample prepared from the smallest value of C (.312) shows spectra
typical of amorphous Fe_2O_3, in the sense that it does not exhibit complete
magnetic splitting even at 4.2 K. At all temperatures there appear spectra
with a quadrupole splitting of about 1 mm/s, which is quite large but nor
abnormal for amorphous materials. The presence of a large QPS in the room
temperature spectra of samples coming from C = 5.0 and lower appears to
confirm our earlier speculation about the partially amorphous nature of the
samples prepared by this method.

3. In all samples showing a QPS doublet, the intensity of the two lines
are unequal. This asymmetry increases with decreasing C and is about 14
per cent in the room temperature spectrum of the "amorphous" sample (C =
.312). Such a Goldanskii - Karyagin asymmetry is not expected in randomly
oriented polycrystalline samples unless the f-factor itself shows an
angular dependence. This can be explained by assuming an anisotropy in
mean-square vibrational amplitudes of the lattice vibration i.e. $\langle x^2 \rangle =
\langle z^2 \rangle$. Such a condition may appear if the surface effects are predominant

since the vibrational amplitudes of surface atoms would be different along
and normal to the surface.

CONCLUSIONS

1. The microemulsion technique shows promise as a general method for
preparing a variety of microcrystalline metals and metal oxides.

2. Under special circumstances it may even be used as a low temperature
technique for producing amorphous materials.

3. The particle size may be controlled by suitable choice of the concentration of the initial aqueous phase. An important advantage of this
method appears to be the narrow particle size distribution of the resulting
microcrystals.

4. The Fe_2O_3 microparticles prepared by this method exhibit interesting
structural changes as a function of the particle size. The detailed mechanism of this phenomenon is not completely understood.

Acknowledgements

     This work has been done in collaboration with Dr. V.R. Palkar and Dr.
M.S. Multani. The author acknowledges the constant encouragement of Prof.
R. Vajayaraghavan.

REFERENCES

1. K. Sattler et al, Phys. Rev. Lett. 45 (1980) 821.
2. Schroeer et al, Phys. Rev. Lett. 19 (1967) 632.
3. Boutonnet et al, Colloids and Surfaces, 5 (1982) 209.

LYOTROPIC MESOPHASES - LIQUID CRYSTALS

STRUCTURAL TRANSFORMATIONS IN LYOTROPIC LIQUID CRYSTALS

J. Charvolin

Laboratoire de Physique des Solides*
Bât. 510, Université Paris-Sud, 91405, Orsay, France
* Laboratoire associé au CNRS (LA 02)

INTRODUCTION

The broad lines of the polymorphism of assemblies of amphiphilic
molecules in presence of water are now well drawn. They can be found in
basic review articles about phase diagrams[1,2] and structures[3,4]. These
presentations are limited to descriptions of phase diagrams and structures;
they do not consider the processes by which one structure transforms into
another when the thermodynamical parameters of the phase diagram, water
content and temperature, vary. In order to approach this problem we shall
first analyze the structures described in the above classical works in
purely geometrical terms. We shall consider them as organizations of space
in two media, aqueous and paraffinic, of various connectivities. This will
lead us to distinguish two classes of structural transformations: one
corresponding to transformations without change of connectivity in the two
media; the other corresponding to transformations with changes of connectivity. In the first case there is only growth, deformation and ordering
of aggregates of amphiphilic molecules, without topological change. Thermodynamical models of aggregation and ordering, inspired from those
developed for micellar aggregation and ordering of molecular crystals and
thermotropic liquid crystals can give account, and in a few examples predict, some aspects of the phase transformations. In the second case the
appearance of new connections between the aggregates when the transformation takes place changes the topology of the system. Such changes
cannot be predicted by the above models and require specific treatments.
This imposed the search for a more general frame within which the essential
features of these models might be used. We shall propose a geometrical
approach which consists in looking at the various structures as solutions
to "frustrations" in the bilayers of the lamellar phases. Finally, we shall
describe a few recent experimental studies which lead to the revision of
some aspects of the classical descriptions, in agreement with the above
proposition.

STRUCTURES AND CONNECTIVITY

We describe the structures as partitions of space in two media, aqueous (a) and paraffinic (p), organized in connected subspaces of various
numbers (one subspace is said connected when it is possible to find a path
connecting any two points of it which does not cross a subspace of dif-

ferent chemical nature). We classify them according to their numbers of
(a) and (p) subspaces. For the so-called "direct" structures, most often
obtained with mono-alkyl amphiphiles, three cases can be distinguished:

i)   (a) is one infinite subspace and (p) is made of an infinite number of
     finite or infinite subspaces. This is the case of the micellar,
     nematic and cylindrical phases, examples of which are pictured in
     Fig. 1.

ii) (a) is one infinite subspace and (p) is made of two infinite sub-
     spaces. This is the case of the cubic phases, one example is shown in
     Fig. 2.

iii) (a) and (p) are made of an infinite number of infinite subspaces.
     This is the case of the lamellar phase shown in Fig. 3.

     (A quite symmetric classification, where (p) and (a) are exchanged,
holds also for the so-called "inverse" phases, most often obtained with
di-alkyl amphiphiles).

     It is clear that transformations between structures in each of these
three cases (for instance miceller-nematic, nematic-cylindrical, flat-
lamellar corrugated-lamellar) are different from these between structures
in different cases (for instance cylindrical-lamellar, lamellar-cubic) as
they do not imply a change in the connectivities of the subspaces whereas
the latter do. We therefore distinguish two classes of transformations and
shall illustrate them with a few typical examples.

TRANSFORMATION WITHOUT CHANGE OF CONNECTIVITY

     We choose our examples in the ternary phase diagram sodium decyl
sulfate/decanol/water, the relevant part of which is sketched in Fig. 4.
Two types of structural evolutions can be found along two lines of this
phase diagram. Along line (A), where the water content varies while the
soap to alcohol ratio keeps constant, the degree of order of the system
increases as the water content decreases. Along line B, where the water
content keeps constant but the soap to alcohol ratio varies, changes of
symmetry in ordered phases occur.

<u>Onset of Order in a Micellar Solution</u>

     Along line A the degree of order increases by steps as the concen-
tration of amphiphilic molecules increases, in a manner somewhat remi-
niscent of that observed in thermotropic liquid crystals[5]. First, at
very low concentration, the amphiphilic molecules are dispersed as monomers
in the aqueous solution. At the critical micellar concentration they start
aggregating in micelles which are dispersed in the solution without long
range order, - this is the isotropic liquid. As the concentration
increases the aggregates grow very slowly in an anisotropic manner. For
sufficiently high concentration the aggregates become organized in the
solution with long range orientational order and short range translational
order, this is the nematic liquid which is generally known to present three
phases, two uniaxial and one biaxial[6]. Finally, at higher concentration,
the aggregates build a two dimensional lattice corresponding to a long
range translational order of rectangular symmetry, -this is the crystalline
phase. (It must not be forgotten here that the terminology, liquid,
nematic, crystal concerns the ordering of labile aggregates of amphiphilic
molecules, where shape and size change all along the sequence, and not of
rigid molecules, as in ordinary thermotropic liquid crystals).

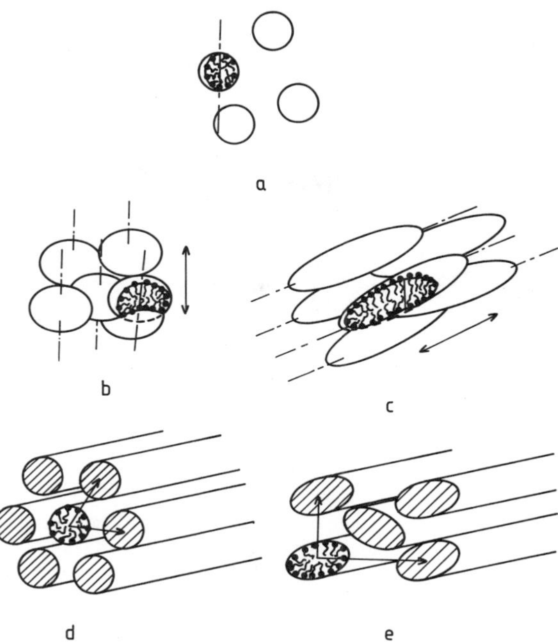

Fig. 1.   Structures with one infinite aqueous subspace and an infinity of
           paraffinic subspaces: (a) micellar phase of spheroidal aggregates,
           (b) and (c) uniaxial nematic phases of anisotropic aggregates (the
           arrows indicate the director of the phases), (d) and (e) hexagonal
           and rectangular pgg phases of cylindrical aggregates (the arrows
           represent the lattice of the structure).

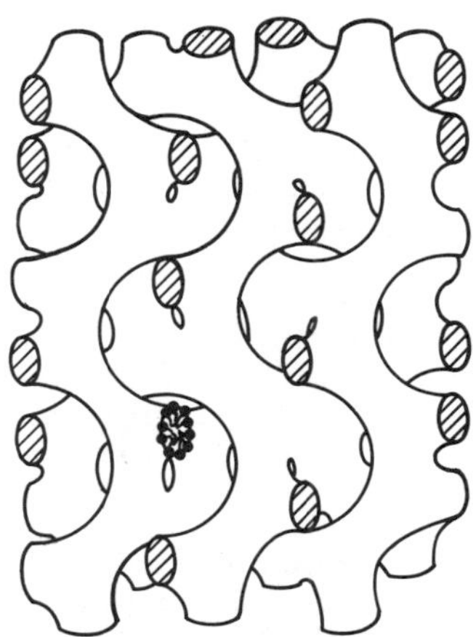

Fig. 2.   Structure with one infinite aqueous subspace and two interwoven
           infinite paraffinic subspaces: a cubic structure Ia3d.

        A thermodynamical modelization of such a sequence is being developed
at the moment[7].   Schematic presentations of its main lines have been
reported recently[8,9]. We shall just sketch a few basic ideas. The
starting point of the aggregation is a small spheroidal aggregate. If the

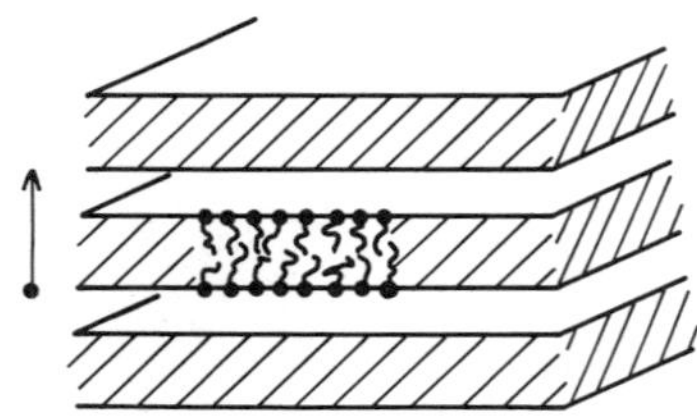

Fig. 3. Structure with an infinity of infinite aqueous and paraffinic subspaces: a lamellar structure (the arrow indicates the period of the structure).

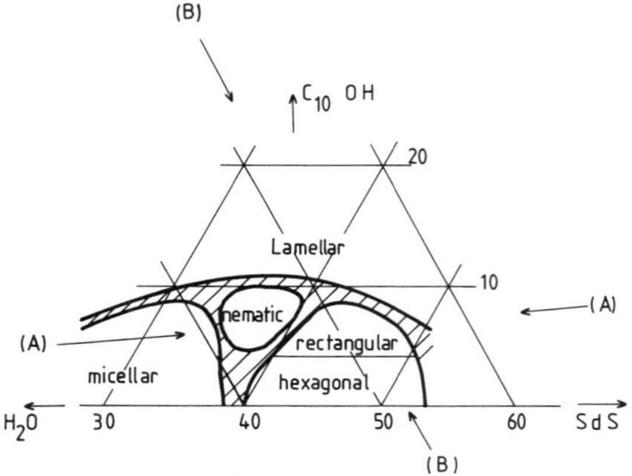

Fig. 4. Schematic representation of part of the sodium decyl sulfate/ decanol/water phase diagram. The concentrations are in weight per cent. The hatched regions are polyphasic domains. The arrows (A) and (B) indicate the direction of the lines discussed in the text.

molecules are not pleased to be in a spherical environment, they build anisotropic aggregates as their concentration increases (for instance, rod-like ones,) in order to decrease the percentage of molecules in this environment. This is obviously a factor of growth as the longer the rod, the smaller the number of molecules in its hemispherical end caps. This factor is counteracted by the dilution entropy of the solution which favors small aggregates. The competition between these two terms controls the size of the aggregates in the ideal micellar solution. As the concentration goes on increasing the micelles start interacting. The first term responsible for the non-ideality of the solution is the entropy associated with the excluded volume. As this volume is smaller for orientationally ordered aggregates than for disordered ones, this term favors the ordering of the system. The appearance of nematic phases of nearly parallel aggregates, or ordered phases of parallel aggregates, can be accounted for by studying the competition between this term and the orientational entropy of the solution. It appears that the existence of nematic phases depends on the fact that the aggregates can be highly concentrated while still being small. This necessary slow growth is ensured by the action of entropic terms which can be strengthened, in the case studied here where two amphiphiles are present, by the heterogeneous distribution of the two molecules within an aggregate[10]. When the nematic order parameter increases the model also shows that the competition between the excluded volume and orientational entropies is at the origin of a very effective

228

size/anisotropy coupling which favors the sudden growth of the aggregates
and their organization is an ordered phase with order parameter 1.

Symmetry Changes in Cylindrical Phases

Line B goes through a domain of the phase diagram where the
amphiphilic molecules build infinitely long cylinders organized along two
dimensional lattices of various symmetries. As the concentration of
decanol increases, the section of the cylinders becomes more anisotropic
and the symmetry changes by steps from hexagonal, p6m, to rectangular, cmm
then pgg[5,11]. These processes are most obviously driven by the introduction of decanol. Its role can be qualitatively understood following a
recent study which shows that sodium decyl sulfate and decanol are not
uniformly distributed within the anisotropic section of the cylinders in a
rectangular phase. Sodium decyl sulfate is preferentially located in the
regions of high interfacial curvature whereas decanol prefers the regions
of low interfacial curvature[12]. Thus decanol makes the formation of flat
interfaces more favorable and its introduction in an aggregate of circular
cross section transforms the latter into an aggregate with anisotropic
cross section of lower average interfacial curvature. When the anisotropy
of the section becomes sufficient, the hexagonal p6m symmetry is deformed
into a rectangular cmm one. The further passage to the pgg symmetry can be
understood considering that the quadrupolar moment of the electric potential of the aggregate section is greatly enhanced by the heterogeneous
distribution of the two amphiphilic species within one aggregate.

This type of structural transformation has not been modelized yet but,
here also, sources of inspiration are certainly to be found in the treatment developed for molecular crystals.

TRANSFORMATIONS WITH CHANGES OF CONNECTIVITY

We choose our example in the binary phase diagram potassium laurate/-
water, which is sketched in Fig. 5. A most spectacular example of transformation with connectivity changes appears along line (A) where the lamellar phase transforms into a cubic one as the water content increases. As
already stated, the lamellar phase is made of infinite numbers of infinite
aqueous and paraffinic sub-spaces whereas the cubic phase is made of tow
infinite paraffinic sub-spaces separated by one infinite aqueous subspace.
During the transformation a rather complex pattern of connections has
therefore to be established between the soap lamellae of the lamallar phase
in order to build the two interwoven structures of the cubic phase. Less
spectacular, but not less typical, is the following transformation of the
cubic phase into a hexagonal phase along the same line (A), or that of the
lamellar phase into intermediate ones of the cylindrical type along line
(B).

A thermodynamical approach of the type of the ones presented above is
of no help here to predict such complex behaviors. Indeed these thermodynamical models are developed on the basis of a priori hypothesis about
the structures of the aggregates. They can be efficient in the case of
structural transformations concerning growth and slight deformations of
aggregates of very simple shapes, seen as spheres and ellipsoids of various
aspect ratio, without topological changes but not in the cases described
just above where so important topological changes take place. Clearly an
approach along a different direction, which provides the topologies of the
structures, is needed as a basis for the application of thermodynamical
consideration.

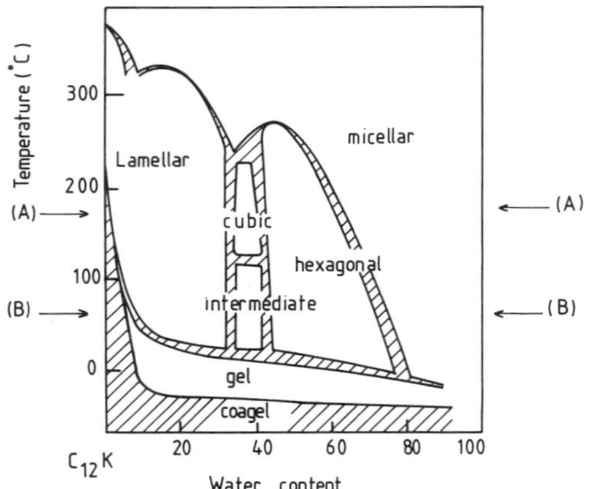

Fig. 5. Schematic representation of the potassium laurate/water phase
diagram. The concentrations are in weight per cent. The hatched
regions are polyphasic domains. The arrows (A) and (B) indicate
the directions of the lines discussed in the text.

The basic idea of our approach to this problem comes from the observation of the existence of internal stresses within bilayers. The areas of
the polar and paraffinic layers constituting a bilayer do not necessarily
vary in concordance when the parameters of the phase diagram are
changed[13]. Differences between these two areas impose symmetric curvatures on the two layers of the bilayer so that, in the absence of any other
reaction of the system, the compactness of the core is no longer preserved,
as shown in Fig. 6. This situation can be described as a frustration
between curvature and thickness which can be solved by a change of structure only[13]. We have examined the possible solutions in developing a
purely geometrical argument following which frustration is solved by the
introduction of defects of rotation, or disclinations[14]. The physical
validity of such an approach lies in the fact that curvature energies
associated with disclinations, are generally much lower in liquid crystals
than those of compression, associated with compacity. The solutions found
that way correspond to organizations of amphiphilic molecules having the
same topologies than those of the liquid crystalline and micellar phases.
Nevertheless geometry alone cannot predict the symmetry of the structures,
or their sequences, and this is where physical factors, other than those of
the origin of the frustration, have to intervene. For instance a given
level of frustration in the bilayers of a lamellar phase can be solved
geometrically by solutions having the topologies of either a cylindrical
phase or a cubic phase. This is illustrated in a certain way by the fact
that in the phase diagram of Fig. 5. cylindrical and cubic phases exist on
the same side of the lamellar phase for the same water content, however
their well-defined locations with temperature is certainly the result of
the action of the physical terms which are not yet in this approach.

RECENT OBSERVATIONS

The preceding paragraphs were written within the frame of the classical descriptions of the structure[1,4] where, particularly, the lamellar
phases are described as being made of infinite bilayers of amphiphilic
molecules. However these descriptions might be modified as recent experimental studies show that bilayers of lamellar phases at the limits of
their domains can be fragmented in a way which can be looked at as a pre-

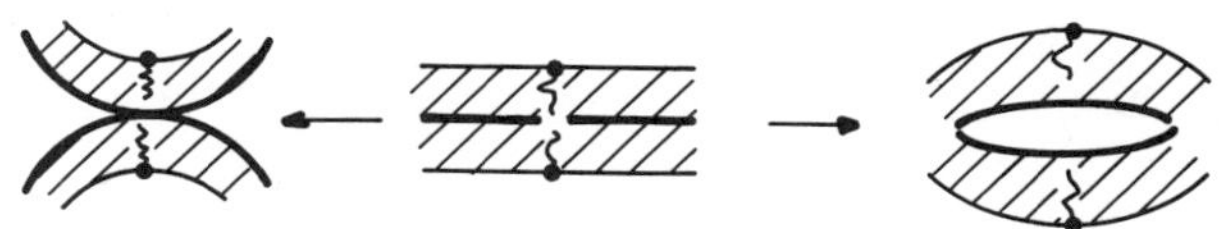

Fig. 6.   Center: a bilayer with equal interfacial and mid areas, left
          (right: the interfacial area is smaller (larger) than the middle
          one.

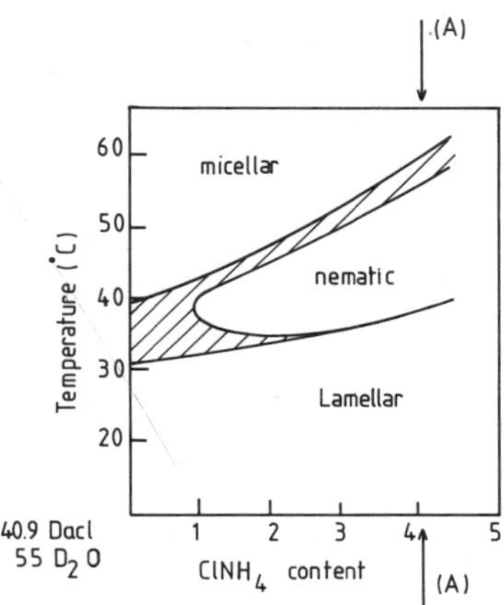

Fig. 7.   Schematic representation of one section of the decyl ammonium
          chloride/water/ammonium chloride phase diagram. The concen-
          trations are in weight per cent. The hatched regions are
          polyphasic domains. The arrows (A) indicate the direction of the
          line discussed in the text.

cursor of the phase transformations. The first example of such a situation
was found in the ternary phase diagram decylammonium chloride/water/-
ammonium chloride, sketched in Fig. 7, where we studied the lamellar-
nematic transition along line (A)[15]. The X-ray pattern of the lamellar
phase in the vicinity of the transition, Fig. 8, not only exhibits meridian
Bragg spots, associate to the lamellar order, but also diffuse vertical
bands, which are to be associated to a fragmentation of the lamallae in
rather mono dispersed aggregates. The transition corresponds to the loss
of the translational ordering of the fragments of bilayers. A second
example was found in the lamellar phase of the sodium dodecyl sulfate/water
system in the vicinity of its transformation into a cylindrical phase of
oblique p1 or p2 symmetry[16]. Here also the X-ray pattern of the lamellar
phase presents diffuse scatterings which are to be associated to a fragmentation of the lamellae. The organization of the fragments presents some
relation with that of the ribbons in the oblique phase. A similar
situation prevails in the lamellar phase of the sodium decyl sulfate/
decanol/water system, whose phase diagram was discussed previously, Fig. 4,
in the vicinity of its transformation into the rectangular phase pgg along
line (B)[11].

    Such examples of fragmentation of bilayers in lamellar phases can be
interpreted within the frame of the frustration model as solutions found by
the system to accommodate very low levels of frustration. They also show
that lamellar-nematic or lamellar-cylindrical transformations which might
have been thought of being the second class, with connectivity changes, may
belong to the first class, without connectivity change. This suggests that

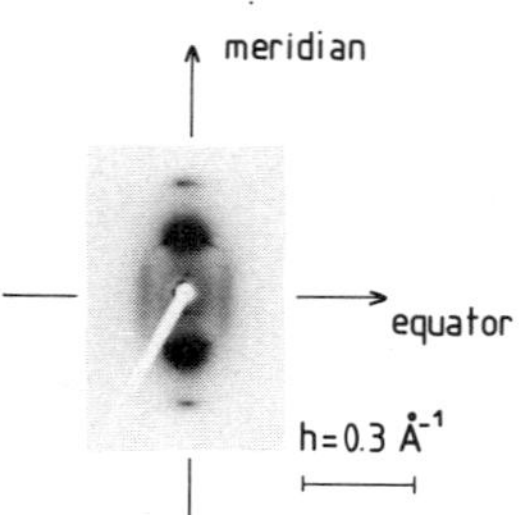

Fig. 8.   X-ray diagram of an oriented lamellar phase of decylammonium chloride/water/ammonium chloride (line (A), 36°C, on Fig. 7). The lamellae are horizontal. (The small meridian spots, close to the beam stop, are artefactual, the Bragg spots are over exposed).

the geometrical change of connectivity and the physical change of order are not necessarily at the same place in the phase diagrams.

CONCLUSION

We have presented a few detailed experimental studies of structural transformations in ordered lyotropic liquid crystals. We have examined some of the models proposed to give account of them and we have stressed upon the importance of geometrical features which might be of help to develop an integrated approach of those complex systems. Among the information obtained some of them might be of interest also for analyzing disordered systems. They concern the process of aggregation, a more accurate description of the growth and its factors, the properties of mixed films, the coupling between relative concentrations and curvature, and the properties of bilayers, notion of frustration and fragmentation. As these information concern anisotropic local properties they are more easily studied by scattering methods in ordered systems than in disordered ones.

REFERENCES

1. P. Ekwall, Composition, properties and structures of liquid crystalline phases in systems of amphiphilic compounds, in "Advances in Liquid Crystals", vol. 1, G. H. Brown ed., Academic Press, New York (1978).
2. A. Skoulios, Amphiphiles: organisation et diagrammes de phases, Ann. Phys. 3, 421 (1978).
3. V. Luzzati, X-ray diffraction studies of lipid-water systems, in "Biological Membranes", vol. 1, D. Chapman ed., Academic Press, New-York (1967).
4. J. Charvolin and Y. Hendrikx, Lyotropic nematic phases of amphiphilic compounds, in "Liquid Crystals of One and Two Dimensional Order", W. Helfrich and G. Heppke ed., Springer-Verlag, Berlin (1980).
5. Y. Hendrikx and J. Charvolin, Structural relations between lyotropic phases in the vicinity of the nematic phases, J. Physique, 42, 1427 (1981).
6. L. J. Yu and A. Saupe, Observation of a biaxial nematic phase, Phys. Rev. Lett. 45, 1000 (1980).
7. W. M. Gelbard, A. Ben Shaul, W. E. McMullen and A. Masters, Micellar growth due to interaggregate interaction, J. Phys. Chem. 88, 861 (1984). W. E. McMullen, A. Ben Shaul and W. M. Gelbart, Rod/disk coexistence in dilute soap solutions, J. Coll. Interface Sci. 98, 523 (1984). W. M. Gelbart, W. E. McMullen, A. Masters and A. Ben Shaul, Partitioning of cosurfactants in mixed micelles: size

enhancement and nematic stability, Langmuir, 1,101 (1985).
W. M. Gelbart, W. E. McMullen and A. Ben Shaul, Nematic stability
and the alignment induced growth of anisotropic micelles, J. Physique, 46, 1137 (1985).

8. W. M. Gelbart, A. Ben Shaul, A. Masters and W. E. McMullen, Effects of
    interaggregate forces on micellar size, in "Physics of Amphiphiles:
    Micelles, Vesicles and Micro-emulsion", V. Degiogio and M. Corti
    ed., North-Holland, Amsterdam (1984).

9. J. Charvolin, From micelles to liquid crystals, Mol. Cryst. Liq.
    Cryst. 113, 1 (1984).

10. Y. Hendrikx, J. Charvolin and M. Rawiso, Segregation of two
    amphiphilic molecules within non-spherical micelles, J. Coll.
    Interface Sci. 100, 597 (1984).

11. S. Alpérine, "Etudes structurales de transformations de phase dans des
    cristaux liquides lyotropes." Thèse de doctorat de troisième cycle,
    Orsay (1985).
    S. Alpérine, Y. Hendrikx and J. Charvolin, in preparation.

12. S. Alpérine, Y. Hendrikx and J. Charvolin, Internal structure of
    aggregates of two amphiphilic molecules in a lyotropic liquid
    crystal, J. Physique Lett. 46, L-27 (1985).

13. J. Charvolin, Crystals of interfaces: the cubic phases of amphiphile/
    water systems, J. Physique, 46, C3-173 (1985).

14. J. F. Sadoc and J. Charvolin, Frustrations in bilayers, J. Physique,
    47, 683 (1986). J. Charvolin and J.F. Sadoc, Bicontinuous cubic
    phases, J. Physique 48, 1559 (1987).

15. M. C. Holmes and J. Charvolin, Smectic-Nematic Transition in a
    lyotropic liquid crystal, J. Phys. Chem. 88, 810 (1984).

16. P. Kékicheff, B. Cabane and M. Rawiso, Structural defects of a lamel-
    lar lyotropic mesophase, J. Physique Lett. 45, L813 (1984).

SOME ASPECTS OF LYOTROPIC LIQUID CRYSTALS

G. Chidichimo, G. A. Ranieri,
F. P. Nicoletta*, M. Meuti* and R. Bartolino*

Department of Chemistry
*Department of Physics
Unical Liquid Crystals Group
University of Calabria
Arcavacata di Rende, 87036 Cosenza, Italy

INTRODUCTION

We present in this review our most recent results in the field of
Lyotropic Liquiid Crystals as well as related literature studies. The
first part deals with diffusion of water in different Lyotropic mesophases
which have been investigated by Nuclear Magnetic Resonance (NMR)
techniques. Data concerning systems such as the L_α and H_α mesophases
obtained from binary mixtures of Potassium Palmitate (KP) and water and
those obtained from a ternary mixture of Potassium Palmitate, Potassium
Laurate (KL) and water, giving Ribbon (RB_α) mesophase, will be presented.
The main purpose of such studies was to provide a deeper understanding of
factors which determine the mobility of water in Lyotropic Liquid
Crystals. Together with such studies of microscopic properties we have
been interested in the investigation of properties at macroscopic level.

In the second part we report the more recent results obtained from
birefringence and light scattering experiments in Nematic Lyotropics.

The systems studied are ternary mixtures of Sodium decyl sulfate
(SdS)-Decanol-water and KL-Decanol-water, which present three different
Nematic mesophases such as the Calamitic (NC), the Discotic (ND) and the
Biaxial one (NbX). The purpose of such research is to study the elastic
properties of these systems, the normal modes of thermal fluctuations and
flow properties, including the pretransitional effects near phase
transitions.

PART 1

Water Diffusion in Lyotropic Liquid Crystal by
Nuclear Magnetic Resonance (NMR) Spectroscopy

Nuclear Magnetic Resonance (NMR) provides a simple and useful
experimental technique for studying and measuring the rate of molecular
motion (e.g., rotational and translational) in liquids as well as in
solids.

There are two rather different approaches to the study of diffusion in liquids by means of NMR.

One method employs measurements of the relaxation times (e.g., T_1 and $T_{1\rho}$) as a function of temperature and/or of the Larmor frequency. This experimental method is usually classified as indirect (or microscopic) since it may allow the determination of the mean atomic jump time (e.g., diffusion correlation time), τ_c, that is a quantity related to the microscopic nature of diffusive motion. In an isotropic system, τ_e can be related to the macroscopic coefficient of diffusion D, as expressed by Fick's law through the Einstein relation:

$$D = \frac{<r^2>}{6\tau_e}$$

where $<r^2>$ is the mean square jump distance.

It should be considered that this technique is successful in giving information on the diffusion process only when the nuclear spin-relaxation is controlled by translational motion. A number of relaxation mechanisms may be in fact competitive over a sizeable temperature range, and the problem exists of separating the contribution to relaxation due to diffusion-dependent and diffusion-independent mechanisms.

The other NMR method for studying diffusion is the spin-echo technique in the presence of a known magnetic field gradient. This is a direct (or macroscopic) method since it gives a direct measurement of the diffusion coefficient D_{NMR}, which gives the macroscopic atomic migration as opposed to the microscopic jump time τ_c, detected in the relaxation times experiments.

The method is equivalent to tracer diffusion experiment, where one monitors the migration of a tagged atom over a given interval of time. Here the tagged atom is the nucleus itself, which will experience different fields and therefore different precessional frequencies, according to its location along the magnetic field gradient applied across the sample. In a spin-echo experiment, in the presence of a steady magnetic gradient field g, following the $\pi/2 - \tau - \pi$ pulse sequence, an echo will appear at time 2τ, with amplitude[1]

$$A(2\tau) = \exp[-(2\tau/T_2 + 2\gamma^2 g^2 D_{NMR}\tau^3/3)]$$

and therefore the macroscopic diffusion coefficient, D_{NMR}, can be measured directly. Here T_2 is the transverse relaxation time, γ the gyromagnetic ratio.

It would appear that D_{NMR} can be measured over a large range by varying g and τ appropriately. However, due to a number of experimental limitations, the continuous field gradient method is restricted to $D_{NMR} > 10^{-7}$ cm^2/sec and to systems with unrestricted diffusion processes.

For slower diffusion rates a modified experiment using a pulsed field gradient (PFG) has been developed by Stejskal and Tanner[2], which extends the range of D_{NMR} up to about 10^{-9} cm^2/sec and can be used also when the atoms are diffusing in a finite volume (restricted diffusion). Here large field gradient, g, are applied for a short time δ, between the $\pi/2$ and π rf pulses, and between the π pulse and the echo signal. Normally the following equation holds:

$$\ln \frac{A(g)}{A(0)} = -\gamma^2\delta^2 g^2 D_{NMR}\left(\Delta - \frac{\delta}{3}\right), \tag{1}$$

where A(g) and A(0) are the amplitudes of the echo signal with and without field gradient, respectively, and Δ is the time separation of the two gradient pulses. The PFG-NMR technique is particularly suitable in the study of water mobility in lyotropic systems where the diffusion coefficient is of the order of 10^{-5} - 10^{-6} cm^2/sec.

However, such experiments have been performed just in a few laboratories, and few papers have been reported in the literature concerning, for example, water diffusion in randomly oriented L_α and H_α mesophases[3,4], stacked bilayers[5,6] and lamellar phase of the binary system AOT/water[7].

The water diffusion in lyotropic liquid crystals depends on the structure of the amphiphilic aggregates, on their spatial orientation and on the composition of samples.

In fact, water molecules cannot travel across the amphiphilic aggregates and for this reason their mobility is lower than that of pure water, depending on the shape and spatial packing of such aggregates which behave as boundary walls enclosing local water domains.

Furthermore, it should be considered that restriction to water mobility exists in the interspace among different aggregates. The aggregate surface can in fact temporarily link the water molecules, and as a consequence a fast equilibrium is established between free and bound water molecules. Since the time scale, over which diffusion is measured in the PFG-NMR experiment is much higher than the mean time a water molecule exchanges between the free and bound state, the experimental value of D_{NMR} will not give information on the details of the exchanging process, and will depend on the population ratio between free and bound water.

It has been shown[8-9] that the experimental value of D_{NMR} can be generally written as:

$$D_{NMR} = f[(1 - P)D_W + PD_b]$$ (2)

where D_W and D_b are the diffusion coefficient of free and bound water respectively, P is the fraction of bound water and f is the structural factor, which account for the constraints to water diffusion due to the spatial boundary effects of the lyotropic aggregates. Its value may range from zero to a pure water value of unity. It has been shown[8] that the structural factor of a given mesophase can be obtained by measuring D_{NMR} in two samples of the same mesophases with a different stoichiometric composition, by using the relation:

$$f = \frac{D_{NMR}^i - K_{iJ}D_{NMR}^J}{(1 - K_{iJ})D_W}$$ (3)

where index i and J refer to two different lyotropic mixtures of the same mesophase and K_{iJ} is the ratio P_i/P_J between the fractions of bound water in the samples i and J respectively. By defining W_i and W_J as the water weight fractions in the samples i and J respectively, the factor K_{iJ} can be obtained through the equation:

$$K_{iJ} = \frac{W_J(1 - W_i)}{W_i(1 - W_J)}$$ (4)

under the reasonable assumption that for a given lyotropic mesophase the chemical bound water per unitary weight of amphiphile does not depend on the composition. By applying Eqs. (2) - (4) it is possible to determine the structural factor and the probability of bound water at the different compositions using D_{NMR} data obtained from two different-composition mixtures of the same lyotropic mesophase. The validity of such an approach has been tested (Refs. [8-9]) in the case of the lamellar and hexagonal mesophases. A short summary of the results is given in the next two paragraphs.

<u>Lemellar L_α Mesophases[8]</u>

The structural factor f in the case of non-aligned L_α mesophases can be obtained by simple theoretical arguments. Let us consider a cluster of lamellar aggregates having a common direction. We can define a reference frame xyz having the xy plane parallel to the lamellar aggregates (see Figure 1). Let D_x, D_y and D_z be the diffusion coefficients along x, y and z respectively.

If the axis of measurement (e.g., the field gradient direction) is oriented in the direction defined by the polar angles θ and ϕ, one can write:

$$D_{NMR} = D_z \cos^2\theta + D_x \sin^2\theta\cos^2\phi + D_y \sin^2\theta\sin^2\phi. \tag{5}$$

In the present case it is evident that:

$$D_z = 0; \quad D_x = D_y = D_\perp$$

where the subscript $\perp$ indicates the direction perpendicular to the local director. Eq. (1) can then be written:

$$A(g)/A(0) = \exp[- KD_\perp \sin^2\theta] \tag{6}$$

where $K = \gamma^2\delta^2 g^2 (\Delta - \frac{\delta}{3})$.

Since a non-aligned L_α mesophase is a polycrystalline sample with a random θ distribution, the measured quantities are averaged over all the possible orientations. Then:

$$<A(g)/A(0)>_\theta = \frac{1}{2} \int_o^\pi \exp[- KD_\perp \sin^2\theta]\sin\theta d\theta. \tag{7}$$

In the condition $KD_\perp < 1$, usually met in our experiment, the above integral can be evaluated giving:

$$\ln <A(g)/A(0)>_\theta = - \gamma^2\delta^2 g^2 (\Delta - \frac{\delta}{3})(\frac{2}{3}D_\perp). \tag{8}$$

The measured quantity D_{NMR} obtained by plotting $\ln[A(g)/A(0)]$ versus Δ (in the approximation $\Delta \gg \delta/3$) will be then $(2/3)D_\perp$. Since $D_\perp$ is only affected by the chemical binding of the water to the aggregate surface, one can write:

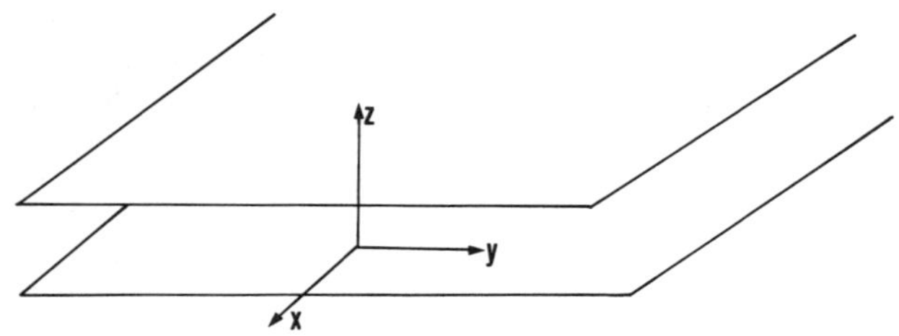

Fig. 1. Lamellar boundary in the magnetic gradient field.

$$D_{NMR} = \frac{2}{3} [(1 - P)D_w + PD_b].\qquad(9)$$

Physically the 2/3 factor accounts for the reduction of the diffusion due to the geometric constraints imposed by the lyotropic aggregates, and therefore it is the structural factor f for lamellar mesophases.

The same conclusion has been reached by Callaghan and Soderman by following a similar procedure{7]. An experimental test of the above conclusion is the measurement of the diffusion coefficient in the L_α mesophase of the binary system Potassium Palmitate (KP) and water[8]. Three different samples of such a mesophase having water weight fractions of 0.28 (sample 1), 0.31 (sample 2) and 0.37 (sample 3) respectively have been analyzed.

The data are shown in Figure 2. The following structural factors have been obtained by using Eqs. (3) and (4) and by pairing the data corresponding to the three different samples as shown in Table 1. The agreement with the predicted 2/3 value appears quite good.

Hexagonal H_α Mesophase[9]

The structural factor f of the hexagonal H_α mesophase can be derived by following a procedure similar to that used for lamellar mesophase. In this case a cluster of cylindrical aggregates such as that shown in Figure 3 must be considered and a local frame having the xy plane containing the cylindrical aggregates, and therefore the director, can be defined.

Here we have:

$$D_x = D_y = D_\| = [(1 - P)D_w + PD_b].\qquad(10)$$

Along the z axes diffusion is limited to the interspace among amphiphilic aggregates. It has been assumed[9] that D_z is proportional to the fraction of empty space existing among the cylindrical aggregates which has been expressed as $(d - d_i)/d$, d being the distance between the axis of neighboring cylinders and d_i the cylinder diameter. It follows:

$$D_z = D_\perp = \frac{d - d_i}{d} [(1 - P)D_w + PD_b]\qquad(11)$$

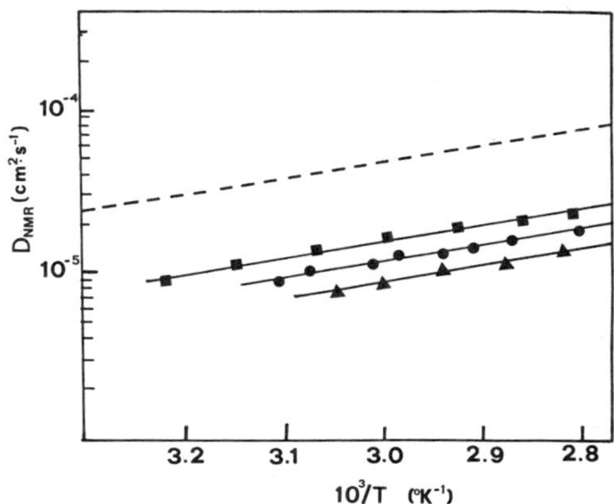

Fig. 2. PGF-NMR diffusion data in lamellar lyotropic system of Potassium Palmitate and water. (■ sample 1; ● sample 2; ▲ sample 3; --- pure water).

Table 1. Values of Structural Factor f
for L_α Mesophase Obtained by
Pairing Data of Figure 2

Sample Pairs	Structural Factor (PFG–NMR Data)	
1 − 2	0.58	± 0.04
1 − 3	0.62	± 0.04
2 − 3	0.64	± 0.04

and

$$<A(g)/A(0)>_\theta =$$

$$= \frac{1}{2} \int_o^\pi \exp[-K(D_\parallel \sin^2\theta + D_\perp \cos^2\theta)]\sin\theta d\theta \tag{12a}$$

$$\simeq \exp[-K(\frac{2}{3} D_\parallel + \frac{1}{3} D_\perp)]. \tag{12b}$$

The D_{NMR} diffusion then takes the form:

$$D_{NMR} = \frac{1}{3} [2 + \frac{d - d_i}{d}]\{(1 - P)D_w + PD_b\} \tag{13a}$$

or, considering that $D_b \ll D_w$:

$$D_{NMR} = \frac{1}{3} [2 + \frac{d - d_i}{d}]. \tag{13b}$$

The structural factor f for H_α mesophase results:

$$f(H_\alpha) = \frac{1}{3} [2 + \frac{d - d_i}{d}]. \tag{14}$$

An experimental study of the water diffusion, in H_α mesophase, has been presented also in [9]. Diffusion data are summarized in Figure 4.

Samples 1, 2, 3, 4 have water weight fractions of 0.425, 0.472, 0.537 and 0.73 respectively. By using Eqs. (3) and (4) the structural factor has been obtained by pairing the data. Results are reported in Table 2.

A slight but significant increment of f with increasing water content can be observed. This can be explained in terms of an increasing of the distance d between the axes of the cylindrical aggregates (see Eq. (14)).

A rather good agreement has been observed[9] for the structural factors as calculated from D_{NMR} data (through Eqs. (3) and (4)) and from X-ray data[11] by using Eq. (14).

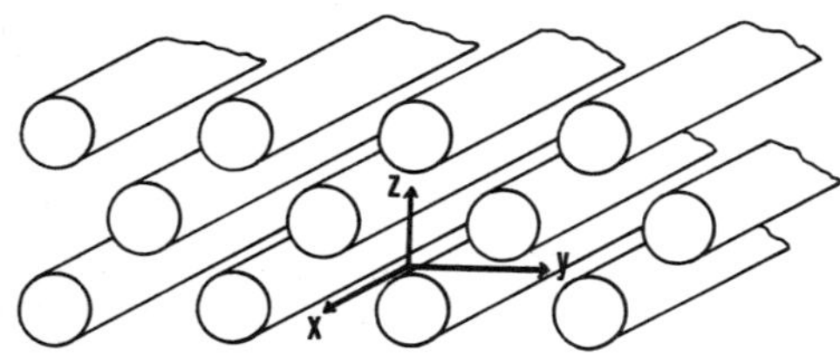

Fig. 3. Spatial organization of the Hα mesophase (notation used in the text).

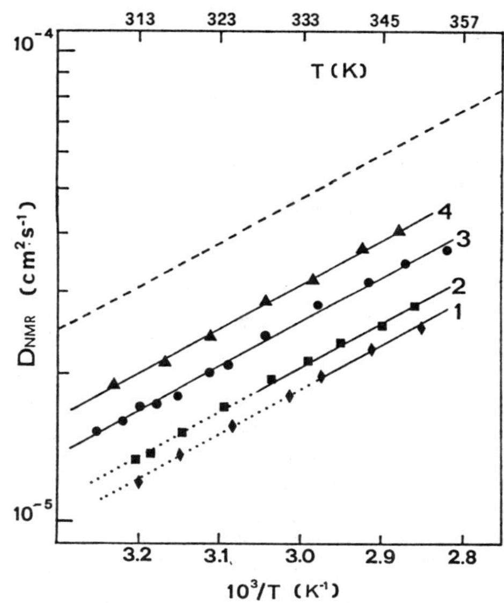

Fig. 4. Temperature dependence of water diffusion coefficient from the lyotropic mixtures 1, 2, 3, 4 and from pure water (dashed lines): (a) hexagonal Hα mesophase (continuous lines); (b) another mesophase sandwiched between Lα and Hα mesophases (dotted lines).

Figure 5 shows the trend of the chemically bound water probability P versus the composition, obtained by Eq. (13b).

Ribbon Mesophase[10]

Recent studies have shown that such a mesophase can be obtained by mixing lipids, having a different chain length, as well as different polar head[12-15]. A systematic study of water diffusion in such a lyotropic system has not yet been done and only few data are available[10]. It is expected that the structural factor value for such a mesophase is similar to that of the H_{α} mesophase, since the shape of ribbon aggregates is not very far from that of a deformed cylinder.

Diffusion data from a ribbon phase obtained by mixing Potassium Palmitate (7.92 m%), Potassium Laurate (1.11 m%) and water (90.97 m%) are shown in Figure 6.

Table 2. Values of Structural Factor f for H_{α} Mesophases Obtained by Pairing Data of Figure 4

Sample Pairs	Structural Factor (PFG–NMR Data)	
1 – 2	0.74	± 0.02
1 – 3	0.75	± 0.02
1 – 4	0.77	± 0.02
2 – 3	0.75	± 0.02
2 – 4	0.78	± 0.02
3 – 4	0.79	± 0.02

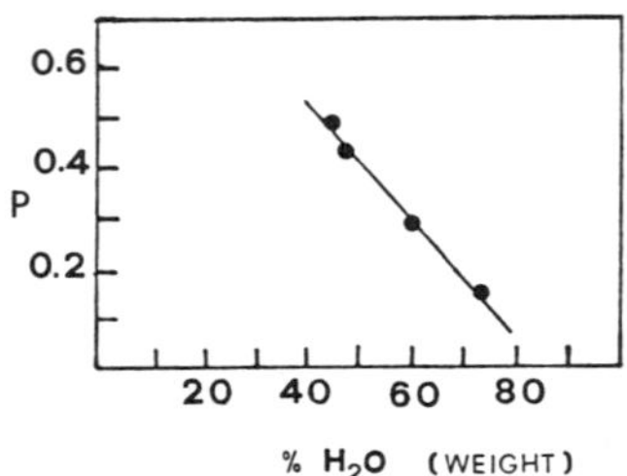

Fig. 5. Bound water fraction P as a function of water weight percentage in
Hα mesophase.

Here the ribbon mesophase corresponds to the temperature range 25 -
60°C, as shown by NMR spectral profiles studies. Below 25°C the ternary
system correspond to a mixture of cylindrical and lamellar (L$_\beta$) aggregates
(the percentage of lamellar mesophase increases with decreasing
temperature). Above 60°C the system becomes a mixture of ribbon and
lamellar L$_\alpha$ aggregates (the ribbons gradually disappear with increasing
temperature). The data obtained from the ternary mixture are compared in
Figure 6 with those obtained from a L$_\alpha$ binary system: KP (12.5 m%) and
water (87.5 m%), and from a H$_\alpha$ mesophase: KP (3.9 m%) and water (96.1 m%).

It is interesting to observe that in monophase systems the slope of
D_{NMR} against T is almost equal to that of pure water.

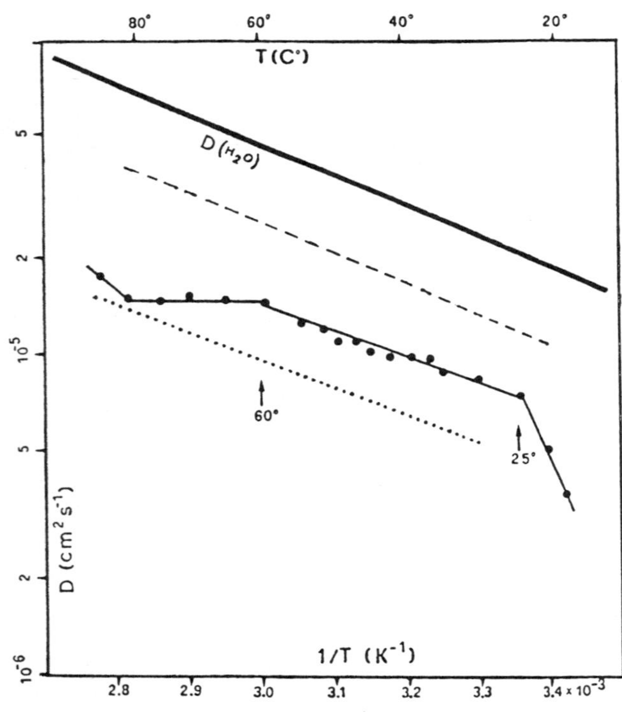

Fig. 6. Temperature dependence of the diffusion constant obtained from the
lyotropic mixtures: ● KP 7.92 m%, KL 1.11 m%, water 90.97 m%;
(dot line) KP 12.5 m%, water 87.5 m%; (dash line) KP 3.9 m%, water
96.1 m%.

PART 2

Lyotropic Nematics

Lyotropic nematics are solutions of amphiphilic aggregates of finite dimensions and anisotropic shape which generally occur in narrow ranges of compositions and temperatures[16]. Their elastic and optical properties are similar to those of thermotropic nematics, both being classes of highly anisotropic liquids[17,18]. Since it is very easy to obtain oriented monodomains of lyotropic nematics[16,19], their physical properties can be described in terms of order parameters. Generally nematic phases lie between the isotropic micellar phase and Lamellar or Hexagonal anisotropic mesophases, having intermediate structures function of concentration, temperature and molecular properties of the surfactant.

Up to now two uniaxial nematic phases, with opposite magnetic susceptibility $\Delta\chi$ and birefringence $\Delta\varepsilon$ [17,20], and a biaxial one have been observed.

When investigated by X-ray and neutron diffusion scattering, discotic nematics Nd appear as small disks randomly distributed in solution with their axes almost parallel to an average direction $\bar{n}_o(\bar{r})$ (the director). Discotic aggregates of sodium decylsulphate-decanol-water have a thickness of 20 Å and a diameter of $\cong$ 57 Å. Calamitic nematics, Nc, are cylinders with axes almost parallel to $\bar{n}_o(\bar{r})$. Their average diameter and length are 26 Å and $\simeq$ 60 Å respectively, in the case of the SdS-decanol-water system. Other systems show similar dimensions.

Biaxial nematics (Figure 8) N_{bx}, are cylinders with a rectangular section[21] (even if micelles have not a rigid structure due to thermal fluctuations). They are characterized by three mutually perpendicular twofold axes whose directions are denoted by the vectors $\bar{n}_o(\bar{r})$, $\bar{m}_o(\bar{r})$ and $\bar{l}_o(\bar{r}) = \bar{m}_o(\bar{r}) \times \bar{n}_o(\bar{r})$.

To define in a quantitative way the order of a mesophase, order parameters have been introduced which vary from 1 (in full ordered phases) to zero (in disordered phases).

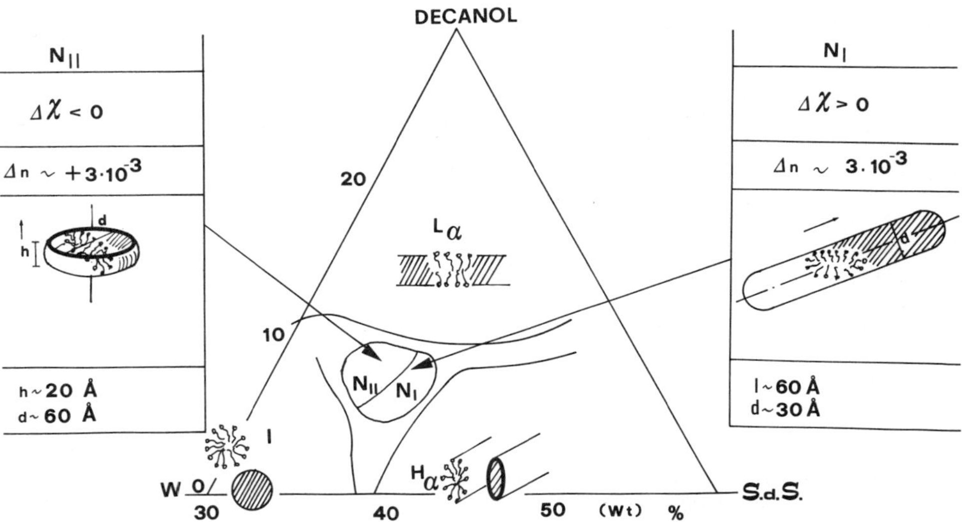

Fig. 7. A schematic phase diagram which shows some mesophases formed in this SdS-dec-water system. The two uniaxial phases are on the left; the disk-like nematic, and on the right, the calamitic one.

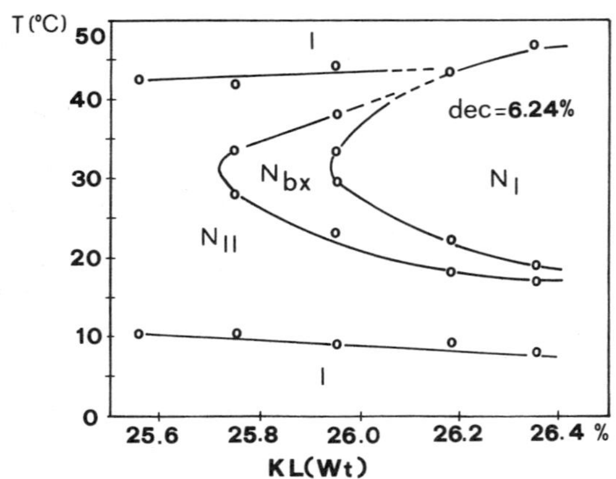

Fig. 8. Biaxial nematic region for: Kl–dec–water system at fixed decanol
concentration.

From a macroscopic point of view the order of an uniaxial nematic is
described by the tensor order parameter[22]:

$$Q_{\alpha\beta} = S(n_\alpha n_\beta - \frac{1}{3}\delta_{\alpha\beta}) \tag{15a}$$

with S normalization constant and $\alpha,\beta = x,y,z$; while for a biaxial nematic
[23]:

$$Q_{\alpha\beta} = S(n_\alpha n_\beta - \frac{1}{3}\delta_{\alpha\beta}) + \frac{1}{3}T(\ell_\alpha \ell_\beta - m_\alpha m_\beta) \tag{15b}$$

where T is another constant, $Q_{\alpha\beta}$ is a physically measurable quantity which
can be, for example, identified with the dielectric or diamagnetic
susceptibility or can be related to the differential cross section in
scattering experiments.

Elasticity

In this section we report our most recent light scattering
experiments. In this case the properties of samples in the hydrodynamic
limit are studied; then liquid crystals can be considered as continuous
media[22,23]. The elastic continuum theory for liquid crystals was
obtained by Frank[24] expanding the free energy density of the liquid
crystal sample, f, around its homogeneous state value, f_o.

In the absence of external field, the difference $f - f_o = f_d$ is due
to fluctuations or distortions of director $\delta\bar{n} \simeq \bar{n} - \bar{n}_o$, and to order
parameter fluctuations. f_d is generally a function of the order
parameters and their spatial derivatives.

Using only relevant tensors, the distortion free energy density of an
uniaxial nematic is:

$$f_d = \frac{1}{2}K_{11}(\nabla\bar{n})^2 + \frac{1}{2}K_{22}(\bar{n}\cdot\nabla\times\bar{n})^2 + \frac{1}{2}K_{33}(\bar{n}\times\nabla\times\bar{n})^2 \tag{16}$$

where K_{ii} are the Frank-Oseen's bulk elastic constants for the residual
three elementary distortions. We neglect the surface terms since
the general formula requires 5 bulk and 2 surface constants.

The intrinsic structure of micelles will give rise to other
contributions in the free energy density expansion, due to micellar

244

flexibility and to local concentration fluctuations which could be
responsible for local phase transition.

The application of an external field gives[22]:

$$f_m = \frac{1}{2\mu_o} \Delta\chi(\bar{H}\cdot\bar{n})^2 \qquad f_e = -\frac{1}{2} E_o \Delta E(\bar{E}\cdot\bar{n})^2, \qquad (17a)$$

$\Delta\chi$ and $\Delta\varepsilon$ being respectively the magnetic and electric susceptibilities.
Recently Govers and Vertogen[25] have derived the expression of the
free energy density also for a biaxial nematic. As predicted by Brand
and Pleiner[26,27] the free energy density needs 12 bulk elastic
constants, 3 surface elastic constants (generally neglected) and 5
twist elastic constants to account for chiral biaxiality. The complete
bulk expression is:

$$
\begin{aligned}
f_d = &\, K_1(\bar{n}\cdot\nabla\times\bar{n}) + K_2(\bar{m}\cdot\nabla\times\bar{m}) + K_3(\bar{m}\times\bar{n})[(\bar{m}\cdot\nabla)\bar{n}] + K_4(\bar{n}\times\bar{m})[(\bar{n}\cdot\nabla)\bar{m}] + \\
&+ K_5[(\bar{m}\times\bar{n})(\bar{m}\times\nabla\times\bar{n}) + (\bar{n}\times\bar{m})(\bar{n}\times\nabla\times\bar{m})] + \tfrac{1}{2}K_1(\nabla\bar{n})^2 + \tfrac{1}{2}K_2(\bar{n}\cdot\nabla\times\bar{n})^2 + \\
&+ \tfrac{1}{2}K_3(\bar{n}\times\nabla\times\bar{m})^2 + \tfrac{1}{2}K_4(\nabla\bar{m})^2 + \tfrac{1}{2}K_5(\bar{m}\cdot\nabla\times\bar{m})^2 + \tfrac{1}{2}K_6(\bar{m}\times\nabla\times\bar{m})^2 + \\
&+ \tfrac{1}{2}K_7[\bar{n}\cdot(\bar{m}\times\nabla\times\bar{m})]^2 + \tfrac{1}{2}K_8[\bar{m}\cdot(\bar{n}\times\nabla\times\bar{n})]^2 + \\
&+ \tfrac{1}{2}K_9[\bar{m}\cdot\nabla\times(\bar{n}\times\bar{m})]^2 + \tfrac{1}{2}K_{10}[\bar{n}\cdot\nabla\times(\bar{m}\times\bar{n})]^2 + K_{11}[\nabla\times(\bar{n}\times\bar{m})]^2 + \\
&+ \tfrac{1}{2}K_{12}[\nabla(\bar{n}\times\bar{m})]^2.
\end{aligned}
\qquad (17b)
$$

The uniaxial director fluctuations, in the isotropic phase of a
Lyotropic Liquid Crystal are normal to the initial alignment and there
will be two components of $\delta n(\delta\bar{n}_\alpha, \alpha = 1,2)$ which will give rise to two
collective normal modes of relaxation whose rates are:

$$\Gamma_\alpha(q) = \frac{1}{\eta_\alpha(\bar{q})} (K_{\alpha\alpha}q_\perp^2 + K_{33}q_\parallel^2) \qquad (18a)$$

$q_\perp$ and $q_\parallel$ are the components of scattering wave vector $\bar{q}$ normal and
parallel to $\bar{n}(\bar{r})$ and η_α is the effective viscosity of α mode. Under
particular geometries it is possible to separate the fluctuation modes and
calculate $\Gamma_\alpha = (K_{\alpha\alpha}/\eta_\alpha)q^2$ by light scattering[22]. It is then possible to
evaluate $\partial_\alpha = K_{\alpha\alpha}/\eta_\alpha$ from a (q^2,Γ) plot.

The two collective normal modes of biaxial fluctuations in uniaxial
phase of lyotropic nematics are of relaxational type with rates

$$\Gamma_S = \frac{a(T_c - T) + cq^2}{\nu} \qquad \Gamma_T = \frac{\nu}{\nu'}\,\Gamma_S, \qquad (18b)$$

where a and c are the coefficients of the Landau expansion of the free
energy density (see below): T_c is the uniaxial-biaxial nematic transition
temperature, ν and ν' are effective viscosities. If one changes the
polarization geometries it is possible to separate the two modes, using
polarized light for mode S and depolarized light for T mode.

Phase Transitions

Phase transitions in Lyotropic Nematics are usually of second order.
They are well explained by Landau phenomenological theory[27], which
assumes that, near the transition, the free energy density is

$$f_d = \frac{1}{2} AQ_{\alpha\beta}Q_{\beta\alpha} + \frac{C}{4}(Q_{\alpha\beta}Q_{\beta\alpha})^2 + \frac{1}{2}L_1\partial_\alpha Q_{\beta\gamma}\partial_\alpha Q_{\beta\gamma} + \frac{1}{2}\partial_\alpha Q_{\alpha\beta}\partial_\beta Q_{\beta\gamma} + \cdots \qquad (19)$$

where $A \simeq a_o(T - T_c)$, C is independent of T, and T_c is the phase transition temperature.

Order parameters have important fluctuations when the temperature T approaches T_c, since small range order of the ordered phase starts to establish in the disordered phase. The length over which an order parameter fluctuation can extend is the coherence length ξ. This length steeply increases while T approaches T_c, causing the divergence of macroscopic quantities according to some critical exponents of the reduced temperature $E = (T - T_c/T_c)$.

Some of the most important critical exponents are[28]:

Exponent	Quantity	Law
α	specific heat	$c_v \sim E^{-\alpha}$
β	order parameter	$S \sim (-E)^{\beta}$
ν	coherence length	$\xi \sim E^{-\nu}$
γ	magnetic susceptibility	$\chi \sim E^{-\gamma}$

These critical exponents are linked together by some relations. By optical experiments it is possible to obtain ν and γ, whose values in a mean field theory (e.g., Landau theory) are respectively $\nu = 1/2$ and $\gamma = 1$.

Recent models have been proposed to study the defect rule in phase transitions (dislocation-mediated phase transitions)[29].

This effect could be quite important, in Lyotropic Nematics, where micellar modifications and order modifications start near the transition.

Allain[30] observed by electron microscopy that in the lamellar phase of the systems $C_{12}EO_5$-water and $C_{12}EO_6$-water (aqueous solutions of pentaethylene glycol and hexaethyleneglycol dodecyl ether) the equilibrium defects increase with temperature. These defects built dislocation loops which cross the layers. (See also J. M. Di Meglio, this meeting.)

Experimental Description

The most studied systems (SdS-decanol-water and KL-decanol-water) were observed by Saupe[31] and Charvolin[32] which characterized these Lyotropic phases. The orientational properties were studied by light microscopy and NMR, while the structure and the dimension of micelles were determined by X-ray and neutron scattering.

The elastic and dynamic properties are generally studied by optical techniques. In particular, conoscopic observations and birefringence versus temperature data[33] have been used in order to determine the invariants of the order parameters[34]. In uniaxial phases the order parameters depend linearly on T and one of the two birefringences is zero; in the biaxial phase both birefringences are different from zero.

Interference measurements show that the birefringence is two orders of magnitude lower than in thermotropic nematics and changes optical sign during the N_d - N_c transition.

Measurements by Galerne et al. (Figure 9) have better quantified these results and extended them to biaxial phases. From these measurements the invariants are in full agreement with theory; the transitions N_d- N_{bx} and N_{bx} - N_c are of second order, while the transition I - N_d is

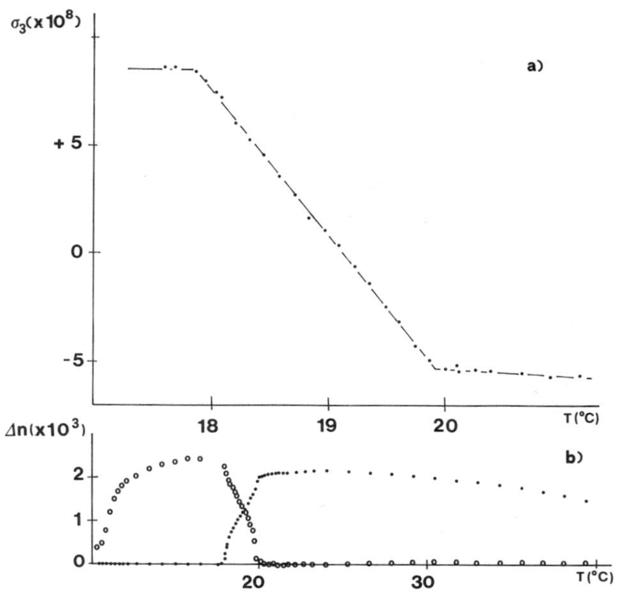

Fig. 9.　(a) Plot of the invariant $\sigma_3 = 4\pi\Omega_{\alpha\alpha}$ versus T showing its linear behavior in the biaxial range. (b) Plot of birefringences versus T: both are different from zero only in the biaxial phase.

slightly of first order[35]. When a magnetic field is applied to an isotropic phase very near to T_c, a birefringence is induced as a pretransitional effect (inducing nematic order[22]), ($\Delta n = \Delta\varepsilon\Delta\chi H^2/9\bar{n}_o \times (T - T_c^*)$) where n is the average refractive index and T_c^* is the temperature for which the transition I – N_d becomes of second order.

From magnetically induced birefringence measurements in isotropic phase Kumar et al. [36] obtained a value of $a_o = 2.9 \times 10$ ergs K^{-1} cm^{-3} and $\nu = 0.64 \pm 0.05$ for the transition N_d – L_α in samples of decylammonium chloride-ammonium, chloride-water ($DAC1-NH_4Cl$-water) [37]. They found, besides, that $\gamma = 1.01 \pm 0.04$ in aqueous solution of cesium perfluoro-octanoate (CsPFO-water) [38].

No absolute values of elastic constants in lyotropic systems have been obtained up to now since measurable quantities are combinations of K_{ii} with other parameters. Haven et al. [39], studying static and dynamic deformations (induced by a magnetic field), in the isotropic phase of $DAC1-NH\ Cl-H_2O$, obtained the birefringences, the ratios $K_{11}/\Delta\chi$ and $K_{33}/\Delta\chi$, and thus the ratios K_{11}/K_{33} versus the temperature (Figure 10).

These data show that the ratio K_{11}/K_{33} is of the same order of magnitude as in thermotropic nematics (in agreement with some of our preliminary results obtained using a different technique[40], see below) and the bend constant has a critical behavior at the nematic-neat soap transition.

Lacerda Santos[41] has found that, in the discotic phase of KL-decanol-water, the ratio $(D_1/D_2) \simeq 7.2$, in contrast with a thermotropic nematic where (D_1/D_2) is equal to 1.4. Such big anisotropy is similar to that obtained for the ratio D_3/D_2 in thermotropic calamitic nematic. From this, the above authors suggest that in lyotropics the ratio (D_3/D_2) is one order of magnitude lower than in thermotropic nematics. Since the ratio D_i/D_J depends on both K_{ii}/K_{JJ} and η_J/η_i and since K_{ii}/K_{JJ} is almost the

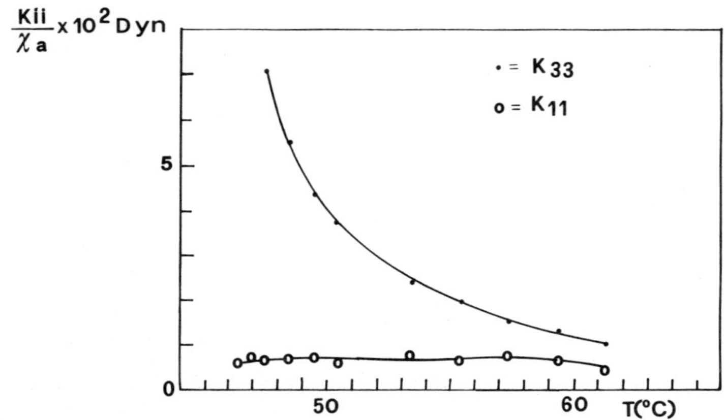

Fig. 10. Behavior of elastic constants K_{33} and K_{11} at the N-L$_\alpha$ phase
transitions.

same in both lyotropics and thermotropics, the observed anisotropy must be
attributed to a viscosity effect due to a backflow of micelles that let be
η_1 one order of magnitude lower than η_2.

Measurements of biaxial fluctuations, thermally excited in the disk
phase[42], show that the line widths of polarized and unpolarized modes
undergo a critical slowing down. In particular, Lacerda Santos[41] found
that Γ' and Γ depend linearly on temperature and that $\nu/\nu' = 4.3$ (see
Figure 11).

Copic[43] has obtained from dynamical measurements that, in contrast
with what happens in thermotropic nematics, the twist elastic constant K_{22}
is bigger than the bend elastic constant, K_{33}. This surprising result
could be due to a particular thermal history of Lyotropic solutions.

It should be possible, on principle, to obtain the absolute value of
elastic constant from total and differential cross section light

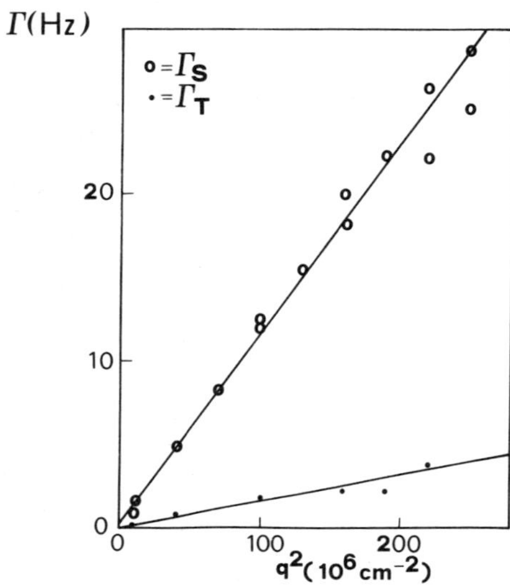

Fig. 11. Relaxation rates versus q^2 of the two modes Γ_S and Γ_T in the
disc-like phase.

measurements[44]. On this particular subject work is in progress in our
laboratory, but many experimental difficulties, such as the poor signal to
noise ratio obtained in oriented Lyotropic nematics, are encountered.

CONCLUSIONS

     Our research activity in these fields will continued in future since
many problems still remain. In particular, the structural and dynamical
role of water in the formation of condensed micellar mesophases must be
clarified. Even from a macroscopic point of view, the peculiar
characteristics of Lyotropic mesophases need to be studied further in
order to elucidate the deep connections among thermodynamical properties
of mesophases and molecular structures.

REFERENCES

1. N. Y. Carr and E. M. Purcell, Phys. Rev., 94:630 (1954).
2. E. D. Stejskal and J. E. Tanner, J. Chem. Phys., 42:288 (1965).
3. R. Blinc, V. Dimic, J. Pirs, M. Vilfan and I. Zumpancic, IIS Report
    R-592 (1970).
4. R. Blinc, K. Easwaran, J. Pirs, M. Vilfan and I. Zumpancic, Phys. Rev.
    Lett., 25:1327 (1970).
5. J. T. Gordon and J. Tiddy, J. Chem. Soc. Faraday Transaction, I:73
    (1973).
6. P. Ukleja and J. W. Doane, in: "Ordering in Two Dimension", Siuha,
    ed., Elsevier North Holland Inc. (1980).
7. P. T. Callaghan and O. Soderman, J. Phys. Chem., 87:1737 (1983).
8. G. Chidichimo, D. De Fazio, G. A. Ranieri and M. Terenzi, Chem. Phys.
    Lett., 117:514 (1985).
9. G. Chidichimo, D. De Fazio, G. A. Ranieri and M. Terenzi, Mol. Cryst.
    Liq. Cryst., 135:223 (1986).
10. G. Chidichimo, J. W. Doane, A. Golemme, G. A. Ranieri and M. Terenzi,
    Mol. Cryst. Liq. Cryst., 132:275 (1986).
11. V. Luzzati, in: "Biological Membranes", Chapman, ed., Vol. 1,
    Academic Press, New York (1968).
12. G. Chidichimo, A. Golenne and J. W. Doane, J. Chem. Phys., 82:4369
    (1985).
13. G. Chidichimo, A. Golenne, P. Westermann andJ. W. Doane, J. Chem.
    Phys., 82:536 (1985).
14. G. Chidichimo, N. A. P. Vaz, Z. Yaniv and J. W. Doane, Phys. Rev.
    Lett., 49:1950 (1983).
15. Y. Hendrikx and J. Charvolin, J. Phys., 42:1427 (1981).
16. K. Radley and L. W. Reeves, Can. J. Cehm., 53:2998 (1975); K. O.
    Lawson and T. J. Flautt, J. Am. Chem. Soc., 89:5489 (1967).
17. L. J. Yu and A. Saupe, Phys. Rev. Lett., 45:1000 (1980).
18. J. Charvolin and Y. Hendrikx, Liquid Crystals of One and Two
    Dimensional Order, in: "Chemical Physics", Springer Verlag (1980).
    Y. Hendrikx, J. Charvolin, M. Rawiso, L. Liebert and M. C. Holmes, J.
    Phys. Chem., 87:3991 (1983).
19. R. Bartolino, T. Chiaranza, M. Meuti and R. Compagnoni, Phys. Rev.,
    A26:1116 (1982).
20. M. Meuti, G. Barbero, R. Bartolino, T. Chiaranza and F. Simoni, Il
    Nuovo Cimento, 3D:30 (1984).
21. A. Saupe, P. Boonbrahm and L. J. Yu, J. Chim. Phys., 80:7 (1983).
22. P. G. de Gennes, "The Physics of Liquid Crystals", Oxford University
    Press (1975).
23. E. A. Jacobsen and J. Swift, Mol. Cryst. Liq. Cryst., 87:29 (1982).
24. F. C. Frank, Discuss. Faraday Soc., 25:19 (1958).
25. E. Govers and G. Vertogen, Phys. Rev., A30:1998 (1984).

26. H. Brand and H. Pliner, Phys. Rev., A24:2777 (1981).
27. L. D. Landau and E. M. Lifshitz, Statistical Physics, Pergamon (1975).
28. H. E. Stanley, "Phase Transition and Critical Phenomena", Oxford University Press (1971).
29. W. Helfrich, in: "Physics of Defects", Les Houches, North Holland, p. 716 (1980).
30. M. Allain, J. Phys., 46:225 (1985).
31. A. Saupe, Il Nuovo Cimento, 3D:16 (1984).
32. J. Charvolin, Il Nuovo Cimento, 3D:3 (1984).
33. M. Meuti, R. Bartolino and T. Chiaranza, Sol. State Comm., 48:751 (1983).
34. Y. Galerne and J. P. Marcerou, Phys. Rev. Lett., 51:2109 (1983).
35. Y. Galerne and J. P. Marcerou, J. Phys., 46:589 (1985).
36. S. Kumar, J. D. Litster and C. Rosemblatt, Phys. Rev., A28:1890 (1983).
37. S. Kumar, L. J. Yu and J. D. Litster, Phys. Rev. Lett., 50:1672 (1983).
38. C. Rosenblatt, S. Kumar and J. D. Litster, Phys. Rev., A29:1010 (1984).
39. T. Haven, D. Harmitage and A. Saupe, J. Chem. Phys., 75:352 (1981).
40. F. P. Nicoletta, Thesis, University of Calabria (1984), unpublished.
41. M. B. Lacerda Santos, Y. Galerne and G. Durand, J. Phys., 46:933 (1985).
42. M. B. Lacerda Santos, Y. Galerne and G. Durand, Phys. Rev. Lett., 53:787 (1984).
43. M. Copic, T. Ovsenik and M. Zganik, Mol. Cryst. Liq. Cryst., 113:77 (1984).
44. E. Miraldi, C. Oldano, L. Trossi and P. Taverna Valabrega, Il Nuovo Cimento, 60B:165 (1980).
     E. Miraldi, C. Oldano, P. Taverna Valabrega and L. Trossi, Mol. Cryst. Liq. Cryst., 82:231 (1982).

STRUCTURAL MODIFICATIONS INDUCED BY EXTERNAL AGENTS ON

MULTILAMELLAR LIPOSOMES

P. Mariani and F. Rustichelli

Istituto di Fisica Medica, Facoltà di Medicina e Chirurgia
Università di Ancona (Italy)

INTRODUCTION

Phospholipids, a principal component of natural membranes together
with proteins, are molecules with a polar head-group, charged or zwitterionic, and two hydrocarbon chains. When placed into an aqueous medium,
they form a variety of lyotropic structures as a function of water content
and temperature, which are disordered at atomic level and yet display a
high degree of long range organization[1]. In particular, multilamellar
liposomes, layer lattices of alternating and closed bimolecular lipid
sheets intercalated by aqueous spaces, have a strong similarity with the
lipid bilayers of biological membranes and can be regarded as models or
prototypes of living cells[2]. Information about functions and properties
of biomembranes can be obtained by studying such relatively simple structures.

Interactions between phospholipid model membranes and exogenic perturbing factors (as small molecules[3,4], proteins[5], drugs[6] and ionizing radiations[7] are widely investigated: incorporation of a solute in a
lipid bilayer as well as ionizing radiation in fact perturbs the phase
equilibrium and then the physical structural and dynamical properties of
the membrane, such as phase transition temperatures, permeability and
protein functions. Different techniques, as Differential Scanning Calorimetry, Nuclear Magnetic Resonance, Electron Spin Resonance, Fluorescence
spectroscopy, X-Ray and Neutron Diffraction, Electron Microscopy have been
employed to investigate such induced modifications. Concerning some of
these techniques, Chapman and Hayward[8] report recent developments and
applications.

In this paper we report some results obtained by using X-ray diffraction in order to describe perturbed lipids: multilamellar liposomes of
synthetic phosphatidylcholines (PC) were employed as a suitable model to
observe the influence of γ-irradiation, thus excluding the effect associated with the presence of proteins, while the interaction effects and
location of extrinsic molecules were considered in both synthetic and
natural lecithin (egg PC) liposomes.

STRUCTURE AND PHASE BEHAVIOR OF PHOSPHATIDYLCHOLINE LIPOSOMES

The structure, phase behavior and thermotropic properties of multi-lamellar PC liposomes in excess water have been the subject of numerous studies[1,9,10].

When lecithins with two identical acyl chains are considered, two phase transitions can be observed on differential scanning calorimetry thermograms: the first, at lower temperature, which corresponds to the so-called pretransition, broad and at low enthalpy, attributed to the transition from L_β, to P_β, phase and the second, sharp and first order, the so-called main transition, corresponding to the transition from P_β, to L_α phase[11]. A typical calorimetric thermogram obtained by heating distearoylphosphatidylcholine (DSPC) liposomes is shown in Fig. 1. Recently a third transition, observable when samples are stored for a long time at low temperature, was found at temperatures below the pretransition: the subtransition corresponds to a bilayer "crystal" $\rightarrow$ bilayer L_β, structural rearrangement of hydrocarbon chain packing[12].

Three typical X-ray diffraction patterns, corresponding to the three different L_β,, P_β, and L_α phases, are observed at different temperatures, as shown in Fig. 2. for dipalmitoylphosphatidylcholine (DPPC) liposomes.

The low angle diffraction peaks indicate for all temperatures a lamellar structure: below the pretransition, several sharp reflections in the ratio of 1:1/2:1/3... arising from a one-dimensional lattice are observed. In the P_β, phase several strong reflections, together with many additional peaks appear in this region. The pretransition in fact is associated with a structural transformation from a one to two-dimensional monoclinic lattice, consisting of stacked lamellae distorted by a periodic ripple in the plane of bilayer. At the main transition, the lattice reverts to one-dimensional lamellar: sharp low angel reflections appear in the X-ray diffraction patterns again in the ratio of 1:1/2:1/3... The interlamellar repeat distance, which can be obtained from the angular position of the low angle peaks by using the Bragg law, shows a characteristic temperature dependence as appears from Fig. 3: the bilayer thickness increases with increasing temperature, reaching a maximum value in the $P_{\beta1}$ phase, and then decreases above the main transition temperature.

Information about the structure inside the layer can be obtained by the analysis of the high angle diffraction patterns. Below the pretransition temperature, a narrow and intense peak together with a broader band appear in this region. This observed profile is associated with hydrocarbon chains rigid and full extended, tilted with respect to the lipid bilayer and packed in a distorted quasi-hexagonal lattice[11]: a sketch of this distorted packing of chains is reported in Fig. 4. From the angular position of the two high angle peaks, one can deduce two interplanar spacings, d_1 and d_2, (where d_2 is the spacing corresponding to the diffraction peak which has twice the intensity of that corresponding to d_1[13]) by applying the Bragg law.

With increasing temperature, the acyl chains become more regularly packed. In the P_β, phase a unique broad reflection appears in the high angle diffraction region: the chains are tilted with respect to the plane of bilayer and are packed in a hexagonal array. At the main transition, the acyl chains of phospholipid melt, assuming a liquid-like arrangement, as can be deduced from the highly diffuse reflection at wide angle in the L_α phase.

The mixed chain PC show, in addition to the subtransition, which again depends sensitively upon storage at low temperature and in some case disap-

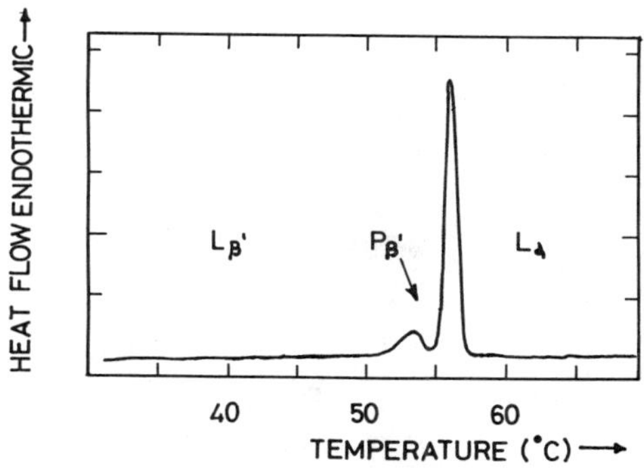

Fig. 1. Characteristic high-sensitivity thermogram obtained by heating
multilamellar liposomes of DSPC in excess water. The nomenclature
used is that of Luzzati[1] and Tardieu et al.[10].

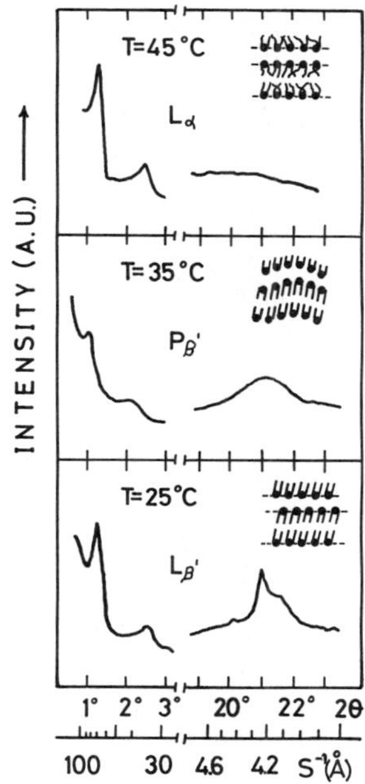

Fig. 2. Characteristic X-ray intensities recorded as a function of
scatering angle 2θ from DPPC liposomes below the pretransition
(T=25°C), intermediate between the two (T=35°C) and above the main
transition (T=45°C). The inserts report the corresponding models
of molecular organization. On the horizontal axes S^{-1} values are
also reported, where $S = 2 \sin \theta / \lambda$. The low-angle region
corresponds to a Bragg spacing >10Å, the high-angle region to a
Bragg spacing <10Å.

pears[9], only the main transition. The X-ray diffraction patterns,
obtained in the different phases, coincide with those found in lecithin
with two identical chains having passed with pretransition[9]. Some difference could be attributed to non-tilted arrangement of acyl chains. The
thermotropic behavior of pure PC liposomes in excess water is graphically
reported in Fig. 5.

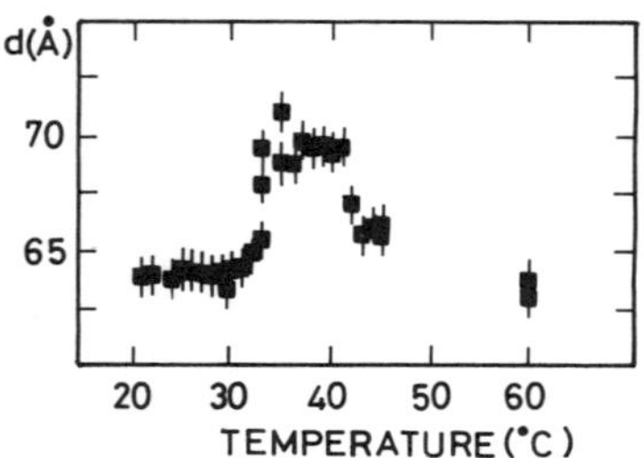

Fig. 3. Typical observed temperature dependence of the layer periodicities d for DPPC liposomes in excess water[16].

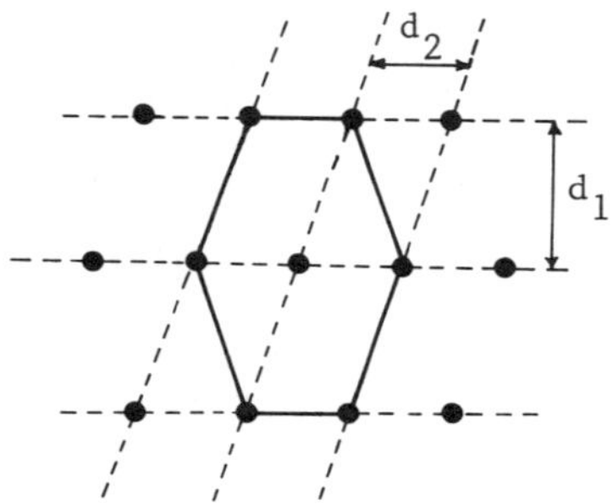

Fig. 4. Sketch of the distorted hexagonal packing in a plane perpendicular to the chain axes in the L_β, phase. The dots represent the mean positions of the acyl chains.

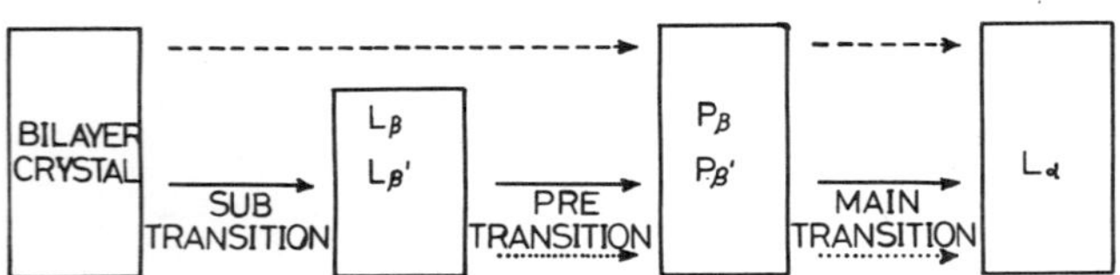

Fig. 5. Scheme of thermotropic behavior of pure PC liposomes in excess water. The full arrows show the phase transitions of PC with identical acyl chains; the dashed arrows indicate the behavior of mixed-chain PC; the exceptional case of dimyristoylphosphatidylcholine is emphasized by the dotted arrows. This general scheme does not consider the tilt of acyl chains: umprimed phase appear at the side of primed phases[1,10].

Phospholipid mixtures, as complex as those obtained from biological membranes, could be expected to lead to intricate phase diagrams when the number of components becomes large[1]. In natural egg-PC, several phases have been observed by X-ray diffraction at high lipid concentration[14], however the low lipid containing phase at room temperature is still lamellar with melted chain (L_α).

STRUCTURAL MODIFICATIONS INDUCED BY γ-RAY RADIATION

The physical modifications induced in biomembranes by γ-ray radiation, which are hypothesized to be important factors in cell death[15], were

254

recently investigated[16,17], on suitable models by using X-ray diffraction
and differential scanning calorimetry. In order to exclude radiation
effects associated with the presence of proteins and peroxidative lipid
damage, multilamellar liposomes of satured PC were employed. Two different
synthetic PC were used to prepare liposomes, namely the dipalmitoylphosphatidylcholine (DPPC) and the distearoylphosphatidylcholine (DSPC), to
detect the eventual influence of chain length variations.

DPPC and DSPC liposomes were irradiated in aerated conditions at room
temperature with γ-rays from a ^{60}Co source at a rate of about 0.5 Gy/s.
High doses were reached in both the experiments to enhance the effects of
irradiation. The two characteristic endothermic peaks, corresponding to
the above mentioned transitions, appear in calorimetric curves, also for
most irradiated samples in spite of previous calorimetric work[18] However
some change can be noticed in calorimetric thermograms when the γ-dose
increases: the pretransition peak broadens and flattens and the main transition endotherm enlarges. In the case of DPPC liposomes, whose calorimetric scans are shown in Fig. 6 for different irradiated samples, the main
transition peak appears to split into two for irradiation times longer than
about 10 hours, while in the case of DSPC liposomes a shoulder is observed
on the high temperature side of the curve above the same value of
irradiation.

The X-ray diffraction experiments performed at different temperatures
show the characteristic patterns of the three phases, still visible also
for DPPC and DSPC samples irradiated for longer times. Some modifications
can be observed: in the case of low angle diffraction peaks in the L_β,
phase, the γ-irradiation induces reduction of the angular position, i.e. an
encrease of layer thickness; Fig. 7 shows the irradiation time dependence
of layer periodicity in the case of DPPC liposomes. As far as the other
phases are considered, changes of lamellar thickness are less easily
detectable: in any case, a shift towards higher values of lamellar distances was observed as a function of γ-dose.

With respect to modifications induced by γ-rays on the structural
organization of acyl chains inside the L_β, phase, Fig. 8 reports for
example the continuous change in the high angle region of the diffraction
pattern of DPPC samples when the dose increases. However, some differences
can be noticed if DPPC or DSPC irradiated liposomes are considered. A
monotonic decrease of the packing periodicities, together with an increase
of the hexagonal symmetry, which are shown, in Figs. 9 and 10, was detected
in DPPC samples as a function of irradiation time. A quite different
dependence of the packing periodicities with the time of irradiation is
observed for DSPC liposomes: a discontinuous behavior occurs. The inter
chain distances, for example, tend to be greater below 10 hours of γ-ir-
radiation, while a decreasing of d_1 and d_2 distances was observed above
this value[17]. In both the experiments, no relevant changes for increasing γ-dose were detected in the high angle region of the X-ray patterns in
P_β, and in L_α phases.

Previous NMR and fluorescence experiments report changes in polarity
and permeability of the membrane, as a function of irradiation, which
appear consistent with penetration of water into the bilayer[18,19], this
can be a consequence of the formation of lysolecithins and free fatty
acids, together with the appearance of new chemical compounds, such as
cross-linked lipids. The observed behavior of the calorimetric and structural data could be attributed to these different mechanisms which, by
considering the significant changes in physical properties at about 10
hours of irradiation expecially evident in DSPC liposomes[17], appear to
act at different values of γ-dose. The abrupt increase of layer thickness
could be due to the penetration of water into the hydrophobic region of

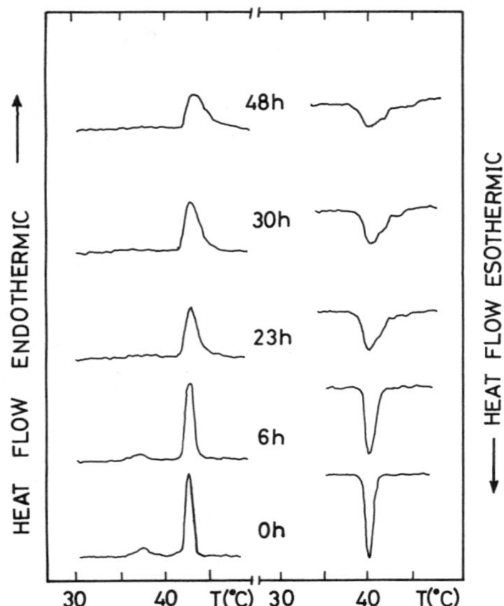

Fig. 6. Thermograms obtained on heating (left) and on cooling (right) different DPPC samples[16]. The γ-irradiation times (hours) are indicated on the corresponding curves: the dose rate was 0.45 Gy/s.

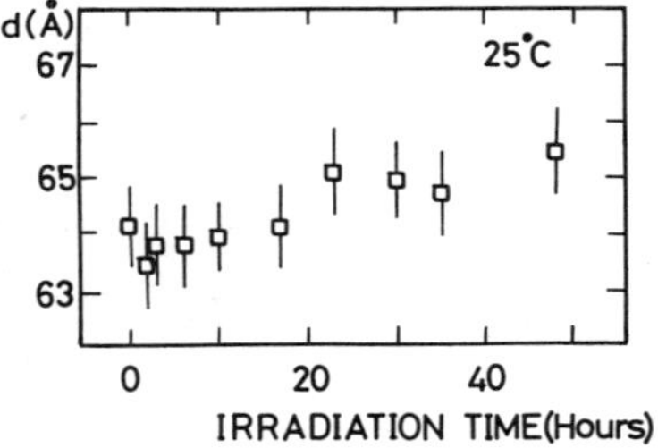

Fig. 7. Observed irradiation time dependence of layer periodicity in the L_β, phase at 25°C in the case of DPPC irradiated liposomes[16].

bilayer for the presence of inverted micelles of lysolecithin; the formation of free fatty acids are cross-linked lipids could explain the reduction of the in-plane intermolecular distances in the L_β, phase and the corresponding increase in hexagonal symmetry and could induce the increasing of the main transition temperature and the decreasing of its co-operativity[3]. Moreover the existence of lipids of different kinds is suggested also by the change in shape of the calorimetric peak associated to the P_β, → L_α transition. Previous calorimetric studies[18] on the same lipid systems report that the pretransition disappears for an absorbed dose of 70 kGy delivered at a rate of 0.05 Gy/s. Albertini et al.[16,17] did not observe the disappearance of the pretransition peak even at the largest doses: the different behavior observed could be related to the difference on the dose rate.

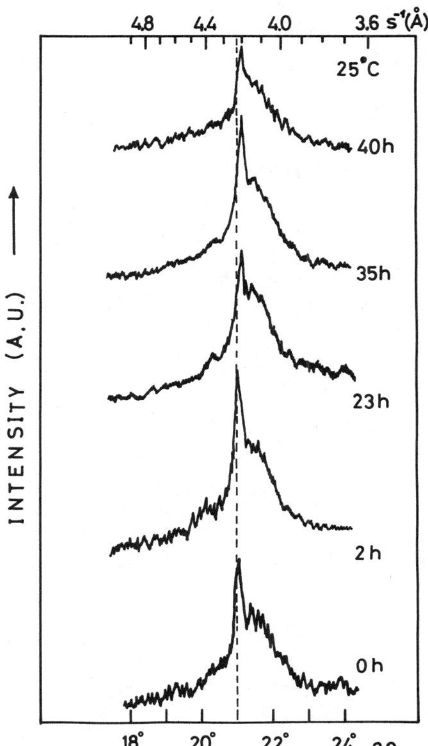

Fig. 8. High angle diffraction patterns in the L_β, phase at 25°C of
different DPPC liposomes[16]. The γ-irradiation times (hours) are
indicated on the corresponding profiles (dose rate of 0.45 Gy/s).

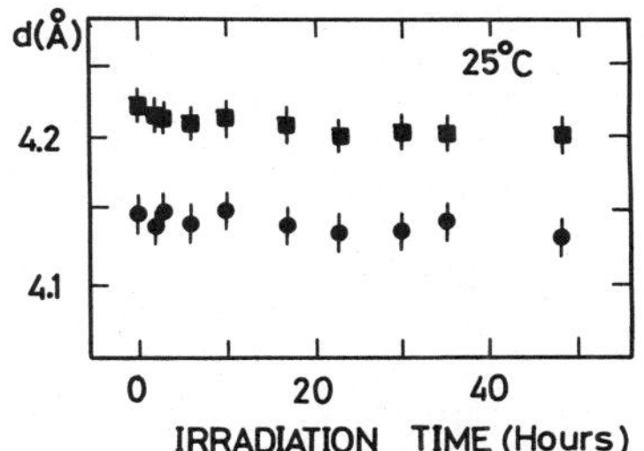

Fig. 9. Observed irradiation time dependence of d_1(■) and d_2(●) spacings
in the L_β, phase at 25°C in the case of DPPC irradiated
liposomes[16].

INTERACTION EFFECTS OF EXTRINSIC MOLECULES

Structural and thermodynamical studies on phase transition behavior
and on molecular organizations can provide information about the nature of
the induced perturbation and suggestions on the location of the additive
guest molecules inside the bilayer.

The effect of drugs on the physical state of membranes are widely
investigated[6]: in particular it was proposed that the change in fluidity[21] and thickness[22] of lipid bilayer is due to the action of

257

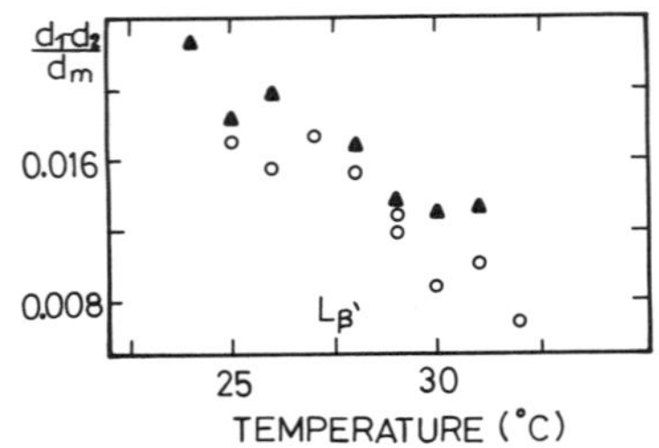

Fig. 10. Temperature dependence of relative difference $2(d_1-d_2)/(d_1+d_2)$ for the DPPC control sample (▲) and 48h DPPC irradiated sample (O) in $L_{\beta'}$ phase[16].

anesthetics. Contradictory results concerning the effect of benzyl alcohol (BA), a local anesthetic, on bilayer thickness have been reported: this molecule appears to increase[23], to decrease[24] and not to change[25] the thickness of membrane with anesthetic concentration. Interesting results have been recently obtained by using X-ray diffraction, in combination with NMR experiment, on egg PC liposomes after the addition of BA[26].

The anesthetic was incorporated at different concentration on egg PC liposomes with different water content: the X-ray diffraction patterns, taken at room temperature, show that all samples are in a lamellar state with liquid-like arrangement of chains, namely in L_α phase. From the measured interlamellar repeat distances, the lipid layer thickness d_ℓ and the surface area S_ℓ of lipid molecules can be calculated[1] if a suitable model concerning the extrinsic molecule location can be postulated. In this case, a portion of the water is assumed to penetrate into the hydrophobic portion of the bilayer, while BA molecules are located between PC molecules. The repeat distance of lamellar phase shows a different behavior as a function of BA/PC molar ratio when the water content changes, as can be seen in Fig. 11. In excess of water, when the concentration of anaesthetic increases, the lamellar thickness and then the lipid bilayer thickness $d\ell$ remain nearly unchanged, while at lower water content the addition of BA appears to decrease both values. NMR experiments show that in any case the rate of molecular motion increases.

The behavior of interlamellar distance, lipid bilayer thickness and surface area when the water content changes at constant BA/PC molar ration, which is shown in Fig. 12, appears more complex. In particular, the lipid thickness d_ℓ for the sample without BA appears to decrease on adding water up to about 46% of water/(water+lipid) (w/w). Above this value, d_ℓ is independent on the water concentration. At a relatively low water content, in fact, the chains in the bilayer are in an extended structure, due to the cohesive force between lipid molecules. When the content of water increases, the forces are weakened and the structure is relaxed to a disordered state.

Interacting with BA, PC bilayers appear to be unusually disordered: at about 30% water, the lipid bilayer thickness reaches the same value as that of the sample without BA and remains nearly constant when the water concentration still increases. The value at maximum hydration then appears to be a lower limit of lipid bilayer thickness: this leads to the difference in the effect of BA at different water contents. When the chains are in an extended structure, namely at lower water content, the hydrocarbon region becomes thinner and more disordered by binding of BA; at excess water, the anaesthetic does not perturb the chain order any further.

258

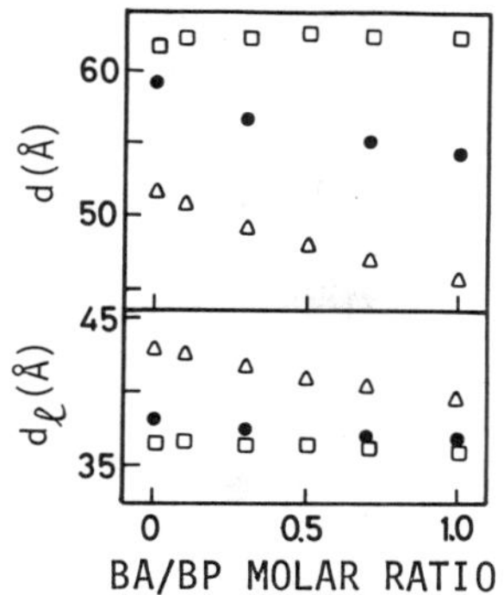

Fig. 11. Observed BA concentration dependence of layer periodicity d and lipid bilayer thickness d_ℓ on egg – PC lamellar liposomes with different content of water[26]: (□) 65, (●) 40 and (△) 20% water/ (water + lipid) (W/W).

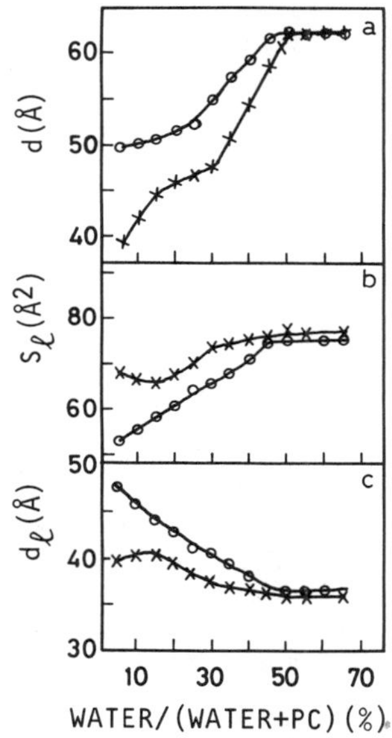

Fig. 12. Layer periodicity d, surface area of lipid molecules S_ℓ and lipid bylayer thickness d_ℓ as a function of water content, for egg – PC liposomes without (○) and with (x) BA (BA/PC molar ratio of 1)[26].

Similar X-ray diffraction experiments were recently performed in order to investigate the effect of tryptophane (TRP) on a model membrane of DPP[27]. This aminoacid, actually used in studying intrinsic protein fluorescence, appears to be a good probe for monitoring lipid-protein interaction. The X-ray diffraction patterns observed at different temperatures indicate that the three phases, $L_{\beta'}$, $P_{\beta'}$ and L_α, are present for all the concentrations of TRP at any DPPC/water weight ratio investigated. An abrupt decreasing of the pretransition temperature appears to be induced by the aminoacid: the calorimetric data confirm this behavior and show that the co-operativity of the main transition gets smaller for TRP content sufficiently high. A strong increase in interlamellar repeat distance appears in all phases at low TRP concentration: this fact could be associated with hydration effects on the bilayer surface, as no modifications in the interchain organization are observed. Moreover, changes near the headgroup region without modification deep in the hydrophobic "core" are detected by EPR measurements. Higher concentration of TRP seems to modify this behavior, expecially in the $P_{\beta'}$ phase: X-ray diffraction and EPR data, together with some fluorescence quenching results, indicate in fact that the acyl chains could also be perturbed when the content of this aminoacid increases sufficiently.

CONCLUSIONS

Structural studies on model membranes are useful for understanding the role of phospholipid molecules in biological membranes as well as the nature of their interaction with exogenic perturbing factors.

The relatively simple chemistry and structure of phosphatidylcholine liposomes in excess water allows interaction effects on these models to be easily investigated, and permits eliminating the contribution due to membrane components other than lipids. X-ray diffraction, in combination with other biophysical methods widely used to study the static and dynamic order of membranes, appears as a suitable technique for studying the perturbed membranes as well. In fact it can provide some information about the effects on lipids of ionizing radiation and about the location of some guest molecules inside the bilayer; moreover it has shown why different effects are obtained when using the same anesthethic in membranes at different water concentrations.

REFERENCES

1. V. Luzzati, X-ray diffraction studies of lipid-water systems, in: "Biological Membrane", D. Chapman Ed. Academic Press, London (1968).
2. C. Tanford, Monolayers, micelles, lipid vescicles and biomembranes, Presented at: International School of Physics "E. Fermi" XC Course, Physics of Amphiphiles: Micelles, Vesiscles and Microemulsions Varenna (1983).
3. M. K. Jain and N.M.Wu, Effect of Small molecules on the dipalmitoyl lecithin liposomal bilayer. III, Phase transition in Lipid Bilayer, J. Membrane Biol. 34: 157 (1977).
4. A. W. Eliasz, D. Chapman and D. F. Ewing, Phospholipid phase transition. Effects on n-alcohols, n-monocarboxylic acids, phenylalkyl alcohols and quaternary ammonium compounds, Biochim. Biophys. Acta, 448: 220 (1976).
5. D. Chapman, J. C. Gomez-Fernandez and F. M. Goni, Intrinsic protein-lipid interactions, FEBS Letters, 98: 211 (1979).
6. B. P. Cater, D. Chapman, S. M. Hawes and J. Savicle, Lipid phase Transitions and drug interactions, Biochim. Biophys. Acta, 363: 54 (1974).
7. J. C. Edwards, D. Chapman, W. A. Cramp and M. B. Yatvin, The effect of ionizing radiation on biomembrane structure and function, Prog. Biophys. Molec. Biol., 43: 71 (1984).
8. D. Chapman and J. A. Hayward, New biophysical techniques and their application to the study of membranes, Biochem. I., 228: 281 (1985).
9. J. Stumpel, H.HIBL and A. Nicksch, X-ray analysis and calorimetry on phosphatidylcholine model membranes: the influence of length and position of acyl chains upon structure phase behaviour, Biochim. Biophys. Acta, 727: 246 (1983).

10. A. Tardieu, V. Luzzati and F. C. Reman, Structure and polymorphism of the hydrocarbon chains of lipids: a study of lecithin water phases, J. Mol. Biol. 75: 711 (1973).

11. M. J. Janik, D. M. Small and G. G. Shipley, Nature of the thermal pretransition of synthetic phospholipids: dimyristoyl- and dipalmitoyllecithin, Biochemistry, 15: 4575 (1976).

12. J. Ruocco and G. G. Shipley, Characterization of the sub-transition of hydrated dipalmitoyl phosphatidylcholine bilayers: X-ray diffraction study, Biochim. Biophys. Acta. 684: 59 (1982).

13. A. Wattes, K. Harlow and D. Marsh, Charge-induced tilt in ordered-phase phosphatidylclycerol bilayers: Evidence from x-ray diffraction, Biochim. Biophys. Acta. 645: 91 (1981).

14. F. Reiss-Husson, The structure of liquid crystalline phases of different phospholipids, monoglicerides, and sphingolipids, anhydrous and in the presence of water, J. Mol. Biol. 25: 363 (1967).

15. T. Alper, The role of membrane damage in radiation induced cell death, Adv. Exp. Med. Biol. 84: 139 (1977).

16. G. Albertini, E. Fanelli, L. Guidoni, F. Ianzini, P. Mariani, F. Rustichelli and V. Viti, X-ray diffractometry and calorimetry studies of structural modifications induced by γ-irradiation in phosphatidylcholine multilamellar liposomes, Int. J. Radiat. Biol. 48: 785 (1985).

17. G. Albertini, E. Fanelli, L. Guidoni, F. Ianzini, P. Mariani, F. Rustichelli and V. Viti, X-ray diffractometry and calorimetry studies of structural modifications induced by x-ray radiation on distearoylphosphatidylcholine liposomes, Int. J. Radiation Biol. 52:145 (1987).

18. G. Erriu, M. Ladu, and G. Meleddu, Modifications induced on phosphatidylcholine multilayer by Co-60 γ-rays, Biophys. J. 35: 799 (1981).

19. L. Guidoni, F. Ianzini, P. L. Indovina, and V. Viti, ^{1}H and ^{2}H NMR studies of water in γ-irradiated phosphatidylcholine multilamellar liposomes, Int. J. Radiat. Biol. 48: 117 (1985).

20. F. Ianzini, L. Guidoni, P. L. Indovina, V. Viti, L. Erriu, S. Onnis, and P. Randaccio, Gamma-irradiation effects on phosphatidylcholine multilayers liposomes: calorimetric, NMR and spectrofluorimetric studies Radiat. Research, 98: 154 (1984).

21. A. G. Lee, Model for action of local anaesthetics, Nature, 262: 545 (1976).

22. D. A. Haydon, B. M. Hendry, S. R. Levinson and J. Requena, Anaesthesia by the n-alkanes: a comparative study of nerve impulse blockage and the properties of black lipid bilayer membranes, Biochim. Biophys. Acta, 470: 17 (1977).

23. R. G. Ashcroff, H. G. L. Coster and J. R. Smith, The molecular organization of bimolecular lipid membranes: the effect of benzyl alcohol on the structure, Biochim. Biophys. Acta, 469: 13 (1977).

24. L. Ebihara, J. E. Hall, R. C. MacDonald, T. J. McIntosh and S. A. Simon, Effect of benzyl alcohol on lipid bilayers: a comparison of bilayer systems, Biophys. J. 28: 185 (1979).

25. J. Leyes and R. Latorre, Effect of anesthetics benzyl alcohol and chloroform on bilayers made from monolayers, Biophys. J., 28: 259 (1979).

26. T. Shibata, Y. Sugiura and S. Iwaxanagi, Effects of benzyl alcohol on phosphatidylcholine lamellar phase with different water contents, Chem. Phys. Lipids, 31: 105 (1982).

27. G. Albertini, G. Curatola, P. Mariani, F. Rustichelli, and G. Zolese, Lipid aminoacid interaction: a study of tryptophane effects on dipalmitoylphosphatidylcholinemultilamellar liposomes, to be submitted to Chemistry and Physics of Lipids.

DETERMINATION OF THE RIGIDITY CONSTANT OF THE AMPHIPHILIC

FILM IN BIREFRINGENT MICROEMULSIONS, SPIN-LABELING AND

QUASIELASTIC LIGHT SCATTERING EXPERIMENTS

J. M. di Meglio, M. Dvolaitzky and C. Taupin

Physique de la Matière Condensée, Collège de France
11 Place Marcelin-Berthelot, 75231 Paris Cedex 05, France

Much activity has been devoted to the intensive study of lyotropic
systems and more specifically of microemulsions[1] because of their practical applications: oil enhanced recovery, cosmetic industry, pharmacology
etc.[2]. From a more fundamental point of view, these systems are very
interesting because of their fascinating properties: critical behavior[3,4], percolation transition[5], ultralow interfacial tensions[6] and
extraordinarily rich phase diagrams[7,8]. The usual application of microemulsions recovers isotropic thermodynamically stable dispersions of water
in oil (or oil in water) whose characteristic size is of order of one
hundred Å and thus are transparent to visible light. These systems are
stabilized by a surfactant forming an interfacial film separating oil and
water often associated to a cosurfactant (usually a short-tail alcohol).

The very first microemulsion model is due to Schulman[9] and only took
into account the interfacial tension between oil and water and the surfactant free energy. In particular, this model was at a very local scale
(i.e. at the scale of the surfactant molecule) and neglected the entropy of
the film (in the case of a randomly bent interface), electrostatic energies
between surfactants, interactions between interfacial films and curvature
energies. The essential difficulty in completing a microemulsion theory is
that all these effects are small, competing together, not independent of
each other, and varying with the system. However, a particular emphasis
has been given to the role of the curvature energy[1,10,11] and in this
paper, we present a detailed study of flexibility.

We have investigated a lamellar system which is very close to isotropic microemulsions in its phase diagram[12]. These systems present
lamellar textures when observed by microscope between crossed polarizers,
and are birefringent, but their structure is easily destroyed through a
gentle stirring revealing an extremely fragile interfacial film. They can
be swollen by oil and thus have reticular distances of up to several hundreds of Å as observed by small angle X-ray scattering. A simplified phase
diagram is represented in Figure 1 and shows that a tiny relative added
amount of cosurfactant leads to the isotropic classical droplet microemulsion. These particular phases are present in some other microemulsion
systems[8,13].

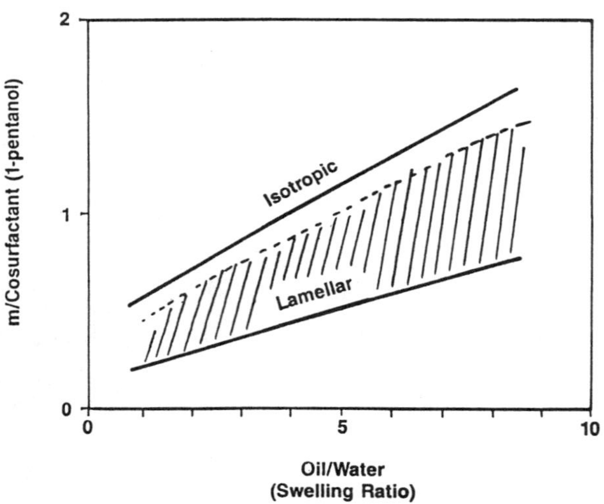

Fig. 1. Schematic phase diagram of the studied region. The volume of
water is 1 ml and the water/surfactant ratio is 2.5. It can be
seen that only a few tenths of ml cosurfactant is needed to go
from the lamellar phase into the isotropic phase.

The curvature energy per unit area is written as[1]:

$$E_c = \frac{1}{2} K \left[\frac{1}{R} - \frac{1}{R_o} \right]^2$$

where K is the rigidity constant of the film, R the radius of curvature and
R_o the spontaneous radius of curvature, i.e. the radius of curvature that
the film would adopt in absence of any interaction[14]. We expect that, in
their domain of existence in the phase diagram, the lamellar phases have a
zero spontaneous curvature; this assertion is supported by the fact that
the phases remain lamellar even for very large oil/water ratios (swelling
ratios). For highly flexible films, one can define a persistence
length:[1]

$$\xi_K = a \, \exp\left[\frac{2\pi K}{kT} \right]$$

where a is a molecular length (typically the minimum radius of curvature,
i.e. the length of the alkyl tail of the tensioactive molecule). Thus a
very small decrease of the rigidity would lead to a drastic decrease of the
persistence length according to the scheme of Figure 2. Schulman had
supposed that one role of the cosurfactant could be to decrease the rigidity of the interfacial film; this would explain the transition that we
observed from the lamellar birefringent phase to the isotropic state.

Our systems were prepared from commercially available compounds. The
surfactant was sodiumdodecylsulfate (SDS) (Serlabo), the cosurfactant was
1-pentanol (Serlabo), oil was cyclohexane (Merck) and water was distilled
three times before use. The water/surfactant ratio was 2.5 (by weight) in
all our samples. The phases were obtained through careful titration by the
cosurfactant pentanol. There is about one molecule of cosurfactant per
surfactant in the interfacial film for the lamellar phase to be compared to
two for the isotropic droplet microemulsion. We have performed two sets of
experiments; one by varying the swelling ratio and staying at the linear

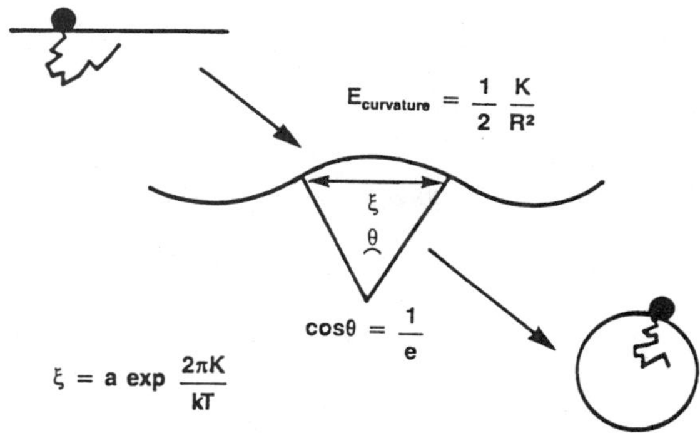

Fig. 2.   As the persistence length of the film varies exponentially with
the rigidity constant, a small decrease of this rigidity (for
instance by adding cosurfactant, as suggested by Schulman) may
explain the transition towards very curved phases.

border where the lamellar phase appears (to ensure that the chemical composition of the interfacial film remains constant) and the other by varying
the amount of cosurfactant as a fixed swelling ratio.

These lamellar phases are easily oriented in parallel wall glass
containers (path length 100µm) by heating them at around 75°C. The
anchoring is of the homeotropic type, and oily streaks are visible but
could almost be all eliminated through the heating process. The fact that
we can orient the samples is very favorable to their study by methods
sensitive to orientation as the spin labeling technique[15]: this permits
us to test the orientation of the lamellae.

Spin labeling consists of studying the electronic paramagnetic resonance of a nitroxide radical probe. This probe is grafted on the alkyl
chain of a surfactant molecule[16] whose alkyl chain length is identical to
the one of the tensioactive molecules of the interfacial film (SDS) (see
scheme below).

$$CH_3 - (CH_2)_{11} - C - (CH_2)_3 - N^+ - CH_3SO_4^-$$

We use one probe molecule per one thousand surfactant molecules so
that we do not perturb the interfacial film; indeed, we do not observe any
modification of the phase diagram of the labeled samples.

The spectra of these peculiar lamellar systems revealed an unusual
feature[17]: the spectra obtained with oriented samples could not be simply
deduced from the "powder spectra" i.e. the spectra obtained in a geometry
where all the orientations have the same occurrence of probabilities
(Figure 3). This is due to the fact that the lamellae exhibit long wave
length undulations (Figure 4). The amplitude of these undulations
increases with the swelling ratio of the samples and tends to a limit
(Figure 5). We will describe this phenomenon in the following through the
theory developed by Helfrich and de Gennes.

The main idea is to say that the interfacial film is undulated because
of its high flexibility, and that the amplitude of the undulations is

limited by the interactions between adjacent lamellae. The free energy per
unit area is thus written as:

$$E = \frac{1}{2} \frac{K}{R^2} + \frac{1}{2} U'' z^2 \text{ with } U'' = \frac{\partial^2 W}{\partial z^2}$$

where W is the interlamellae potential and z is the distance of the lamella
to an ideally plane reference lamella. We have distinguished two types of
interactions:

- attractive interactions (van der Waals)
- repulsive interactions (Helfrich)[18].

These latter are predominant and are steric repulsions due to the
fluctuations of the highly flexible lamellae. We thus have:

$$U'' = 5.04 \frac{(kT)^2}{Kd^4}$$

where d represents the average distance between lamellae, which can be
computed from the relative amounts of oil and water and from the area per
polar head of the surfactant (determined by X-ray scattering). The theory
developed by de Gennes and Taupin allows us to estimate the average θ^2:

$$\theta^2 = \frac{kT}{\pi K} \text{Ln}[\ \frac{\xi u}{a}\ ] \text{ with } \xi u = [\ \frac{K}{U''}\ ]^{1/4}.$$

We have first undertaken a temperature investigation of a sample with
a swelling ratio of 4[19]. This sample presents a lamellar texture as
observed through an optical microscope up to 70°C. Figure 5 represents the
square angular spread of the normal to the lamellae with respect to the
temperature. From the slope, we can deduce an estimation of the rigidity
constant K by assuming that it is fairly independent of the temperature.
We find $K = 6.4 \ 10^{-15}$ erg.

The above behavior ($\theta^2 \propto T$) is not very surprising and just describes
thermal fluctuations. Another more interesting prediction of the de Gennes
model is that:

$$\theta^2 = \frac{kT}{\pi K} \text{Ln } d + \text{constant}$$

where d is the distance between lamellae. This is shown on Figure 7. This
leads to another determination of the rigidity constant[19] which is independent of the constants of the model (Hamaker constant, minimum radius of
curvature a). We find $K = 4.10^{-14}$ erg. The fact that this prediction is
well verified corroborates that the chemical composition of the interfacial
film is the same in all the studied samples. The rigidity of the interfacial film is thus between 10^2 and 10^3 times less than in lecithin systems
($K_{lec} = 2.10^{-12}$ erg)[20,21]. We attribute this difference to the fact that
in lecithin systems there is no cosurfactant.

These undulations have been also observed by us in non-ionic binary
lamellar systems[22], where a transition of flexibility has been revealed.

Our aim as we began this study was of course to investigate the influence of the cosurfactant alcohol on the rigidity of the interface. We have
thus performed spin labeling experiments on a sample with a swelling ratio
of four varying the amount of cosurfactant starting from the minimum needed
to get the phase. All the samples studied present characteristic lamellar
structures. The amplitude of the undulations increases with the cosurfactant, and as the added amount is relatively small with respect to the

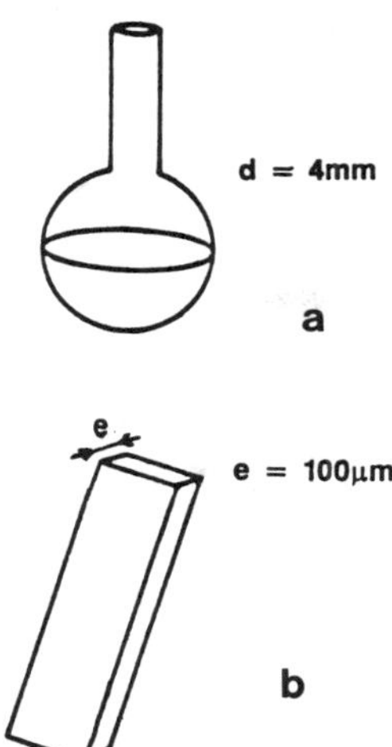

Fig. 3. Different cells used in our experiments. The cell (a) gives an
isotropic distribution of the orientation of the lamellae through
the anchoring of the layers to the wall. The samples were
oriented by heating in cell (b) and their orientation was checked
by observation by microscope through crossed polarizers.

volume of the sample we expect that the reticular distance remains the
same. Thus the observed increase of the undulation amplitude corresponds
to a decrease of the rigidity constant as represented in Figure 8.

The classical spin labeling technique is not useful for investigating
dynamics at times longer that 10^{-6}s. The fact that we find $\theta^2 \propto T$ indicates
that the undulations we see are the thermal fluctuations of the lamellae.
In order to investigate the dynamics of these fluctuations, we have performed a quasielastic light scattering experiment, this technique being
sensitive to very long time constants ($>10^{-4}$s) for the apparatus we set
up[23].

We wanted to observe undulations of the lamellae out of a ideally flat
reference plane, so the scattering vector $\vec{q}$ has to be parallel to the
lamellae; this condition is not so easy to fulfil and may explain some of
the difficulties that we encountered in obtaining a signal. We have used
the same sample containers as in the spin labeling study (2 mm wide, around
5 cm long and 100 μm path) and the samples were oriented in the same way.
The incident beam was from an ionized krypton laser (wave length = 5309 Å)
and we have analyzed the scattered intensity for an angle of diffusion
lying between 9° and 60°. This corresponds to q^{-1} between 845 and 5345 Å;
thus we mainly expect to see collective modes (120 Å$<d_{ret}<$360 Å).

By careful analysis of the theoretical predictions available for this
scattering geometry[24], it appears that the pure undulation mode is the
most probable. Its dispersion relation is:

$$\omega = i\,\frac{K}{\eta d}\,q^2$$

where η is the viscosity of the interlayer solvent, d the reticular distance and q the module of the scattering vector.

In all the experiments, the dynamic/static signal ratio is small and
decreases strongly with the swelling ration of the phases. We attribute

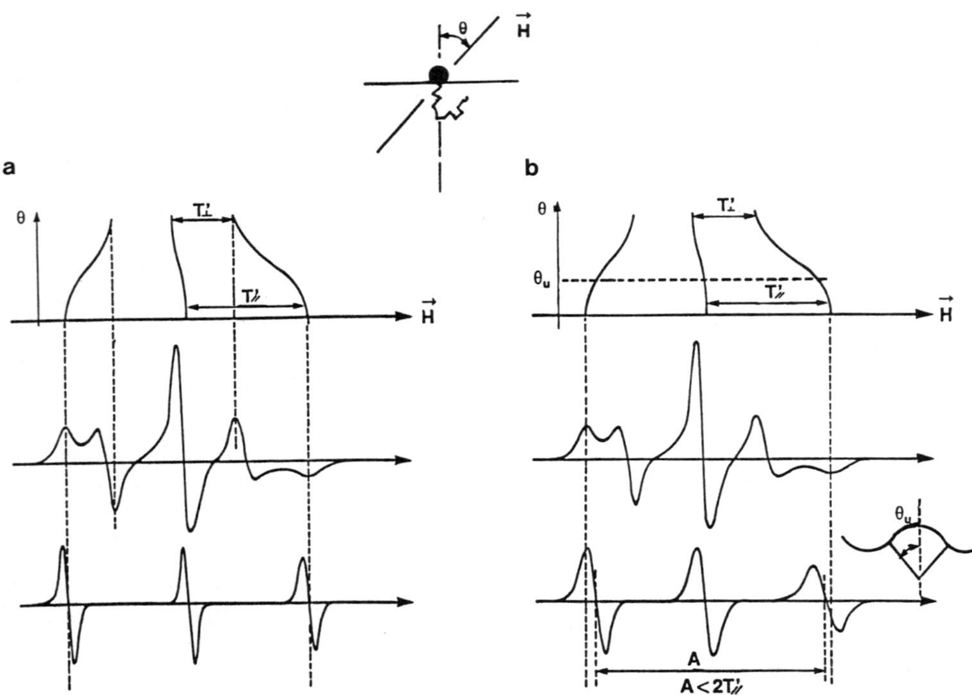

Fig. 4. Schematic spectra obtained with: (a) a "good" lamellar phase such as a smectic A liquid crystal. The isotropic spectrum consists of the superposition of the resonance peaks of all the orientations and can be easily simulated by means of standard procedures using lorentzian line shapes. The oriented spectrum can be deduced simply from the characteristics of the powder spectrum. (b) A birefringent microemulsion phase. The isotropic spectrum has the same shape but the characteristics of the oriented spectrum are different by the shift and the line widths. In order to simulate this oriented spectrum, we have to make an integration up to a certain angle θ_u which reflects that the lamellae are undulated.

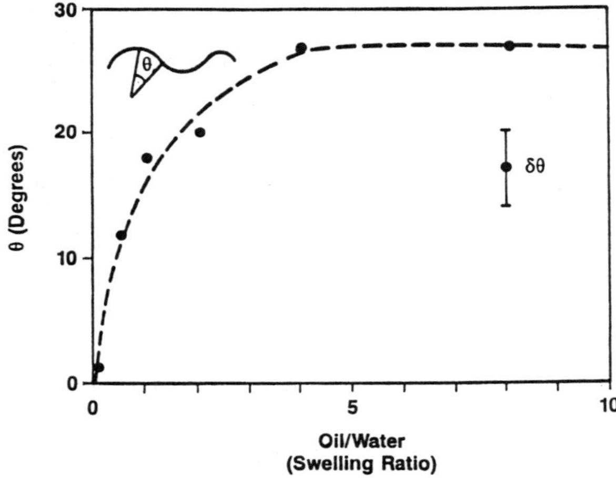

Fig. 5. Amplitude of the undulations as a function of the swelling ratio.

268

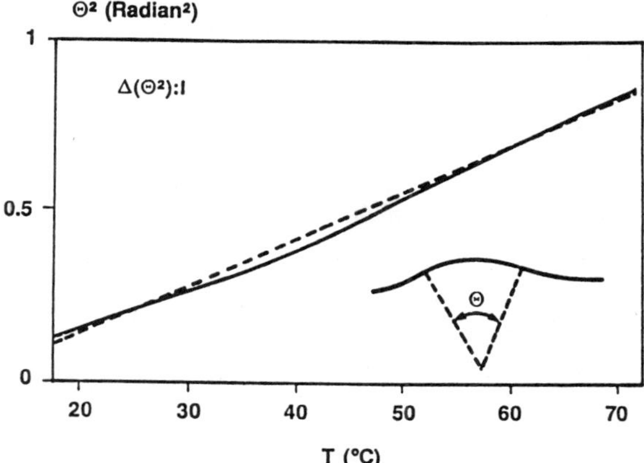

Fig. 6. Amplitude of the undulations as a function of the temperature for a 4 oil/water ratio (full line). The dashed line corresponds to $\theta^2 \propto 10^{-2}\ T$.

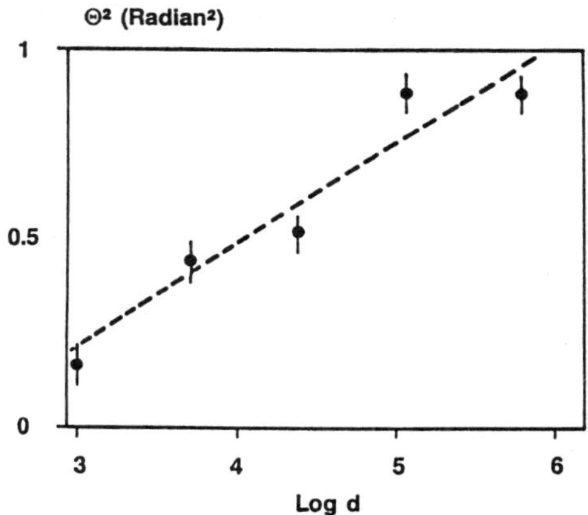

Fig. 7. Amplitude of the undulations as a function of the reticular distance computed from the swelling ratio and the mean area per polar head obtained through X-ray experiments.

this to the presence of textural defects in the samples, mainly induced by manipulation in placing them in the spectrometer. These defects are more likely present for the samples with a large swelling ratio. These defects lead to a large static scattering.

We have analyzed the autocorrelation of the scattered intensity and found that the correlation time τ $(=\omega^{-1})$ is proportional to q^{-2} according to the theoretical predictions. The results are plotted in Figure 9 for three different swelling ratios (2, 4 and 8). From the slopes, we may deduce the rigidity of the interfacial film. The results are reported in the table below:

Oil/water	K/ηd (cgs)	d (Å)	K (erg)
2	$2.6.10^{-7}$	120	$(3.1 \pm 0.5).10^{-15}$
4	$9.2.10^{-8}$	200	$(1.9 \pm 0.3).10^{-15}$
8	3.10^{-9}	360	10^{-16}

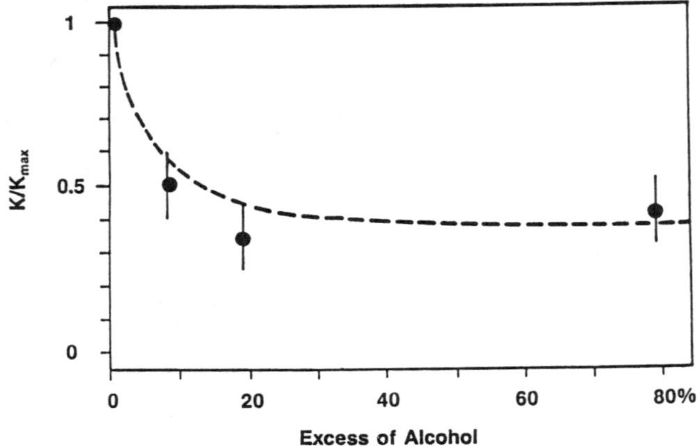

Fig. 8. Relative rigidity constant as a function of excess cosurfactant for a 4 swelling ratio phase. K_{max} is the rigidity of the phase with the minimum cosurfactant amount (0.43 ml).

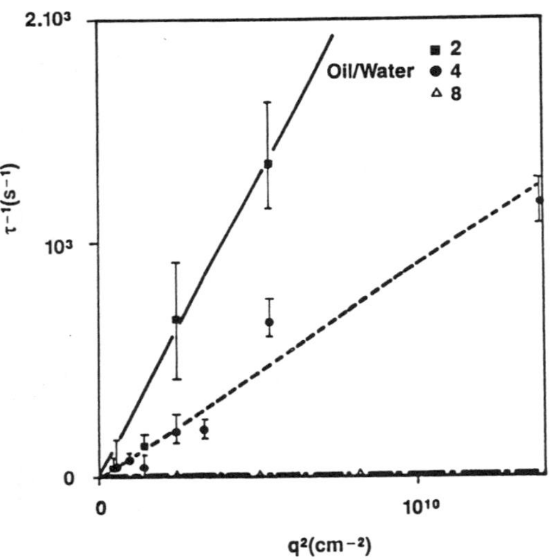

Fig. 9. Relaxation times of the pure undulation hydrodynamical mode as a function of the square of the scattering vector for three different swelling ratios.

We find that the values of K are roughly the same for 2 and 4 and this in good agreement with the spin labeling experiments. In contrast, the 8 sample has a rigidity which is lower by one order of magnitude; for this very swollen sample, the very high fluidity should lead to the presence of many structural defects, and give a very high static scattered intensity hindering the analysis of the spectra. The obtained values for 2 and 4 are in relative good agreement with the ones found previously by ESR.

The effect of the cosurfactant has been studied for a sample with a swelling ratio of 4 by the same technique. We find again that the role of the cosurfactant is indeed to lower the rigidity of the interfacial film. The results are shown in Figure 10. All the above described experiments have allowed us to measure for the first time the rigidity constant of the

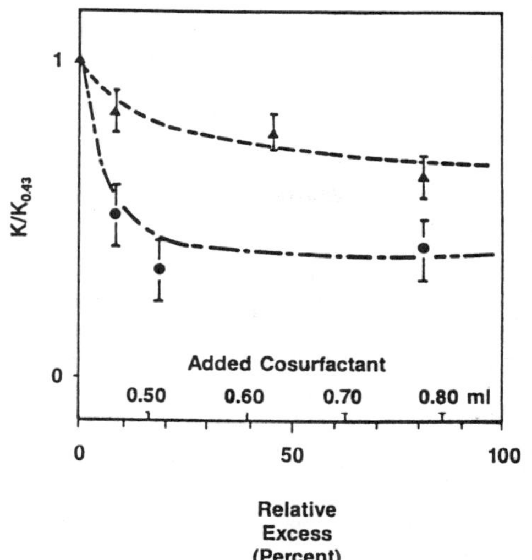

Fig. 10. Relative rigidity constant as a function of excess cosurfactant
for a 4 swelling ratio phase. K_{max} is the rigidity of the phase
with the minimum cosurfactant amount (0.43 ml). Triangles (▲) are
light scattering experiments while stars (*) are the results of
the previous spin labelling experiments.

interfacial film of microemulsion systems. These rigidities are very small
(a little lower than kT) compared to the widely studied lecithin systems.
We have studied lamellar systems, so we expect that the spontaneous curvature should be very small or even zero. We have so mainly concentrated
on the flexibility, and have shown experimentally that the cosurfactant
lowers the rigidity of the interface as indicated by many authors. We do
not exclude that the cosurfactant may affect the spontaneous curvature (as
simple steric models tend to show) but we do not think that this is important in our case as we deal with lamellar phases. We have benefited from
the fact that our system was so close to the isotropic phase that we may
imagine a scenario for the transition to this isotropic phase without
involving the influence of other parameters. Spin labeling experiments
have also revealed that the interfacial film of birefringent microemulsions
include highly curved defects[17] whose number is increased by excess of
cosurfactant. These defects should contribute to the rupture of the film
in the transition to the isotropic phase[25].

Acknowledgements

This work has received partial financial support from P.I.R.S.E.M.
(C.N.R.S.) under A.I.P. n° 2004.

REFERENCES

1. P. G. de Gennes and C. Taupin, J. Phys. Chem., 86 (1982) 2294.
2. A. M. Cazabat and M. Veyssie, Editors, Colloïdes et Interfaces, Summer
   School of the CNRS, Aussois 6-17 Sept. 1983, Editions de Physique,
   Paris 1984.
3. J. S. Huang and M. W. Kim, Phys. Rev. Letters, 47 (1981) 418.
4. A. M. Cazabat, D. Langevin, J. Meunier and A. Pouchelon, J. Phys.
   Letters, 39 (1982) 89.
5. M. Lagues, R. Ober and C. Taupin, J. Phys. Lettres, 39 (1978) L-487.

6. A. M. Bellocq, J. Biais, P. Bothorel, B. Clin, G. Fourche, P. Lalanne, B. Lemaire, B. Lenanceau and D. Roux, a recent review, Adv. Colloid Interface Sci., 20 (1984) 167.

7. A. Pouchelon, D. Chatenay, J. Meunier and D. Langevin, J. Coll. Int. Sci., 82 (1981) 418.

8. D. Roux, Thesis of Doctorat d'Etat, Bordeaux 1984.

9. J. H. Schulman and J. B. Montagne, Ann. N.Y. Acad. Sci., 92 (1961) 366.

10. S. A. Safran, L. A. Turkevich and P. A. Pincus, J. Phys. Letters, 45 (1984) L-69.

11. C. A. Miller, J. Dispersion Science and Technology, 2 (1985) 159.

12. M. Dvolaitzky, R. Ober, J. Billard, C. Taupin, J. Charvolin, Y. Hendricks, Comptes Rendus Acad. Sci., B 295 (1981) 45.

13. S. E. Friberg and C. S. Wohn, Colloid and Polymer Sci., 263, (1985) 156.

14. W. Helfrich, Z. Naturforschung, 28c (1973) 693.

15. L. S. Berliner Ed., "Spin Labeling: theory and applications", Academic Press, New York 1976.

16. M. Dvolaitzky and C. Taupin, Nouv. J. Chim., 1 (1977) 355.

17. J. M. di Meglio, M. Dvolaitzky, R. Ober and C. Taupin, J. Phys. Letters, 44 (1983) L-234.

18. W. Helfrich, Z. Naturforschung, 33a (1978) 305.

19. J. M. di Meglio, M. Dvolaitzky and C. Taupin, J. Phys. Chem., 89 (1985) 871.

20. R. M. Servuss, W. Harbich and W. Helfrich, Biochim. Biophys. Acta, 436 (1976) 900.

21. M. B. Schneider, J. T. Jenkins and W. W. Webb, Biophys. J., 45 (1984) 891.

22. J. M. di Meglio, L. Paz, M. Dvolaitzky and C. Taupin, J. Phys. Chem., 88 (1984) 6036.

23. J. M. di Meglio, M. Dvolaitzky, L. Léger and C. Taupin, Phys. Rev. Lett., 54 (1985) 1686.

24. F. Brochard and P. G. de Gennes, Pramana, 1 (1975) 1.

25. L. Paz, J. M. di Meglio, M. Dvolaitzky, R. Ober and C. Taupin, J. Phys. Chem., 88 (1984) 3415.

HIGHLY CURVED DEFECTS IN LAMELLAR PHASES: NON-IONIC LAMELLAR

PHASES AND BIREFRINGENT MICROEMULSIONS

J. M. di Meglio, M. Dvolaitzky, R. Ober,
L. Paz and C. Taupin

Laboratorie de Physique de la Matière Condensée
Collège de France 11, place Marcelin-Berthelot
75231 Paris Cedex 05, France

INTRODUCTION

In the past few years, new efforts have been devoted to the study of
the interfacial film of amphiphilic molecules because of the increasing
interest for new phases: microemulsions[1] and lyotropic nematics[2].
Among the amphiphiles, non-ionic surfactants[3] have been particularly
studied because of several special features:

- they are commercially available in a very pure form
- their aqueous micellar solutions exhibit a lower consolute point
  whose nature is not completely understood[4-7].
- they are able to form microemulsions without cosurfactant[8] in
  contrast to ionic surfactants. This capability seems to be
  correlated with the extension of the lamellar phase in the phase
  composition diagram.

In this paper, we show by using the spin labelling technique that the
interfacial film of these lamellar phases of non-ionic surfactants is
perturbated by very curved defects which appear to be cores of screw dislocations as described through observation of freeze fractures replica by
electron microscopy.

EXPERIMENTAL SECTION

The pure non-ionic surfactants, pentaethyleneglycol and hexaethylene-
glycol-dodecyl ethers ($C_{12}(EO)_5$ and $C_{12}(EO)_6$), were purchased from Nikko
Chemicals (Tokyo). Their phase diagrams have been published by Tiddy[9]
(Fig. 1). The samples (73% w/w of the surfactant in distilled water) were
observed under the microscope between crossed polarizers and verified to be
lamellar and to exhibit a transition towards a non-birefringent isotropic
phase at 73°C for $C_{12}(EO)_5$ and 66° for $C_{12}(EO)_6$, in good agreement with
Tiddy's diagrams.

The industrial non-ionic surfactant UK 36 and the polyethyleneglycols
(PEG) added to the samples were a gift of Mr. Tilguin (Centre de Recherche
PCUK, Levallois). A given weight of PEG was added to the lamellar phase

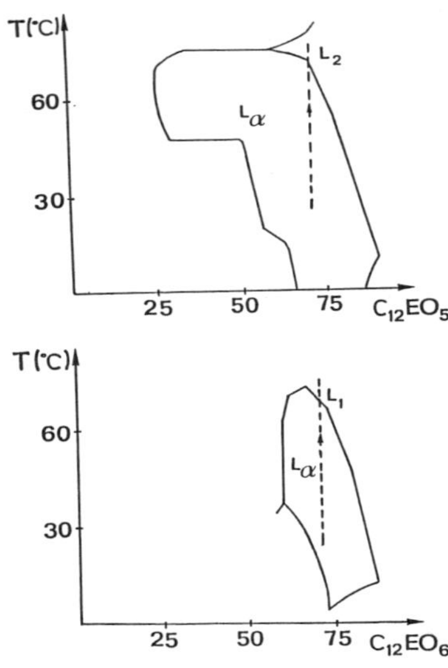

Fig. 1. Schematic representation of the domain of existence of the
lamellar phase L_α for the $C_{12}(EO)_5$-H_2O and $C_{12}(EO)_6$-H_2O systems.
L_1 and L_2 refer to the direct and high-temperature micellar
phases, respectively.

and the mixture was then heated in the non-viscous L_2 phase in order to
increase the solubilization rate.

The spin-label measurements were performed on a Varian E-9 EPR spectrometer. Either an isotropic or a parallel distribution of lamellae could
be obtained by using the orienting effect of walls of different glass
containers[10]. The spin-labelled amphiphile 1, which

resides exclusively in the interfacial film[11], is used at a molar concentration of 0.1% with respect to the non-ionic surfactant. This well-known
technique[12] givens information about the local state of the labelled
carbon of the chain: the order parameter and the correlation times of the
molecular motions.

The hydration rate of the lamellar phase (73% w/w in surfactant) was
determined as follows. A 0.4-ml sample of the lamellar phase was carefully
deposited on 1.1 ml of water in a graduated tube. The equilibrium state of
the final system is located in the L_1 domain (20% w/w in surfactant).
Because of the birefringence of the lamellar phase, the interface between
the two phases can be easily observed between crossed polarizers. The
samples were then thermostated in a water bath at 25°C, well below the
corresponding cloud points (T_c > 40°C), and the decrease of the birefringent volume was recorded as a function of time.

The X-ray experiments were performed on the scattering apparatus
equipped with a rotating copper anode generator, located at the "Centre de
Génétique Moléculaire" in Gif-sur-Yvette. The wavelength was 1.54 A and
the distance between the sample and the counter (of the position-sensitive

proportional type) was varied from 25 cm to 60 cm in order to observe
either diffraction peaks or small-angle scattering.

Electron microscopy of the freeze-fractured replica was performed at
the Centre d'Etudes Nucléaires of Fontenay aux Roses by Manuel Allain.

RESULTS

X-Ray Diffraction Study

The X-ray diffraction patterns of the different samples were recorded
at several temperatures. Fig. 2 shows a typical spectrum. Table 1 gives
the lamellar repeat distance values, in excellent agreement with[13].

Several points should be noticed. The repeat distances, and thus the
molecular areas, are almost insensitive to the temperature increase (2% for
30°C, to be compared to 6% for classical soaps[14]). Addition of 1.3% PEG
600 slightly increases the repeat distance (4%).

The addition of an ethylene oxide group increases the molecular area
(Table 1). This increase is associated with a decrease of the order parameter as observed by Mely[15] in soap systems.

At room temperature the samples exhibit no small-angle scattering and
the diffraction line widths are equal to the width of the incident beam.
This is the classical behavior of well-organized lyotropic lamellar phases
[16].

Local State of the Lamellar Film as Seen by Spin Labelling

Fig. 3 shows a typical spectrum obtained with an unoriented sample.
It is in fact the superposition of two spectra: one component is associated
with a very long angular correlation time and characterizes the lamellar
order[17], while the three line other component (M) is associated with an
almost completely averaged motion of the spin labels at the experimental
time scale (of the order of 10^8 sec). This effect was first observed by
Seelig[18] in hexagonal phases. The M spectrum may be due to the presence

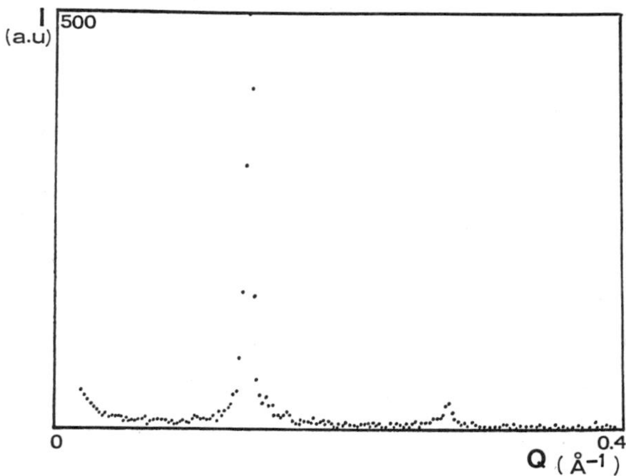

Fig. 2.   X-rays diffraction pattern of the $C_{12}(EO)_5$ lamellar phase at room
temperature. Two orders of diffraction are clearly visible.

Table 1. Lamellar Repeat Distances and Molecular areas Measured by X-Ray
Diffraction for 73% w/w Various Non-Ionic Surfactant-Water
Systems at 20°C

Products	$C_{12}(EO)_5$	$C_{12}(EO)_6$	$C_{12}(EO)_5$ + 1.3 % PEG 600
d(Å)	44.6	42	46.5
A(Å²)	42.9	50	–

of highly curved regions where, because of the fast lateral diffusion
inside the film, the labels undergo an almost complete reorientation (Fig.
4).

The spectra were recorded at different temperatures in the large
domain of existence of these lamellar phases, from 26 to 72°C. Two phenomena occur as the temperature increases:

1)   The order parameter decreases linearly and the slope is very similar
to that measured by Seelig on different lyotropic systems[19]. This effect
is due to the increasing rotation of the methylene groups, which promotes a
more disordered conformation of the chain.

2)   The relative intensity of the (M) spectrum increases. It is difficult
to make a quantitative evaluation of (M) intensity in the unoriented sample
spectrum of Fig. 3 because of the overlap of two types of spectra. It is
easier to do so for samples that are oriented by the walls of flat glass
cells, which give well-separated spectra as shown in Fig. 5. It is then
possible to evaluate the label proportion N which gives rise to an (M)-type
spectrum.

Fig. 6 represents the variation of N with temperature for various
systems. N first increases slowly, then very rapidly a few degrees before
the transition from the lamellar L_α to the high-temperature micellar phase.

Interpretation of the (M) Spectrum

The characteristics of the (M) spectrum give some insight into its
origin:

1)   If the translational diffusion constant D and the correlation time τ
of the spectrum are known, the radius of curvature can be calculated with
the relation $a^2 = 4D\tau$, which is valid for two-dimensional diffusion. The
lateral diffusion coefficient was measured using the method of Träuble[20],
which uses the broadening of the lines due to spin exchange when the label
concentration increases. A value of the order of $(5 \pm 1) \times 10^{-7} cm^2/s$ was
found in agreement with the value of D in cubic soap systems[21]. The
correlation time was determined by solving the Bloch equations modified by
an extra term for rotational Brownian diffusion as described in[22]. The
rigid limit was assumed to be the lamellar phase (order parameter 0.45)
itself. We found $\tau = (1 \pm 0.3) \times 10^{-8}$ s leading to a radius of the order
of 15 ± 5 A.

2)   A question arises about the nature of these curved objects or defects.
Are they intrinsic to the lamellar phase or the result of a phase that has
separated consequently of the presence of the first spin labels?

276

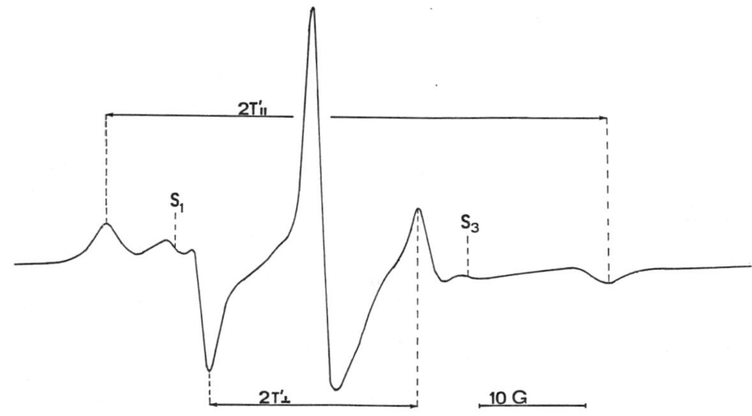

Fig. 3. Lamellar powder spectrum of the $C_{12}(EO)_5$ sample (73% w/w in
surfactant). M_1 and M_3 indicate the first and third line of the M
spectrum, respectively; the splittings $2T'_{//}$ and $2T'_{\perp}$ are the
averaged hyperfine tensor parameters of the L spectrum.

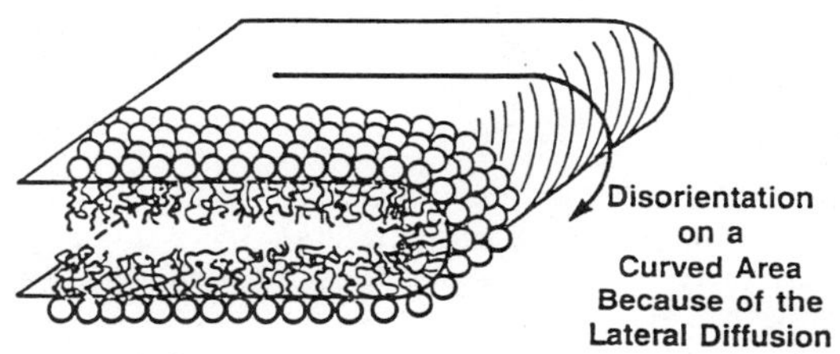

Fig. 4. Example of accident in a lamella.

In order to prove that the label concentration is the same in the two
environments, lamellar and curved, we varied the label concentration from
0.1 to 0.6 mole %. It is known from the diffusion coefficient determination that in these systems at 0.7 mole per cent, the width of the lines
begins to increase due to spin exchange between the nitroxide radicals. We
observed no variation of the line widths with the label concentration.
This fact proves that the curved regions are not specially concentrated in
labeled surfactant; as a consequence the N proportion of labels which
exhibit a (M) spectrum effectively reflects the proportion of curved
regions. Furthermore, the relative intensity of the (M) spectrum is independent of the label concentration, which gives the first indication that
no demixing has been induced in the sample. Moreover, both the L_α - L_2
transition temperature and the cloud point are unchanged by the presence of
0.1 mole per cent of the label. Finally, a microscopic observation under
phase-contrast confirms the phase homogeneity even when N is around 40%. A
demixing of droplets is visible only in the immediate vicinity (1°C) of the
transition. We conclude then that the curved regions are intrinsic to the
lamellar phase.

3)   The activation energy for creation of defects as calculated from the
low temperature slope of Fig. 6, is 12 Kcal/mole associated with the exact
reversibility of the thermal curve at a time scale of a few minutes. In
the non-ionic systems two types of defects can exist: unstable mechanically
induced ones and stable thermally created ones.

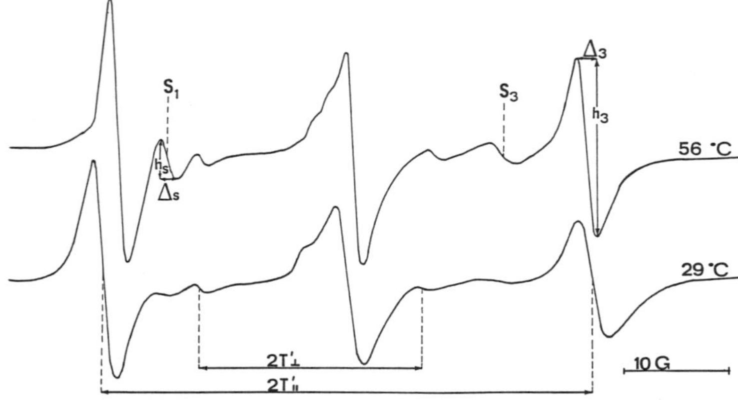

Fig. 5. Oriented spectra with the magnetic field corresponding to the
maximum line splitting (2T') for an homeotropic $C_{12}(EO)_5$ sample
oriented by the glass sides of a 0.1 mm flat tube at 29 and 56°C.
2T' indicates a small contribution of the molecules oriented
perpendicular to the field H by the small sides of the cell.
Taking into account the two successive derivatives of the signal
with respect to the magnetic field, the fraction N of defects is
deduced from the following expression:

$$N = (h_M\Delta_M{}^2)/(h_3\Delta_3{}^2 + h_M\Delta_M{}^2).$$

The shape of the lines (M_1 and L_3) has been verified to be
lorentzian. Furthermore we checked by synthesis and double
integration of the two components (L and M) that overlap is not
a serious problem. The relative accuracy on N is about 30%.

4)   Another interesting feature of such motion-averaged spectra is the
polarity, which is measured by the splitting of the lines and gives information about the local environment of the labelled carbon. First, one can
remark that the spectrum of the high-temperature isotropic phase, which is
characterized by a short correlation time due to curvature and a lower
polarity than L_α (14.4 instead of 15G), appears exactly at the same place
as the (M) spectrum a few degrees below the transition. From this result,
one can infer that the defects might be the high-temperature micellar phase
precursors. Finally it is interesting to consider the change in polarity
with temperature. At low temperature, the polarity is high (16 G), revealing an aqueous environment. The polarity drops very sharply a few degrees
below the transition, corresponding to the increase in the number of defects. These facts suggest that at low temperature the defects behave like
layer edges or pores in the lamellae[24] and that they undergo an important
change just before the transition. At that point, there are two possibilities. The change in polarity is either due to a curvature inversion
of the interfacial film or to the dehydration of the ethylene-oxide part of
the molecule. In a previous study on microemulsion system it was shown
that the structural inversion of the system (w/o → o/w) is associated with
polarity changes. Such an interpretation would indicate the presence of an
inverted micellar phase (of the L_2 type) at high-temperature.

These non-ionic lamellar phases were also observed using the freeze
fracture technique. Screw dislocations have been revealed and part of the
curved area detected by spin labelling can be attributed to the cores of
these dislocations. It is remarkable that he activation energy deduced

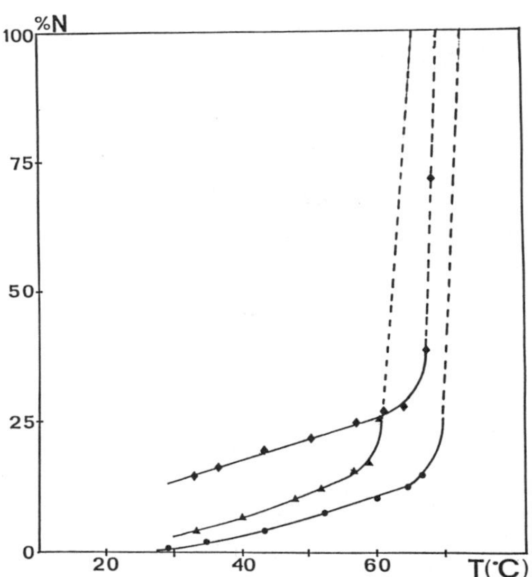

Fig. 6. Plot of the proportion N as a function of temperature for a 73% w/w
surfactant-water system: ($\bullet$) $C_{12}(EO)_5$, ($\blacktriangle$) $C_{12}(EO)_6$ and ($\blacklozenge$)
equimolecular mixture of $C_{12}(EO)_5$ and $C_{12}(EO)_6$.

from different quenchings at different temperatures give the same activation energy as the one deduced from the spin labelling experiments[25].

Effect of Various Additives on the Number of Defects

Two different kinds of additives were used: low-molecular-weight
polyethyleneglycols and molecules with different numbers of ethyleneoxide
groups. These additives are normally present in ethoxylated alcohols, as
impurities that result from the chemical preparation process[3b].

In both cases, the number of defects N increases. Fig. 6 shows the
thermal variation of N for an equimolecular mixture of $C_{12}(EO)_5$ and C_{12}
$(EO)_6$, as compared to the variation in the pure substances. The number of
defects outside the L_α - L_2 transition region is several times larger than
that in the pure compounds. Similar results for various concentrations of
PEG 600 added to the phase are given in Fig. 7. One can calculate that one
molecule of PEG 600 modifies the state of 12 surfactant molecules. Apart
from the increase of the number of defects at low temperature, an addition
of PEG decreases the L_α - L_2 transition temperature, reducing the domain of
the existence of the lamellar phase (Table 2). On the contrary, at room
temperature the limit between L_α and L_1 phase (56% in amphiphile) is unchanged by the addition of 1% of PEG 4000 or 4% of PEG 600.

The industrial product UK 36 which, as regards the cloud point, is
equivalent to the $C_{12}(EO)_5$ compound, but contains the two types of impurities, was also studied. As expected, the number of defects at low
temperature, in this case, is almost ten times larger than it is in the
pure substances.

Role of Defects in the Hydration Rate of the Lamellar Phase

Fig. 8 shows the displacement of the L_α - L_1 phase boundary with time
during the hydration process. As the L_α - L_1 limit is independent of the
presence of the additives, these curves represent the rate of hydration of

Table 2. Transition Temperature as Function of Concentration and
Molecular Weight of Added PEG

Product		PEG 600			PEG 2100	PEG 4000
% (w/w)	0	0.85	1.3	3.8	1.3	1.3
T (°C)	74	65	58.5	45	53.8	44

the L_α phase. One observes that the hydration rate increases with N. At
low temperature, the plot of the half-time of hydration $t_{1/2}$ versus N shows
a clear correlation between the two phenomena. Nevertheless, the effect of
the polymer molecular weight is also a very important factor since the
three different PEG's which were used induce approximately the same number
of defects but lead to very different kinetics of hydration.

These defects were in fact observed by us for the first time in "birefringent" microemulsions[10]. They were also detected by spin labelling.
The number of defects is increased by addition of the cosurfactant and thus
may explain the transition from lamellar order phases towards isotropic
disordered phases. We also have in this case distinguished between two
kinds of defects: textural defects related to the "mechanical history" of
the sample which annealed very well with time and structural defects
thermodynamically stable. We have determined an activation energy for the
latter of 15 kcal/mol.

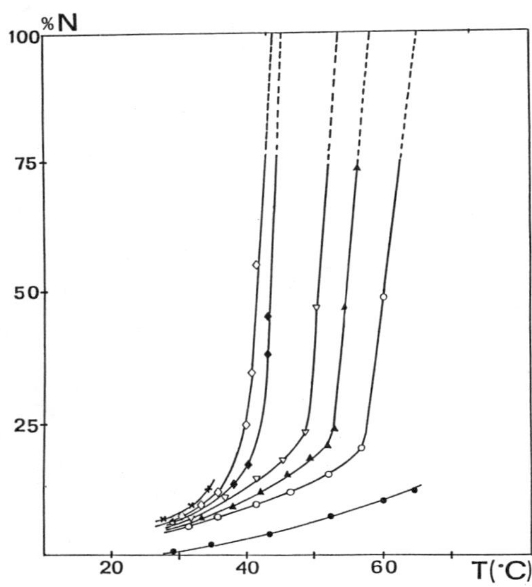

Fig. 7. Plot of the percentage of defects N versus temperature for a 73%
w/w surfactant-water system: ($\bullet$) $C_{12}(EO)_5$, ($\circ$) $C_{12}(EO)_5$ + 0.85 w/w
% PEG (M_w = 600), ($\blacktriangle$) $C_{12}(EO)_5$ + 1.3 w/w % PEG (M_w = 600), ($\triangledown$)
$C_{12}(EO)_5$ + 1.3 w/w % PEG (M_w = 2100), ($\diamond$) $C_{12}(EO)_5$ + 1.3 w/w % PEG
(M_w = 4000), ($\blacklozenge$) $C_{12}(EO)_5$ + 3.8 w/w % PEG (M_w = 600), (x) UKANIL
36.

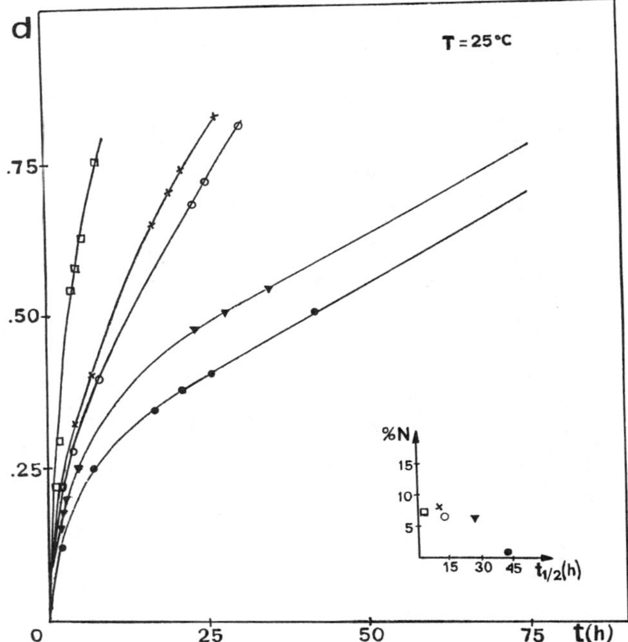

Fig. 8. Variation with time (in hours) of the relative decrease of the volume d equal to $(v_0 - v_{(t)})/v_0$ of the birefringent lamellar phase in contact with water for a 73% w/w surfactant-water system: (●) $C_{12}(EO)_5$, (▼) $C_{12}(EO)_5$ + 1.3 w/w % PEG 600, (○) $C_{12}(EO)_5$ + 1.3 w/w % PEG 2100, (x) UKANIL 36, (□) $C_{12}(EO)_5$ + 1.3 w.w % PEG 4000. Inset: plot of the half-time of hydration $t_{1/2}$ (in hours) versus N.

We can also make a rough estimation of the rigidity constant of the interfacial film from the activation energy value. By assuming that the defects have the topology of a pore (which is close to the one of a core of a screw dislocation) and by taking the radius of that pore equal to the width of a lamella, we find that the curvature energy is equal to 10.8 times the rigidity constant. The activation energy of 15 kcal/mol would thus give a rigidity constant of 9.6×10^{-14} erg in relatively good agreement with a mean value of 2.5×10^{-14} erg obtained from the study of the undulations of the interfacial film by spin labelling.

CONCLUSION

The lamellar phase of the aqueous $C_{12}(EO)_5$ and $C_{12}(EO)_6$ non-ionic surfactants, which has a large domain of existence in temperature (70°C), behaves quite classically as regards x-ray diffraction diagrams, mean areas per molecule, and order paramters. Spin label measurements, on the other hand, demonstrate the presence of highly curved regions which have the following features:

They are thermally nucleated and their number increases enormously a few degrees below the L_α - L_1 transition temperature. Their activation energy is remarkably low (around 12 kcal/mole).

Their number is larger for the $C_{12}(EO)_6$, which has a smaller lamellar domain.

Chemical inhomogeneity or additives increase the number of highly curved regions and promote the $L_\alpha - L_2$ transition, thus reducing the extension in temperature of the L_α phase.

The presence of such defects is clearly correlated with the hydration rate of the lamellar phase. These highly curved regions can be considered as defects of the lamellar structure; we attempted to detect an increase with temperature of the intensity of the small angle scattering. Although the observed effect was very small, it may perhaps indicate the presence of scattering objects of a few tens of $\mathring{A}$. The very efficient role of these defects in the hydration process proves that their formation induces the breaking of the layers. Beyond a screw dislocation[26], one can think of more localized defects such as pores which could explain the increased effect of the highest molecular weight PEG.

Finally, these results raise a problem. Such an effect has never been noticed before in the numerous EPR[17-20] studies of lyotropic conditions that must be fulfilled: there must be a sufficient number of defects (a few per cent) and the lateral diffusion must be fast. The lateral diffusion constant in phospholipidic systems is more than ten times smaller than that in the systems we have studied[20], thereby making it impossible to observe such an effect. The same argument is not a priori valid for ionic surfactant systems. Therefore our experimental observations may be a step in understanding the well-known different behavior of the ionic and non-ionic films.

Acknowledgements

We would like to thank M. Hermant and M. Tilquin (Centre de Recherche PCUK, Levallois) who drew out attention to the properties of non-ionic systems. We have benefited from the hospitality of Vittorio Luzzati and from many discussions about x-ray diagrams with Annette Tardieu at the "Centre de Génétique Moléculaire" (Gif-sur-Yvette).

REFERENCES

1. L. M. Prince, "Microemulsions Theory and Practice," Academic Press, New York, (1977).
2. J. Charvolin, A. M. Levelut and E. T. Samulski, <u>J.Phys.Lett.</u>, 40:L-587 (1973).
3. a) M. J. Schick, "Non Ionic Surfactants," Marcel Dekker, New York (1967).
    b) Ibid. p. 86.
4. M. Corti, V. Degiorgio and M. Zulauf, <u>Phys.Rev.Lett.</u>, 48:1617 (1982).
5. M. Zulauf and J. P. Rosenbusch, <u>J.Phys.Chem.</u>, 87:856 (1983).
6. P. G. Nilsson, H. Wennerstrom and B. Lindman, <u>J.Phys.Chem.</u>, 87:1377 (1983).
7. J. C. Ravey, <u>J.Coll.Int.Sci.</u>, 94:289 (1983).
8. K. Shinoda and H. Kunieda, <u>J.Coll.Int.Sci.</u>, 42:381 (1973).
9. D. J. Mitchell, G. J. T. Tiddy, L. Waring, T. Bostock and M. P. McDonald, <u>J.Chem.Soc.</u>, Faraday Trans. I:79:975 (1983).
10. J. M. Di Meglio, M. Dvolaitzky, R. Ober and C. Taupin, <u>J.Phys.Lett.</u>, 44:L-229 (1983).
11. M. Dvolaitzky and C. Taupin, <u>Nouv.J.Chim.</u>, 1:355 (1977).
12. L. J. Berliner, "Spin Labeling: Theory and Applications," <u>Academic Press Inc.</u>, New York (1976).
13. J. S. Clunie, J. F. Goodman and p. C. Symons, <u>J.Chem.Soc.Trans.Faraday Soc.</u>, 65:287 (1969).
14. B. Gallot and A. Skoulios, <u>Kolloid Z.U.Z. Polymer</u>, 208:37 (1966).

15. B. Mely, J. Charvolin and P. Keller, Chem.and Phys.Lipids., 15:161
    (1975).
16. P. Ekwall, in "Advances in Liquid Crystals," Vol. 1, G. Brown, ed.,
    Academic Press, New York, p. 1 (1971).
17. W. L. Hubbel and A. M. McConnell, J.Amer.Chem.Soc., 93:314 (1971).
18. J. Seelig and H. Limacher, Mol.Cryst.Liq.Cryst., 25:1051 (1974).
19. J. Seelig, J.Amer.Chem.Soc., 93:5017 (1971).
20. H. Trauble and E. Sackman, J.Amer.Chem.Soc., 94:4482 (1972).
21. J. Charvolin and A. Tardieu, in: "Liquid Crystals," Solid State
    Physics, Sup. 14, L. Liébert ed., Academic Press (1978).
22. R. C. McCalley, E. J. Shimshick and H. M. McConnell, Chem.Phys.
    Letters, 13:115 (1972).
23. M. Dvolaitzky, R. Ober and C. Taupin, C.R. Hebd.Acad.Sci., B 293:27
    (1982).
24. W. Helfrich, in: "Physics of Defects," R. Balian ed., North Holland,
    Amsterdam, p. 713 (1981).
25. M. Allain and J. M. Di Meglio, Mol.Cryst.Liq.Cryst., 124:115 (1985).
26. M. Kleman, in: "Springer series in Chemical Physics," Vol. 11, p. 97
    (1980).